Carbon Dioxide Utilisation

P.W.Sh

Nripathi

Also of interest

Carbon Dioxide Utilisation
Volume 2: Transformation
North, Styring (Eds.), 2019
ISBN 978-3-11-066503-1
e-ISBN 978-3-11-066514-7

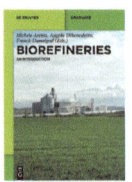

Biorefineries.
An Introduction
Aresta, Dibenedetto, Dumeignil, 2015
ISBN 978-3-11-033153-0
e-ISBN 978-3-11-033158-5

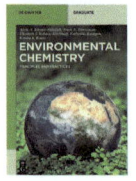

Environmental Chemistry.
Principles and Practices
Rihana-Abdallah, Benvenuto, Roberts-Kirchhoff,
Lanigan, Evans, 2019
ISBN 978-3-11-044330-1
e-ISBN 978-3-11-044331-8

Integrated Bioprocess Engineering.
Posten, 2018
ISBN 978-3-11-031538-7
e-ISBN 978-3-11-031539-4

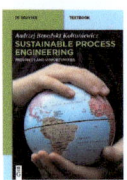

Sustainable Process Engineering.
Prospects and Opportunities
Koltuniewicz, 2014
ISBN 978-3-11-030875-4
e-ISBN 978-3-11-030876-1

Carbon Dioxide Utilisation

Fundamentals

Edited by
Michael North and Peter Styring

Volume 1

DE GRUYTER

Editors
Professor Michael North
University of York
Dept. of Chemistry
Heslington
York YO10 5DD
United Kingdom
michael.north@york.ac.uk

Professor Peter Styring
UK Centre for Carbon Dioxide Utilisation
Chemical & Biological Engineering
Sir Robert Hadfield Building
The University of Sheffield
Sheffield S1 3JD
United Kingdom
p.styring@sheffield.ac.uk

ISBN 978-3-11-056309-2
e-ISBN (PDF) 978-3-11-056319-1
e-ISBN (EPUB) 978-3-11-056352-8

Library of Congress Control Number: 2019941342

Bibliographic information published by the Deutsche Nationalbibliothek
The Deutsche Nationalbibliothek lists this publication in the Deutsche Nationalbibliografie;
detailed bibliographic data are available on the Internet at http://dnb.dnb.de.

© 2019 Walter de Gruyter GmbH, Berlin/Boston
Cover image: Yuji Sakai / Photolibrary / Getty Images
Typesetting: Integra Software Services Pvt. Ltd.
Printing and binding: CPI books GmbH, Leck

www.degruyter.com

About the editors

Michael North is professor of green chemistry in the Green Chemistry Centre of Excellence (GCCE) at the University of York. He obtained a first-class honours degree in chemistry from Durham University and a D.Phil. from the University of Oxford. Prior to his appointment at York, he held organic chemistry academic positions at the Universities of Newcastle, London and Wales. Michael has published more than 200 papers and has an h-index of 50. He was awarded the 2001 Descartes Prize by the European Commission and the 2014 Green Chemistry Award by the Royal Society of Chemistry. He is the deputy director of CO2Chem, the largest international network to bring together academics, industrialists and other parties interested in developing carbon dioxide utilisation and contributed to the 2017 Royal Society Policy Briefing Paper on "The Potential and Limitations of Carbon Dioxide Utilisation". His research interests are focussed on carbon dioxide utilisation, catalysis by the abundant metals in the Earth's crust, chemistry in green solvents and synthesis of polymers from sustainable feedstocks.

Peter Styring is professor of Chemical Engineering and Chemistry at the University of Sheffield. Based in the Department of Chemical and Biological Engineering, he is a former head of Department of Chemistry at Sheffield. He is a fellow of both the Royal Society of Chemistry (FRSC) and the Institution of Chemical Engineers (FIChemE) and is a Chartered Engineer (CEng) and Chemist (CChem). In 2006, he was awarded a prestigious EPSRC Senior Media Fellowship to work with TV, radio and the printed media for promoting chemical engineering and chemistry to the public. Peter is also an associate fellow in the Understanding of Politics at the University of Sheffield.

He is the director of the UK Centre for Carbon Dioxide Utilisation and Director of the CO2Chem Network. Peter sits on the Editorial Boards of Frontier in Energy Research (CCUS) and *Journal of CO$_2$ Utilisation*. He is a member of the Scientific Advisory Board of Carbon XPRIZE and the Board of the Global CO$_2$ Initiative at the University of Michigan. Peter was awarded the 2007 Hanson Medal by the IChemE and was a member of the 2015 Royal Society MP Pairing Scheme, spending time in the Foreign & Commonwealth Office. Peter was panel lead at Mission Innovation on Accelerating CCUS in 2017 and was one of the lead authors of the report published in May 2018. He also contributed to the 2017 Royal Society Policy Briefing Paper on "The Potential and Limitations of Carbon Dioxide Utilisation" and is a member of the Royal Society Steering Group on "Synthetic Fuels".

Peter has a background in chemistry, gaining a first-class honours degree (1985) and PhD in 1989, both from the University of Sheffield. He completed his PDRA position (EU-ESPRIT) at the University of Hull (1989–1990) after which he was appointed Thorn EMI Lecturer in Chemistry in 1990–2000. He became the DERA lecturer in 2004–2008. He left Hull to join the Department of Chemical Engineering at the University of Sheffield as senior lecturer in 2000 and was awarded Personal Chair in 2007.

https://doi.org/10.1515/9783110563191-202

He has been PI on a number of EU grants as UK Lead (WasteKit, SCOT, Artificial Photosynthesis, CarbonNext, Utilising Gaseous Industrial Emissions, Interreg Electrons 2 Chemicals, COZMOS), with total grants approaching 2 M€. He was a Co-I on the 4CU Programme Grant (£4.5M) and PI on numerous UKRI awards [C-Cycle, CO2Chem (three awards), Sustainable Fertilisers, Liquid Fuel and bioEnergy Supply from CO2 Reduction]. Recently, he was PI on a project funded by Costa Coffee, which successfully identified routes to the sustainable recycling of paper coffee cups. This led to him being Co-I on a recent awards Plastics Recycling grant at Sheffield in response to the Attenborough work on single-use plastics highlighted in BBC's "Blue Planet 2". He has published over 110 scientific and numerous policy papers and has an h-index of 30. His co-authored book "Carbon Capture and Utilisation in the Green Economy" has over 25k online views and new work with the Global CO_2 Initiative has resulted in the publication of a set of extensively peer-reviewed standardisation guidelines for Life Cycle and Techno-Economic Assessment of CO_2 Utilisation.

Contents

List of contributing authors

Chapter 1
Michael North (Ed.)
University of York
Dept. of Chemistry
Heslington
York YO10 5DD
United Kingdom
michael.north@york.ac.uk

Peter Styring (Ed.)
UK Centre for Carbon Dioxide Utilisation
Chemical & Biological Engineering
Sir Robert Hadfield Building
The University of Sheffield
Sheffield S1 3JD
United Kingdom

Chapter 2
Michael Carus
nova-Institut GmbH
Chemiepark Knapsack
Industriestraße 300
50354 Hürth
Germany
michael.carus@nova-institut.de

Christopher vom Berg
nova-Institut GmbH
Chemiepark Knapsack
Industriestraße 300
50354 Hürth
Germany

Elke Breitmayer
nova-Institut GmbH
Chemiepark Knapsack
Industriestraße 300
50354 Hürth
Germany

Chapter 3
Katy Armstrong
UK Centre for Carbon Dioxide Utilisation
Chemical & Biological Engineering
Sir Robert Hadfield Building
The University of Sheffield
Sheffield S1 3JD
United Kingdom

Chapter 4
Katy Armstrong
UK Centre for Carbon Dioxide Utilisation
Chemical & Biological Engineering
Sir Robert Hadfield Building
The University of Sheffield
Sheffield S1 3JD
United Kingdom

Peter Sanderson
UK Centre for Carbon Dioxide Utilisation
Chemical & Biological Engineering
Sir Robert Hadfield Building
The University of Sheffield
Sheffield S1 3JD
United Kingdom

Peter Styring (Ed.)
UK Centre for Carbon Dioxide Utilisation
Chemical & Biological Engineering
Sir Robert Hadfield Building
The University of Sheffield
Sheffield S1 3JD
United Kingdom

Chapter 5
Katy Armstrong
UK Centre for Carbon Dioxide Utilisation
Chemical & Biological Engineering
Sir Robert Hadfield Building
The University of Sheffield
Sheffield S1 3JD
United Kingdom

Arno Zimmermann
Technische Universität Berlin
Institute of Chemistry
Straße des 17. Juni 124
10623 Berlin
Germany

Leonard Müller
RWTH Aachen University
Institute of Technical Thermodynamics
Schinkelstraße 8
52062 Aachen
Germany

https://doi.org/10.1515/9783110563191-204

Johannes Wunderlich
TU Berlin
Institute for Advanced Sustainability
Studies e.V. (IASS)
Berliner Straße 130
14467 Potsdam
Germany

Georg Buchner
TU Berlin
Institute for Advanced Sustainability
Studies e.V. (IASS)
Berliner Straße 130
14467 Potsdam
Germany

Annika Marxen
Institute for Advanced Sustainability
Studies e.V. (IASS)
Berliner Straße 130
14467 Potsdam
Germany

Stavros Michailos
UK Centre for Carbon Dioxide Utilisation
Chemical & Biological Engineering
Sir Robert Hadfield Building
The University of Sheffield
Sheffield S1 3JD
United Kingdom

Peter Sanderson
UK Centre for Carbon Dioxide Utilisation
Chemical & Biological Engineering
Sir Robert Hadfield Building
The University of Sheffield
Sheffield S1 3JD
United Kingdom

Stephen McCord
UK Centre for Carbon Dioxide Utilisation
Chemical & Biological Engineering
Sir Robert Hadfield Building
The University of Sheffield
Sheffield S1 3JD
United Kingdom

Henriette Naims
Institute for Advanced Sustainability
Studies e.V. (IASS)
Berliner Straße 130
14467 Potsdam
Germany

André Bardow
RWTH Aachen University
Institute of Technical Thermodynamics
Schinkelstraße 8
52062 Aachen
Germany

Peter Styring (Ed.)
UK Centre for Carbon Dioxide Utilisation
Chemical & Biological Engineering
Sir Robert Hadfield Building
The University of Sheffield
Sheffield S1 3JD
United Kingdom

Reinhard Schomäcker
Technische Universität Berlin
Institute of Chemistry
Straße des 17. Juni 124
10623 Berlin
Germany

Chapter 6
Farnaz Sotoodeh
FeyeCon Development and Implementation
Rijnkade 17A
1382 GS Weesp
The Netherlands
farnaz.sotoodeh@feyecon.com

Tjerk J. de Vries
FeyeCon Development and Implementation
Rijnkade 17A
1382 GS Weesp
The Netherlands

Geert F. Woerlee
FeyeCon Development and Implementation
Rijnkade 17A
1382 GS Weesp
The Netherlands

Chapter 7
Muhammad Akram
Department of Mechanical Engineering
The University of Sheffield
Sheffield S1 3JD
United Kingdom
m.akram@sheffield.ac.uk

Chapter 8
Melis S. Duyar
Chemical and Process Engineering
University of Surrey
Guildford, Surrey, GU2 7XH, United Kingdom
m.duyar@surrey.ac.uk

Shuoxun Wang
Department of Earth and Environmental
Engineering
Columbia University in the City of New York
500 W 120th St.
New York, NY 10027
USA

Martha A. Arellano-Treviño
Department of Earth and Environmental
Engineering
Columbia University in the City of New York
500 W 120th St.
New York, NY 10027
USA

Robert J. Farrauto
Department of Earth and Environmental
Engineering
Columbia University in the City of New York
500 W 120th St.
New York, NY 10027
USA

Chapter 9
Matthew O'Brien
Lennard-Jones Laboratories
Keele University
Keele, Borough of Newcastle-under-Lyme
Staffordshire
United Kingdom
m.obrien@keele.ac.uk

Chapter 10
CD Hills
Engineering Science
University of Greenwich
Chatham Maritime
Kent ME44TB
United Kingdom
C.D.Hills@greenwich.ac.uk

N Tripathi
Engineering Science
University of Greenwich
Chatham Maritime
Kent ME44TB
United Kingdom

C Lake
Civil and Resource
Engineering
Dalhousie University
Halifax, NS B3H 4R2
Canada

PJ Carey
Carbon8 Systems Ltd. Medway Enterprise
Hub
Chatham Maritime
Kent ME44TB
United Kingdom
paulacarey@c8s.co.uk

D Heap
Carbon8 Systems Ltd. Medway Enterprise
Hub
Chatham Maritime
Kent ME44TB
United Kingdom

AT Hills
Engineering Science
University of Greenwich
Chatham Maritime
Kent ME44TB
United Kingdom

Chapter 11
Adrienne Macartney
University of St Andrews
Earth and Environmental Sciences
Department
United Kingdom
am475@st-andrews.ac.uk

Pol Knops
SCW Systems
Diamantweg 36
NL 1812 RC Alkmaar
www.scwsystems.com
pol.knops@scwsytems.com

Chapter 12
Peter Styring (Ed.)
UK Centre for Carbon Dioxide Utilisation
Chemical & Biological Engineering
Sir Robert Hadfield Building
The University of Sheffield
Sheffield S1 3JD
United Kingdom
p.styring@sheffield.ac.uk

Chapter 13
Shankara Gayathri Radhakrishnan
Chemistry Department
University of Pretoria
Pretoria 0002
Republic of South Africa
Shankara.Radhakrishnan@up.ac.za

Emil Roduner
Chemistry Department
University of Pretoria
Pretoria 0002
Republic of South Africa
And

Institute of Physical Chemistry
University of Stuttgart
Pfaffenwaldring 55
70569 Stuttgart
Germany

Chapter 14
L. Pastor-Pérez
Department of Chemical and Process Engineering
University of Surrey
Guildford, GU2 7XH
United Kingdom

E. le Saché
Department of Chemical and Process Engineering
University of Surrey
Guildford, GU2 7XH
United Kingdom
e.lesache@surrey.ac.uk

T.R. Reina
Department of Chemical and Process Engineering
University of Surrey
Guildford, GU2 7XH
United Kingdom
t.ramirezreina@surrey.ac.uk

Part I: **Introductory concepts**

Michael North and Peter Styring

1 Introduction

Welcome to *Carbon Dioxide Utilisation*: from fundamental discoveries to production processes. This book is aimed at advanced undergraduates and recent graduates studying carbon dioxide utilisation. Carbon dioxide utilisation is a highly interdisciplinary topic rooted not only in both chemistry and chemical engineering, but also impacting on physics, biological sciences and environmental science. This text assumes that the reader has a graduate-level knowledge of chemical sciences and in particular is familiar with chemical structures and thermodynamic concepts. Each chapter of this book has been written by one or more authors, at least one of whom is a member of CO2Chem. CO2Chem is a UK-based, global network of academics, industrialists and policy makers with an interest in carbon dioxide utilisation (CDU). It has over 1,000 members, covering not just the sciences but also social sciences such as sociology, economics and politics.

In planning this book we settled on the term carbon dioxide utilisation (often abbreviated as CDU) to most clearly describe the contents, but the topic is also often referred to as carbon capture and utilisation (CCU), which is related to carbon capture and storage (CCS). CDU refers to any technology that can take carbon dioxide and convert it into a more valuable chemical. Carbon dioxide is the end product of all combustion processes such as the burning of fossil fuels (Scheme 1.1) or biomass; it is produced as a by-product in many chemical processes [1] (Table 1.1) and it occurs naturally in the Earth's atmosphere. Prior to the industrial revolution, the concentration of carbon dioxide in the Earth's atmosphere was around 250 ppm, and by 2019 that had increased to over 410 ppm [2], mostly as a result of the global combustion of fossil fuels (coal, oil, gas) that provide the energy to support the lifestyle that we humans enjoy. It is now widely acknowledged that this increase in the atmospheric concentration of carbon dioxide is directly responsible for the global warming that has become apparent since the later decades of the twentieth century and for the associated climate changes that could make large parts of the currently inhabited regions of our planet unfit for human existence or for the production of food crops.

As Scheme 1.1 illustrates, not all fossil fuels produce the same amount of energy per mole of carbon dioxide emitted. Coal is the worst fuel in this respect, with liquid fuels (such as diesel, kerosene and petrol) next and natural gas producing almost twice as much energy per mole of carbon dioxide emitted as coal. Thus, by switching from coal to gas-fuelled power stations it is possible to significantly reduce the carbon dioxide emissions associated with electricity production. This is exactly the approach being taken by many countries (including the UK that has mandated that all coal burning power stations must close by 2025). Unfortunately, however, the known reserves of coal are far greater than those of oil and gas. The 2018 BP statistical review of world energy suggests that there is enough coal in known

https://doi.org/10.1515/9783110563191-001

Coal: $C + O_2 \longrightarrow CO_2$ $\qquad \Delta H_r = -394 \ kJ/mol$
$\quad$ (s) $\quad$ (g) $\qquad\qquad$ (g)

Petrol: $2C_8H_{18} + 25O_2 \longrightarrow 16CO_2 + 18H_2O$ $\quad \Delta H_r = -010,160 \ kJ/mol$
$\qquad$ (l) $\qquad$ (g) $\qquad\qquad\quad$ (g) $\qquad$ (g) $\qquad$ $(-635 \ kJ/mol \ per \ CO_2 \ emitted)$

Gas: $CH_4 + 2O_2 \longrightarrow CO_2 + 2H_2O$ $\qquad \Delta H_r = -803 \ kJ/mol$
$\quad$ (g) $\quad$ (g) $\qquad\qquad$ (g) $\quad$ (g)

Scheme 1.1: Combustion of fossil fuels.

Table 1.1: Major sources of waste carbon dioxide.

Source	Global CO_2 emissions (10^6 t CO_2/year)	CO_2 purity (volume %)
Coal	14,200	12–15
Natural gas	6,320	3–5
Refineries	850	3–13
Cement production	2,000	14–33
Ethylene production	260	12
Iron and steel production	1,000	15
Natural gas production	50	5–70
Ammonia production	150	100

reserves to last for 134 years at the current rate of consumption, but known reserves of oil and gas will be consumed in 50–53 years at current consumption rates [3]. These figures are an oversimplification as new reserves of fossil fuels are being discovered, consumption is not constant (it has increased in all but one year since 1982 and is likely to keep doing so as the human population increases and becomes more affluent) and non-conventional reserves such as gas obtained by fracking are not included. However, the BP figures do illustrate the nature of the problem we face, and give an indication of the timescale: this is not a problem that can just be left for future generations; most readers of this book will expect to be alive in 50 years time! Scheme 1.1 also illustrates that combustion of any fossil fuel still produces carbon dioxide and whilst switching from coal to gas can roughly halve the carbon dioxide emissions associated with production of a fixed amount of electricity, this is not sufficient to meet the level of carbon dioxide emissions reduction required if we are to limit global warming and avoid the worst climate change effects as recommended by the United Nations International Committee on Climate Change.

Of course, there are ways of generating electricity that do not involve the combustion of fossil fuels. One option is to burn freshly grown biomass rather fossilised biomass. This is usually considered to be carbon neutral as the biomass has absorbed carbon dioxide from the atmosphere during its growth and this is simply returned to the atmosphere during its combustion. However, transportation of the biomass to the power station (which may be on a different continent) must be considered and the land use change needed to allow biomass burning on a global scale would be significant, resulting in competition for arable land between food and fuel crops. Other well-established electricity production methods do not involve combustion and so directly generate no carbon dioxide. Examples include nuclear and hydroelectric power schemes. However, both of these require large quantities of cement/concrete during their construction, and this results in carbon dioxide emissions (Table 1.1). Nuclear energy has its own safety and political problems and is also very expensive compared to the combustion of fossil fuels or biomass. Hydroelectric power generation involves significant land use change often associated with loss of habitat for possibly endangered species and loss of arable and habitable land. Over the last 10–20 years, there has been a significant shift in electricity generation towards the so-called renewable energy. This involves installations that can capture the energy from sunlight (photovoltaics), wind or wave power. Each individual renewable unit (wind turbine, photovoltaic cell, etc.) generates a tiny amount of electricity compared to the amounts needed globally or even nationally, but these technologies are easily scalable by numbers rather than by size. The effects can be dramatic; at one point in 2017, the UK produced over half its electricity from renewable sources. Throughout 2017, coal burning accounted for less than 7% of all UK electricity production and in 2018, the UK managed to generate all the electricity it needed for three consecutive days without burning any coal [4].

So in the long term, it should be possible to generate all the electricity needed for human activities from non-carbon dioxide-producing sources, but the pathway to get there will be a lengthy one. In addition, transportation fuel is difficult to decarbonise. Electric trains are already well established and electric cars are becoming more common, but the lithium ion battery technology used for electric cars is not viable for larger vehicles such as trucks or ships and air transport is another major unsolved problem. The scale of the air transport problem should not be underestimated. International aviation produces almost as much carbon dioxide as the whole of the UK emissions and more than a number of other countries including Australia, Italy and France [5]. Thus for the foreseeable future, human beings will continue to burn fossil fuels to generate at least some of the energy needed to support our civilisation. Furthermore, as Table 1.1 shows, many chemical processes that produce substances that are essential for modern life are also major producers of waste carbon dioxide. Thus, we could not construct towns and cities without iron, steel and cement and the ammonia produced is almost all used to prepare urea and ammonium nitrate that are used as fertilisers to grow the crops needed to support a population of 7.5 billion human beings. Given that the human population is predicted to reach 10 billion [6] as early as 2050 (just 31 years from the time this book was published), we will need to expand the production of these chemicals (and others) to cope with the 25% increase in the number of human beings on planet Earth.

The preceding paragraphs set the scene for the need for carbon dioxide utilisation. To be able to sustain (and further enhance) human lifestyle whilst coping with a rapidly increasing population, we need to keep generating waste carbon dioxide and to avoid the damaging effects of climate change this carbon dioxide cannot simply be dumped into the Earth's atmosphere. There are two technologies that have been proposed to allow this: carbon dioxide utilisation (i.e., carbon capture and utilisation) and carbon capture and storage. Carbon capture and storage involves capturing carbon dioxide (from point sources or the atmosphere), purifying and pressurising it, transporting the pressurised gas to suitable disposal site and then burying it underground or under water [7]. This is a very energy-intensive process; it has been estimated that for every three coal burning power stations that CCS is fitted too, a fourth power station of comparable specifications would be needed just to supply the energy needed to power the CCS units. Whilst this already seems highly undesirable, the switch from coal-fuelled power stations to gas-fuelled ones makes the situation even worse. The flue gas from a coal-burning power station contains 12–15% carbon dioxide whilst that produced from a gas-burning power station contains only 3–5% carbon dioxide. The reason for this difference can be seen in Scheme 1.1. For combustion of coal, carbon dioxide is (ideally) the only product, and so if all the oxygen was combusted, the flue gas would contain 80% nitrogen and 20% carbon dioxide. In reality, not all the oxygen is combusted to avoid forming highly poisonous carbon monoxide, and so the flue gas from a coal burning power station contains about 12–15% carbon dioxide. However, combustion of natural gas produces two water molecules (i.e. steam) for every carbon dioxide molecule and so the carbon dioxide concentration in the flue gas will be one-third that from burning coal, that is, 3–5%. The consequence of this is that whilst the carbon capture unit attached to a coal burning power station needs to concentrate the carbon dioxide six- to seven-fold, that attached to a gas burning power station would need to concentrate the carbon dioxide by about 20-fold. This requires far more energy to achieve.

One part of carbon capture and storage that is often overlooked is the transportation. This requires a pipeline from each carbon dioxide producer to the disposal site. Coal-fuelled power stations are much larger than gas-powered ones, and so fewer coal-burning power stations are needed. A typical coal-burning power station might produce 2GW of electricity, whilst a gas turbine-based power station typically produces just 50 MW. So around 40 gas-fuelled power stations would be needed to replace just one coal-burning power station. Coal-burning power stations are often located close to one another on coal fields to minimise the costs associated with transporting coal. Hence, a single pipeline could transport the carbon dioxide from multiple coal burning power stations. In contrast, gas is easily transported through the existing natural gas grids (western Europe already obtains gas in a pipeline that comes from Siberia), and so as coal burning-power stations are phased out and replaced with gas-fuelled ones, these are likely to be far more widely distributed in order to minimise the losses associated with transport of the electricity they produce. As a result, even if each gas-burning power station is fitted with a carbon capture unit, the transportation of the carbon dioxide to a storage site becomes prohibitively expensive.

The reader should also be aware of another issue with many proposed carbon capture and storage schemes. They are not really carbon capture and storage schemes at all, but rather enhanced oil recovery [8] schemes being passed off as carbon capture and storage. The difference is illustrated in Figure 1.1. Both processes involve capture and transport of waste carbon dioxide. Although carbon capture and storage transports the carbon dioxide to a storage site where it can simply be pumped underground or under water, enhanced oil recovery transports the carbon dioxide to a partly depleted oil field. The carbon dioxide is then pumped down an oil well to force more crude oil out of the well. This crude oil would not be recoverable by conventional means. The additional crude oil generated in this way is subsequently treated in the same way as any other crude oil, 95% of it will be converted into fuel and burned to produce more carbon dioxide. Clearly if more carbon dioxide is produced when this fuel is combusted than was originally pumped into the partly depleted oil well, then overall the enhanced oil recovery process will generate more carbon dioxide than it consumes and will be a dangerous pyramid scheme. The attractiveness of enhanced oil recovery is of course that it produces more oil and hence more profit for oil companies.

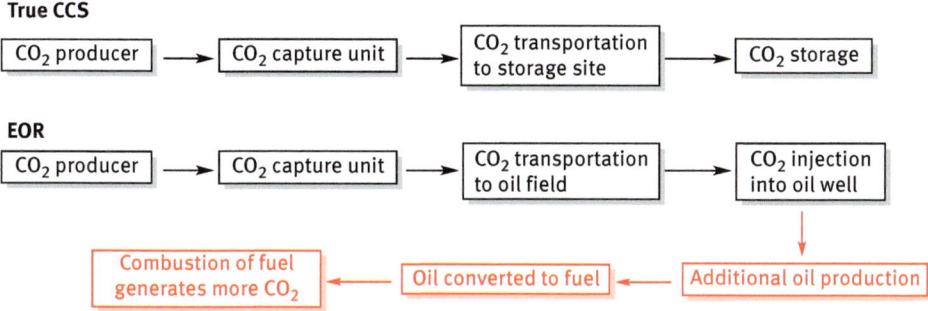

Figure 1.1: Comparative illustration of carbon capture and storage and enhanced oil recovery.

Another issue with carbon capture and storage that is apparent from Figure 1.1 is that it is a linear process that treats carbon dioxide as a waste to be disposed of. This is a continuation of the unsustainable "use once and dispose of" approach that is all pervasive in twenty-first century human society. Other examples include single use plastic and non-recyclable electronic goods that contain rare chemical elements. In contrast, this book is concerned with carbon dioxide utilisation, which treats carbon dioxide not as a waste to be disposed of, but rather as a valuable resource to be recycled and reused as illustrated in Figure 1.2. According to the Lansink Hierarchy of Waste Management [9], the first approach should always be to avoid producing waste in the first place. Thus, if new materials can be produced from carbon dioxide, then we avoid new virgin fossil carbon entering the supply chain. The least preferred option is landfill, which is essentially the fate of carbon dioxide in CCS. Intermediate options include reuse and recycle and recovery of energy: the cornerstone of CDU technologies. As such, implementation of carbon dioxide utilisation represents a shift from an unsustainable linear economy to a sustainable circular economy [10]. As shown in Figure 1.2, the combustion of fuel or production of chemicals generates carbon dioxide (Table 1.1). To close the cycle and convert the carbon dioxide back into fuel requires hydrogen. At present, almost all hydrogen produced commercially is obtained by steam reforming of methane and the water gas shift reaction (Scheme 1.2), which consumes a non-renewable resource (methane) and generates carbon dioxide [11]. As such this is not a sustainable source of hydrogen. An alternative is to obtain hydrogen by the splitting of water into hydrogen and oxygen. This is not currently cost competitive with steam reforming, but is the topic of much research. For the synthesis of chemicals from carbon dioxide, other renewable reactants are required. These should be derived from biomass and an example of such a process, the synthesis of ethylene carbonate from bioethanol and carbon dioxide, is illustrated in Scheme 1.3. The synthesis of ethene from sugar cane via bioethanol is already commercialised in

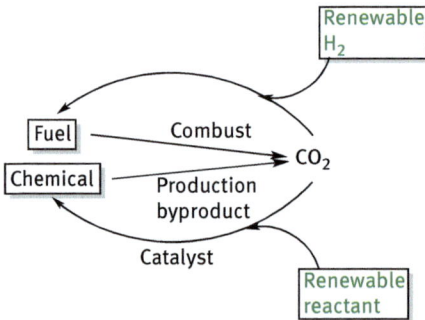

Figure 1.2: Carbon dioxide utilisation.

$$CH_4 + H_2O \xrightarrow{\text{Steam reforming}} CO + 3H_2$$

$$H_2O \downarrow \text{Water gas shift reaction}$$

$$CO_2 + H_2$$

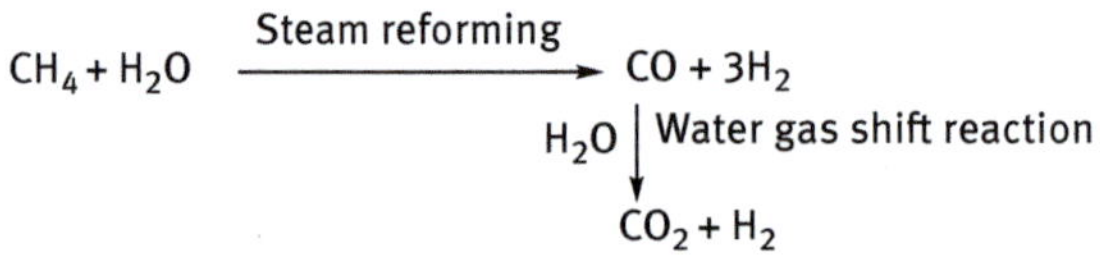

Scheme 1.2: Industrial synthesis of hydrogen.

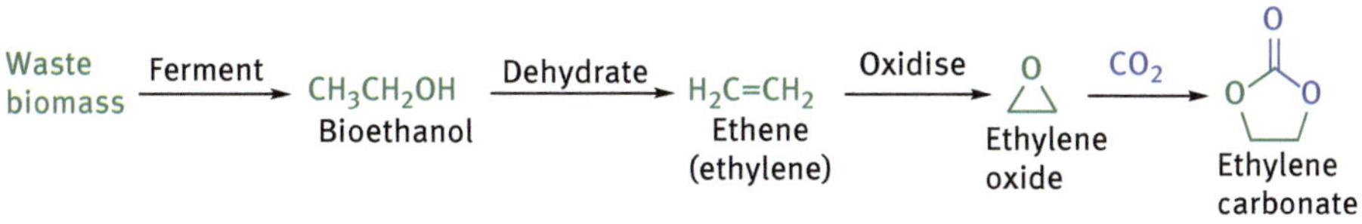

Waste biomass →(Ferment) CH$_3$CH$_2$OH (Bioethanol) →(Dehydrate) H$_2$C=CH$_2$ (Ethene (ethylene)) →(Oxidise) Ethylene oxide →(CO$_2$) Ethylene carbonate

Scheme 1.3: A sustainable route to ethylene carbonate.

Brazil [12], and in the USA, Croda have built a plant for the synthesis of ethylene oxide from bioethanol sourced from corn [13].

One criticism that is sometimes used to belittle carbon dioxide utilisation is the difference in scales of the energy and chemicals sectors. This difference in scale certainly exists: power generation involves the production of gigatonnes (Gt) of carbon dioxide as a by-product whilst even the largest-scale chemical processes only produce about 150 megatonnes (Mt) of product. However, as Figure 1.2 illustrates, carbon dioxide utilisation involves the conversion of carbon dioxide into both fuel and chemicals. The top circle in Figure 1.2 has the potential to be operated on the same scale as energy generation from fossil fuels and hence to make the combustion of fossil fuels obsolete. The lower circle will always be operated on a smaller scale, but here the analogy with a conventional oil refinery and with a biorefinery is informative. Figure 1.2 essentially shows a carbon dioxide refinery and Figure 1.3 compares all three types of refinery. Of these, only the oil refinery is currently operated commercially: 96% of the output of an oil refinery is fuel, the remaining 4% being chemicals that provide the basis for the current global chemicals industry. However, the 4% of the refinery output that goes to the chemicals industry generates over 40% of the profit for the oil refinery! [14] This is why it is important to consider both cycles in Figure 1.2; the route to chemicals may not consume large amounts of carbon dioxide, but it can make the whole carbon dioxide refinery commercially viable. It should also be realised that carbon dioxide mitigation can be achieved not only by capture, but also by avoidance. Under the European Union Renewable Energy Directive (RED) 2, avoidance of carbon dioxide generation is given equal status in emissions legislation to carbon dioxide captured and

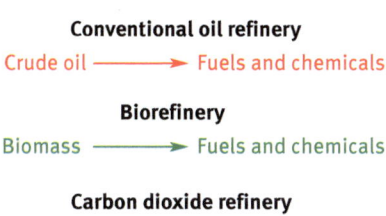

Conventional oil refinery
Crude oil ⟶ Fuels and chemicals

Biorefinery
Biomass ⟶ Fuels and chemicals

Carbon dioxide refinery
CO$_2$ ⟶ fuels and chemicals

Figure 1.3: Comparison of three types of refinery.

permanently stored [15]. This is a progressive piece of legislation that recognises the precepts of the Lansink Hierarchy.

The need for a more sustainable approach to next-generation synthetic fuels and petrochemicals has been highlighted by many governments; particularly in Germany who have invested heavily in CDU, with success, through the BMBF funding instrument [16]. In May 2018, Mission Innovation published a report on Accelerating CCUS [17, 18]. This was agreed through the G20 nations and the European Union as a block. A number of "Priority Research Directions" were proposed and accepted. These include the use of carbon dioxide to make, as the EU defines them, "Synthetic Fuels of Non-Biological Origin" or "e-Fuels". These are fuels that not only replace primarily gasoline and diesel, but which also seek to replace Jet fuel (kerosene), which is more problematic due to strict regulations on its composition. The resulting fuels, including methanol, dimethyl ether and oxymethylene ethers, have been shown to be far superior in their reduced emissions than conventional hydrocarbon fuels. The Royal Society of London have also published an excellent Policy Briefing document [19] on the Potential and Limitations of Carbon Dioxide Utilization (2017) and are due to publish a similar document on Synthetic Fuels in 2019.

There is another advantage associated with a carbon dioxide refinery: point sources of waste carbon dioxide are widely distributed around the planet and atmospheric carbon dioxide occurs at the same concentration everywhere. In contrast, large reserves of oil and gas are localised in often politically unstable parts of planet Earth and suitable land for growing biomass for a biorefinery is also not equally distributed. Thus, carbon dioxide refineries have the potential to be operated anywhere on the planet without the need to source and transport the crude oil or biomass long distances.

The importance of a carbon cycle cannot be underestimated. Nature has for millions of years prior to the industrial revolution managed its own carbon cycle. Carbon dioxide produced from combustion and natural phenomena has been used in photosynthesis to produce the energy for a plant: carbohydrate. When this is combusted or metabolised, the carbon dioxide is re-emitted and the cycle continued. However, an imbalance in the cycle was caused by the industrial revolution and the anthropogenic use of carbon-based materials as a source of fuel. It should be noted that fossil fuels originated from the decomposition of fauna and flora over many millions of years. These natural precursors came from the photosynthetic process in the case of flora and from animal metabolism in the case of fauna. Carbon dioxide in the atmosphere was the source of this carbon. We could therefore consider that what the scientists and engineers are doing in CDU is simply accelerating the fossilisation process using catalysts and process development. It has been argued that reusing carbon dioxide that has been emitted in a combustion process is still fossil carbon and therefore the product is fossil based. This neglects the fact that it is second-life carbon, re-used to produce a product that would otherwise have required virgin fossil oil. If we take the approach that products made from emitted carbon dioxide are still fossil based, then we could equally say that biomass produced in the current environment is also fossil based! Clearly there needs to be a logical disconnect between virgin fossil carbon and second and subsequent life carbon.

What is very clear is that we cannot simply look at a reaction or process in isolation. We need to consider the environmental and economic impact of the whole process across the supply chain surrounding the CDU technology. That is why this book has chapters looking at the techno-, environmental and economic impact of any process that is proposed. The public also needs to be aware of the technologies at an early stage so they are not seen as a problem. This is why this book considers the whole system, including the societal impact. While many of the technologies seem expensive at this time in comparison to fossil-based routes, one must remember that these are nascent technologies. Costs will fall as scale increases and commercial reality kicks in. Televisions and computer monitors transitioned from cathode ray tubes to liquid crystal displays. Liquid crystal display televisions cost thousands of pounds on first release (actually millions of pounds during the development phase), but can now be purchased for under 100 pounds. We should not be put off by high costs during the research and development phase.

This book has been divided into six parts. Part I introduces topics that are essential when considering carbon dioxide utilisation. These include the concept of sustainable development, social acceptance and lifecycle and technoeconomic analysis. The use of carbon dioxide as a solvent is also included in this part. Part II deals with methodologies for separating carbon dioxide from other gases, whilst Part III covers general aspects of carbon dioxide chemistry including its activation and mineralisation. Parts IV–VI contain a series of chapters each of which focusses on one way of converting carbon dioxide into a fuel or commercially important chemical. Part IV considers processes that occur with the aid of a catalyst, Part V electrochemical approaches to carbon dioxide utilisation and Part VI photochemical and plasma-induced reactions of carbon dioxide. The intention is to give the reader a flavour of what can be achieved using each of these methodologies rather than to give a comprehensive coverage. Authors were asked to concentrate the material in their chapter on processes that are already commercial, or which are closest to commercialisation. Each chapter ends with a bibliography, which the reader can use to access additional information. All chapters were written during 2018, and so the literature covered will be that published up to the end of 2017.

References

[1]   J. Wilcox, *Carbon Capture*, Springer, New York, 2012.
[2]   https://www.scientificamerican.com/article/we-just-breached-the-410-ppm-threshold-for-co2/ (Last accessed 20 January 2019).

[3] https://www.bp.com/en/global/corporate/energy-economics/statistical-review-of-world-energy.html (Last accessed 20 January 2019).

[4] https://www.bbc.co.uk/news/business-43879564 (Last accessed 20 January 2019).

[5] https://en.wikipedia.org/wiki/List_of_countries_by_carbon_dioxide_emissions (Last accessed 20 January 2019).

[6] https://www.un.org/development/desa/en/news/population/world-population-prospects-2017.html (Last accessed 20 January 2019).

[7] https://en.wikipedia.org/wiki/Carbon_capture_and_storage (Last accessed 20 January 2019).

[8] https://en.wikipedia.org/wiki/Enhanced_oil_recovery (Last accessed 20 January 2019).

[9] https://www.recycling.com/downloads/waste-hierarchy-lansinks-ladder/ (Last accessed 20 January 2019).

[10] https://www.ellenmacarthurfoundation.org/circular-economy/concept (Last accessed 20 January 2019).

[11] https://www.planete-energies.com/en/medias/close/hydrogen-production (Last accessed 20 January 2019).

[12] http://plasticoverde.braskem.com.br/site.aspx/Im-greenTM-Polyethylene (Last accessed 20 January 2019).

[13] https://www.crodapersonalcare.com/en-gb/discovery-zone/learn/sustainability/sustainable-manufacturing/eco-plant (Last accessed 20 January 2019).

[14] N. Hernández, R. C. Williams, E. W. Cochran, *Org. Biomol. Chem.* 2014, *12*, 2834–2849.

[15] https://www.theicct.org/publications/final-recast-renewable-energy-directive-2021-2030-european-union (Last accessed 20 January 2019).

[16] https://www.bmbf.de/en/index.html (Last accessed 20 January 2019).

[17] Accelerating CCUS, Mission Innovation, DoE, Washington, 2018 (https://bit.ly/2UE82An). (Last accessed 20 January 2019).

[18] http://mission-innovation.net/ (Last accessed 20 January 2019).

[19] https://royalsociety.org/topics-policy/projects/low-carbon-energy-programme/potential-limitations-carbon-dioxide/ (Last accessed 20 January 2019).

Michael Carus, Christopher vom Berg and Elke Breitmayer

2 CO$_2$ utilisation and sustainability

2.1 Summary

The direct utilisation of CO$_2$ as a raw material clearly benefits the environment and comes with several specific advantages. Compared to other carbon sources, fuels and chemicals based on CO$_2$ generally cause the lowest GHG emissions. Compared to the utilisation of biomass, land use efficiency of solar CCU outperforms biomass by a factor of 50–150.

A particular advantage that comes with the utilisation of CO$_2$ is the unlimited availability of the carbon source, both in terms of time and volume. Moreover, renewable energies also offer almost unlimited potential when it comes to meeting the human demand.

CCU thus allows for the "democratisation" of access to raw materials – CO$_2$ is ubiquitous, as this resource can be found everywhere. In future, basically everybody, no matter where they are on this planet, will have the opportunity to harvest carbon, thanks to renewable energies and CCU, and to produce fuels, chemicals and plastics out of it. These products allow solar energy to be stored over a long period of time without any losses. A guaranteed raw materials access by CCU enables a stable economy.

CO$_2$ utilisation in combination with renewable energies is the most sustainable path to fuels, chemicals and plastics from an ecological point of view and can explicitly and substantially contribute to a sustainable economy. Nine out of the 17 UN Sustainable Development Goals are directly addressed through CO$_2$ utilisation in combination with renewable energies.

The utilisation of CO$_2$ is in line with efficiency and consistency strategies, without however requiring a more stringent sufficiency strategy.

2.2 What is the meaning of sustainability?

The concept of "sustainability" or "Nachhaltigkeit" in German can be traced back to Hans Carl von Carlowitz (1645–1714), who managed mining on behalf of the Saxon court in Freiberg (Germany). Despite the court's forest regulations, the impact of timber shortages on Saxony's silver mining and metallurgy industries was devastating. In his work *Sylvicultura Oeconomica* or the *Instructions for Wild Tree Cultivation* [1], Carlowitz formulated ideas for the "sustainable use" of the forest. He raised the question of "how can one conserve and grow forests in a way that

Michael Carus, Christopher vom Berg, Elke Breitmayer, nova-Institut GmbH, Hürth, Germany

https://doi.org/10.1515/9783110563191-002

allows their continuous and sustainable use, because this is indispensable to pre-serve the essence of the land" [2]. His view, that only so much wood should be cut as can be regrown through planned reforestation projects, became an important guiding principle not only of modern forestry but also increasingly for the entire world economy [3].

The term "sustainability" was more widely used once the Brundtland Commission (headed by Gro Harlem Brundtland, the then prime minister of Norway) took up its work in 1983 and found its way into the 1987 Brundtland Report: " 'Sustainable development' is a kind of development that meets the needs of the present without compromising the ability of future generations to meet their own needs and make their own lifestyle choices." At the RIO+10 Summit held in Johannesburg in 2002, the African delegation reduced the concept of sustainability to a simple common denominator – "Enough – for all – forever".

And finally, the German Federal Government used the term sustainability in its "Sustainable Development in Germany" report, published in 2014, adding both a temporal and spatial dimension: "Today and here we want to live neither at the expense of other people in other regions of the world nor at the expense of future generations."

While everybody largely agrees on the general meaning of "sustainability", there are different interpretations when it comes to the concrete application of the term. Very often, the "three pillars model" is used which unites the environmental, social and economic dimensions under the umbrella of "sustainability" as shown in Figure 2.1.

At the Rio-20 conference, the United Nations defined "17 Sustainable Development Goals" as shown in Figure 2.2. The figure shows how the targets are directly connected to CCU.

In 1995, the Wuppertal Institute together with Bund and Misereor also pub-lished an interesting paper outlining the three most important strategies for Germany titled "Sustainable Germany. Thoughts on global sustainable develop-ment": efficiency, consistency and sufficiency [4]. The authors state that all three aspects are essential for an environmentally healthy and just society. Efficiency means an increase in resource productivity to utilise limited resources in the best way. Consistency aims to avoid waste through recyclable resources. Today, this is what we call the "circular economy". However, these two alone will not enable a life without scarce resources; therefore, they added the concept "sufficiency", which refers to a kind of modest life or "living a better life instead of consuming more", which can be achieved by decoupling economic wealth from resource con-sumption. Whether the latter finds global acceptance is doubtful.

Recently, a new public view on the term sustainability became apparent. Johan Rockström of the Stockholm Resilience Centre in Sweden said "The whole concept of sustainability has tipped. Until very recently, the environmental agenda was largely a question of ethics and morality. It was as sacrifice." Now, he says,

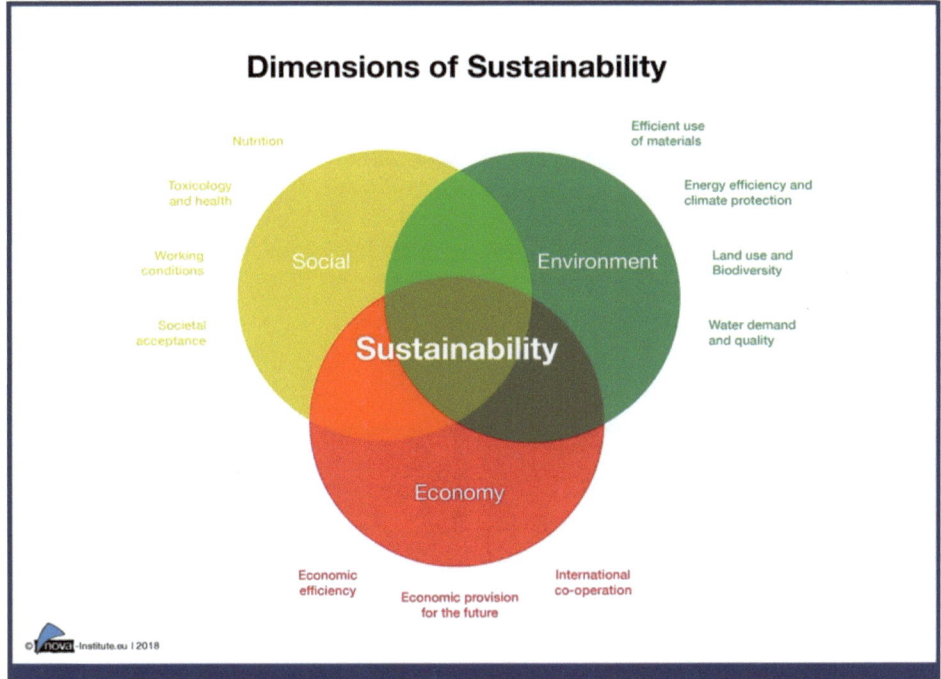

Figure 2.1: The three pillars of sustainability.

"sustainability is seen as the only way to deliver a stable economy. We are into a completely new paradigm" [5].

This will suffice as a brief introduction into the complex issue of sustainability. The following sections will shed some light on this issue and analyse not only the environmental pillar of sustainability, but also efficiency and consistency strategies as well as the concomitant sustainable development goals of the United Nations in regards to the utilisation of CO$_2$.

2.3 CCU and sustainability

Table 2.1 evaluates fuels and chemicals derived from fossil energy sources, biomass and CO$_2$ regarding their impact on important environmental categories, paving the way for an initial assessment of the environmental footprint of CO$_2$ utilisation. It is always assumed that the energy to reduce CO$_2$ is either provided by the emission gas itself (synthesis gas, e.g. from the steel industry) or is generated using renewable resources such as the sun, wind, water or geothermal heat. The use of fossil energy sources to reduce CO$_2$ would lead to very unfavourable energy and CO$_2$ balances and as a consequence is neglected here.

Table 2.1 clearly shows the environmental benefits and the specific advantages of CO_2 utilisation, which will be discussed in more detail in the following paragraphs. Compared to other carbon sources, fuels and chemicals derived from CO_2 generally produce the lowest greenhouse gas emissions (cf. Section 2.3.2) because they do not release any additional fossil carbon and renewable energies harvest solar radiation much more efficiently than natural photosynthesis, the standard process for the production of any biomass. The latter is due to the high level of solar efficiency and the low land use compared to the production of biomass. Searchinger et al. discussed this aspect in a 2017 publication [6]. The authors show that the bioenergy conversion efficiency is a mere 0.1–0.2%, that is, only 0.1 or 0.2% of the solar energy makes its way to the final product (e.g. ethanol). The net solar conversion efficiency (PV) however ranges between 11% and 44% and thus outperforms biomass utilisation by a factor of 122–295. When solar energy is converted into liquid fuels harnessing CCU technologies, efficiency levels of above 50% are feasible even today, that is, the conversion efficiency of solar CCU to liquid fuels beats biomass by a factor of 50–150.

The increasing practicality of solar energy tilts the land use equation further against bioenergy whenever it can be used. Even if 100 ha of good land were to become theoretically available for climate mitigation, they could generally provide at least as much energy and at least 100 times more carbon mitigation if 1 ha were used for solar and 99 to restore forests.

A specific advantage of CO/CO_2 utilisation is the fact that this carbon source is infinitely available without any time- and/or volume-related limitations. On the one hand,

Table 2.1: Evaluation of fuels and chemicals from various carbon sources against the background of environmental sustainability criteria.

C source	Fuels and chemicals		
	Fossil-based (crude oil, natural gas and coal)	**Bio-based (all types of biomass)**	**CO/CO$_2$-based in combination with renewable energies (sun, wind, water, geothermal heat)**
Greenhouse gas emissions	High: in particular due to the release of fossil carbon into the atmosphere	Low to medium: biogenic carbon is recycled. Emissions are caused by fertilisers, pesticides and herbicides	Low: carbon is recycled
Land use	Low	High	Low: per surface area, PV uses solar radiation 50 times more efficiently than the best-performing plants
Water use	Normally low, exception: pollution by accidents	Low to high: depends on the region and the crop	Low
Availability	Finite, only local occurrence	Limiting factor: the land required; generally globally available, but of different local qualities	No limiting factors: both time- and volume-wise to cover the entire fuels and chemicals demand, covering ca. 10% of the deserts with PV systems would be sufficient;[*] globally available anywhere
Contribution to circular value creation	No	Yes	Yes
Repercussions on the food supply, land, water and biodiversity	Normally low, exception: accidents can highly pollute the oceans with immense ecological, economic and social consequences for an affected region	Medium to high: depends on local procedures	Low

Table 2.1 (continued)

C source	Fuels and chemicals		
	Fossil-based (crude oil, natural gas and coal)	**Bio-based (all types of biomass)**	**CO/CO_2-based in combination with renewable energies (sun, wind, water, geothermal heat)**
Solar efficiency (conversion of solar radiation into the final product)	–	< 0.5%	> 10%
Storage of solar energy	Long-term storage of solar energy from millions of years	Agriculture: seasonal, forest: over decades	Fossil or biogenic long-term storages remain intact (avoiding depletion), potential creation of new, loss-free storage in fuels and chemicals such as organic chemicals and carbonates e.g. $CaCO_3$ or $MgCO_3$ via mineralisation
Stability in supply	Depends on geopolitical situation	Crop dependent; can vary due to natural conditions	Stable supply is theoretically possible
Specific risks	Risk of accidents such as oil spills, shipping accidents and other disasters severely impacting the marine environment as well as the local economy	Long-term damage by inappropriate management, for example, biodiversity loss, groundwater depletion, soil erosion, nitrate leaching; risk from climate change	Usually high-energy demand to utilise CO_2. Required energy needs to be "clean" (renewable) to not cause more emissions than are avoided

*Own calculations including losses from conduction and storage. Other sources estimate even lower land requirements (e.g. "DESERTEC, 2014: The DESERTEC foundation, Global "ENERGIEWENDE", The DESERTEC Concept for Climate Protection and Development, May 2014, avaliable online at: https://issuu.com/global_marshall_plan/docs/desertec_atlas_global_energiewende"

there will be enough CO_2 in the atmosphere in the long run, while on the other hand carbon is recycled, that is, it is pulled from the atmosphere or industrial sources and while it is being utilised (or after utilisation), it is released again, replenishing the CO_2 storage which is the atmosphere. The potential of renewable energies to cover human demand is almost unlimited. According to calculations by the nova-Institute, the entire

2050 demand in fuels and chemicals may be met by harnessing PV systems covering less than 10% of the world's deserts combined with CCU technologies.

Another two aspects merit our attention. The access to raw materials is "democratised". In future, everybody, no matter where he or she is on the planet, will generally have the opportunity to harvest carbon using renewable energies and CCU technologies and to produce fuels, chemicals or plastics out of it. These products allow solar energy to be stored over a long period of time without any losses. This point is of particular importance, as carbon already accounts for approximately 60% of resources according to our own estimates; the trend is rising.

Even present analysis shows that the utilisation of CO/CO$_2$ in combination with renewable energies is the most sustainable path to fuels, chemicals and plastics and, from an ecological point of view, can explicitly and substantially contribute to a sustainable economy. Nine out of the 17 Sustainable Development Goals of the United Nations (Figure 2.2) are directly addressed through CO$_2$ utilisation in combination with renewable energies.

- #2 **Zero hunger** – CO$_2$ for proteins as an alternative protein supply, either for feed or even for food.
- #7 **Affordable and clean energy** – CO$_2$-based economy can help to facilitate an energy transition, e.g. when CCU fuels and chemicals provide clean energy storage (if based on renewable energy) that helps to balance renewable energy supply fluctuations and supports an expansion of renewable energy.
- #8 **Decent work and economic growth** – CO$_2$ utilisation can become one of the major growth areas in a low-carbon circular economy.
- #9 **Industry, innovation and infrastructure** – CO$_2$ utilisation is a growing industry field with enormous potential → New and innovative biotechnological and chemical ways of using CO$_2$ and the use of non-purified CO$_2$ will open the door for more applications.
- #10 **Reduced inequalities** – CO$_2$ is a ubiquitous resource, as this resource can be found everywhere; same accounts for renewable energy. An impact to reduce inequality will be determined by a cost-efficient exploitation of renewable energy and the development of know-how, such as efficient capture, catalysts and conversion. A supporting political framework is therefore essential.
- #11 **Sustainable cities and communities** – Local production possibilities of CO$_2$-based fuels and chemicals allow cities and communities to become more sustainable and independent.
- #12 **Responsible consumption and production** – CCU is based on reusing carbon in a circular economy, which aims to increase responsible consumption and production.
- #13 **Climate action** – CCU can contribute to decreased CO$_2$ emissions, by substituting fossil carbon in fuels and chemicals via recycling carbon from fossil and biogenic point sources or directly from the atmosphere.

– #15 **Life on land** – CCU requires much less space than the use of biomass as a renewable feedstock In contrast to biomass, CCU does not require arable land.

CO/CO_2 utilisation is in line with efficiency and consistency strategies, without however requiring a more stringent sufficiency strategy. Imminent raw material bottlenecks, known from fossil and bio-based systems, are largely a matter of the past, thanks to the utilisation of CO/CO_2 in combination with renewable energies.

2.4 CCU and Greenhous Gas emissions (GHG emissions)

But what about the concrete greenhouse gas balances of CO_2 utilisation in combination with renewable energies? Are there any figures available at this time?

Greenhouse gas balances are an essential component of comprehensive environmental footprint calculations, focusing on balancing gases that absorb and emit radiant energy, ultimately contributing to climate change. Comprehensive environmental footprint calculations include further impact categories such as energy requirements, eutrophication, acidification, photosmog potential and so on.

Over the past few years, experts have already determined the environmental footprint of CO_2-based methanol, kerosene and other fuels as well as that of CO_2-based plastic components such as polyols and polyurethanes. The results are promising, depending on their respective product or process path, GHG savings amounted to 20–90% compared to their fossil counterparts. Unfortunately, the results cannot be directly compared, as there are methodological differences between the individual environmental footprint calculations.

For years now there have been intensive discussions on how CCU processes should be treated methodically in LCA. Various international organisations and projects have published specific recommendations in late summer 2018 that recommend to always include "system expansion" as the preferred method of choice.[1] LCA standards such as ISO[2] and ILCD[3] deliberately leave considerable room for manoeuvre,

---

1 Global CO_2 Initiative: Techno-Economic Assessment & Life Cycle Assessment Guidelines for CO2 Utilization. Published August 2018, available under DOI: 10.3998/2027.42/145436, http://hdl.han dle.net/2027.42/145436, by CO2Chem Media and Publishing Ltd, ISBN 978-1-9164639-0-5.
2 ISO 14040:2006 – Environmental management – Life cycle assessment – Principles and framework (2006) and ISO 14044:2006 – Environmental management – Life cycle assessment – Requirements and guidelines by ISO.
3 JRC (2010): International Reference Life Cycle Data System (ILCD) Hand-book – General guide for Life Cycle Assessment – Detailed guidance. First edition March 2010. EUR 24708 EN. Luxembourg. Publications Office of the European Union; 2010.

depending primarily on the goal and scope. This means that the appropriate method has to be selected according to the question the LCA intends to answer.

When utilising CO$_2$, there are some particularities that lead to a reignition of the old discussion about choosing the correct method (IFEU[4]):
- CO$_2$ is both a raw material and an impact category.
- CO$_2$ can come from different sources, fossil or biogenic point sources, the atmosphere or even from natural gas (NG).
- The various methods have considerable implications for the CO$_2$ supplier and also for the CO$_2$ user. Depending on the method chosen, CO$_2$ use can become attractive or unattractive for the supplier or the user.

Table 2.2 below shows which methods are available in order to balance the ecological load that the CO$_2$ brings into the further utilisation process for different CO$_2$ sources.

In order to compare entire *production systems*, for example, comparing fossil-based traditional production with CO$_2$-based production, system expansion[5] is the ideal solution. With this method, you can identify whether a production system is better or worse for different impact categories without distortion through allocations or credits. Moreover, a cross-sectoral shift of burden is not possible here. In addition, it avoids the risk that each sector, supplier and user, will claim an arbitrarily calculated CCU bonus for itself. System expansion is particularly important on a political level, and the results are solid and hardly assailable.

However, when it comes to the balance of specific CO$_2$-based products and their comparison to other products, system expansion is of little help. In these cases, allocation methods for the CO$_2$ and subsequent processes have to be selected. There are several sensible methods and all of them have their specific advantages and disadvantages. As mentioned earlier, the choice of the method is a result of the goal and scope of the LCA and also of considerations such as data availability.
- Substitution and crediting: This method tries to break down the system expansion to the product level. In practice, this is difficult to achieve in complex systems. Especially assigning the credit for an equivalent product can involve considerable uncertainty. There is also uncertainty in the choice of allocation methods. Both are based on uncertain assumptions that strongly influence the results in each case
- Allocation: In principle, an economic allocation is always possible. For CO$_2$, both the ETS price and the market price (if the CO$_2$ has been purified accordingly) can be used. Allocation by mass or energy/exergy can be a good choice, but in some process chains these physical allocation methods make no sense (see table).

---

4 Fehrenbach et al. (2017): Bilanzierung von CO2 für Prozesse in der chemischen Industrie – eine methodische Handreichung. Ifeu – Institut für Energie- und Umweltforschung Heidelberg GmbH.
5 This means expanding the product system to include the additional functions related to the coproducts.

Table 2.2: LCA methodologies for different CO$_2$ feedstocks (nova-Institute 2018).

| CO$_2$ source | CO$_2$ as a product | | | | | CO$_2$ as waste | | |
| | System evaluation | Product evaluation | | | | | | |
	System expansion	Substitution/ credit	Allocation by mass	Allocation by energy/ exergy	Allocation by economy	Cut-off A (at point of separation)	Cut-off B (virtual uptake)	50/50
Direct air capture	–	–	–	–	–	–	–	–
Biogenic point sources (biogas)	✓	✓	✓	✓	✓	✓	✓	✓
Fossil point sources (coal power plant)	✓	✓	–	✓	✓	✓	(✓)	✓
Fossil point sources (chemistry)	✓	✓	✓	✓	✓	–*	–*	–*
Natural gas (water shift reaction)	✓	✓	✓	✓	✓	✓	(✓)	✓
Mineral processes (cement)	✓	✓	✓	–	✓	✓	–	✓
Waste (waste incineration)	✓	✓	–	✓	✓	✓	(✓)	✓
Syngas (steel industries)	✓	✓	✓	–	✓	✓	(✓)	✓

* The entire CO$_2$ currently used (e.g., for carbonating drinks, greenhouses, fire extinguishers) to a large extent comes from this process

- If CO$_2$ is regarded as a waste, a cut-off approach can be used in the same way as in recycling, that is, the impacts of the upstream processes will not be attributed to the waste. In the case of cut-off B, fossil CO$_2$ emissions stay with the CO$_2$ emitter; CO$_2$ using system, however, virtually takes up CO$_2$ from the atmosphere. This approach is fundamentally possible, but methodologically vulnerable, due to decoupling the calculated impacts from the actual CO$_2$ flows. As a consequence, subsequent products can be thought of as "greenwashed". For this reason, we recommend cut-off B only in combination with a preliminary test that compares the entire system via system expansion.
- CO$_2$ from the atmosphere (direct air capture) is a rather simple situation: here, the entire energy and material consumption can be directly and entirely assigned to CO$_2$ capture. Allocations, credits or cut-offs are therefore not applicable, as the process is mono-functional and fully intended to "produce" CO$_2$.

Worldwide efforts towards a uniform methodology for CCU are desired to evaluate the use of CO$_2$ in fuels, chemicals and materials in principle and to steer strategically in the right direction. However, methodical approaches should be used consciously, depending on politically and societal contexts. In this way, the necessary flexibility to tailor the LCA to a specific goal is maintained.

For example, consider the case of a coal-fired power station. If you capture the emissions of the power plant and convert them into fuels, is its fossil CO$_2$ footprint reduced to zero as the CO$_2$ emissions are captured and utilised and therefore not emitted? However, as a consequence the fuel made of the captured carbon gets the full CO$_2$ emissions assigned. Can one reasonably market such a fuel? Contrarily, if the CO$_2$ emissions were to remain with the power plant, the fuel would be almost CO$_2$ free – what is the benefit for the power plant to have the carbon captured then? Some environmental footprint calculations use system expansions to avoid allocation problems of that nature. Without digging too deep here, you can easily imagine the challenges that come with these new technologies.

Still, when taking a systemic perspective, CCU often leads to environmental benefits. Let us take a look at the coal power plant and synthetic natural gas (SNG) produced from CO$_2$. The following set of illustrations are designed to help explain and visualise the possible benefits of reusing CO$_2$. The figures are theoretical examples and model ideal conditions with a 100% process efficiency, that is, they contain no losses. This idealisation means that some other factors are not displayed in the illustrations, although they will have an impact in reality, for example, the energy requirements for separation and purification of CO$_2$.

Starting with Figure 2.3, a typical production and emission process of electricity generation from a NG-fired power plant and a separate fuel generation for ship transport based on NG are depicted. In this "business-as-usual" system, emissions for both the electricity generation and the burning of the ship fuel enter the atmosphere, and both the times, fossil carbon from NG has been utilised for energy generation.

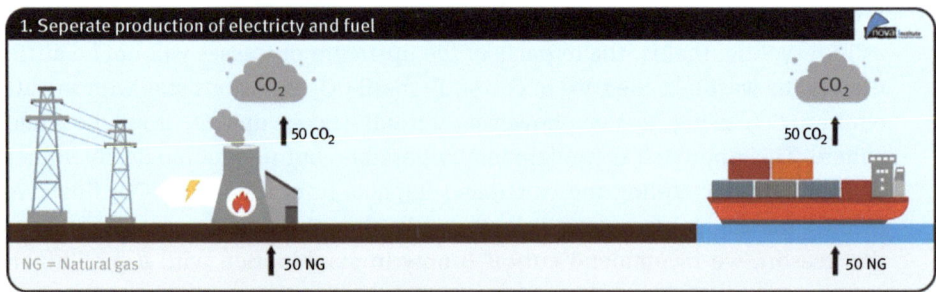

Figure 2.3: Independent production of electricity and fuel.

In Figure 2.4, the electricity production process has been coupled with a geo-storage system for the emissions of the NG plant (CCS). As a consequence of the CCS technology, the emissions of the plant are avoided and stored underground. At the same time, the ship fuel process remains unchanged. In total, and not considering emissions caused by separation and purification of the gas stream, the CCS process reduces the total GHG emissions to the atmosphere by 50%.

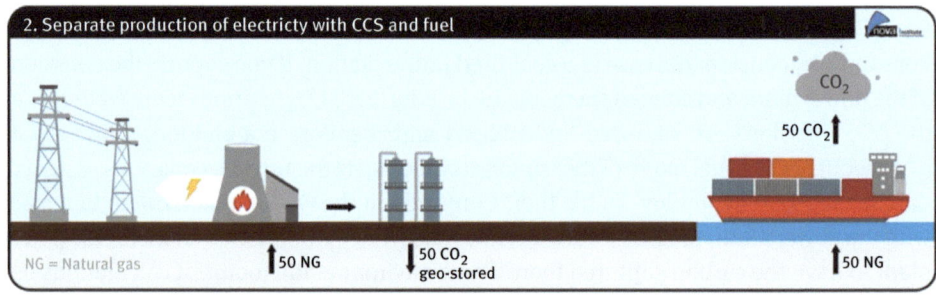

Figure 2.4: Separate production of electricity with CCS and fuel.

Figure 2.5 introduces a CCU process to the system. Here, the emissions from the power plant are captured and transferred to a fuel production plant, which produces SNG under the use of renewable energy. The resulting SNG can be used as a substitute for the NG required as a ship fuel. In the end, the CCU process reduces the overall GHG emissions by 50%.

In Figure 2.6, the same CCU process is used again. But this time, the electricity generation in the power plant is based on renewable biomass (biogas), which is assumed to regrow and bind the same amount of CO_2 that gets released in the process of the electricity generation. Again, the resulting emissions are captured and transferred to the CCU process, which generates SNG as a substitute for standard ship fuel. Factoring in the CO_2 that was taken up by the biomass, overall emissions are reduced by 100%. The amount that was released due to ship transport is covered by the biomass CO_2 uptake.

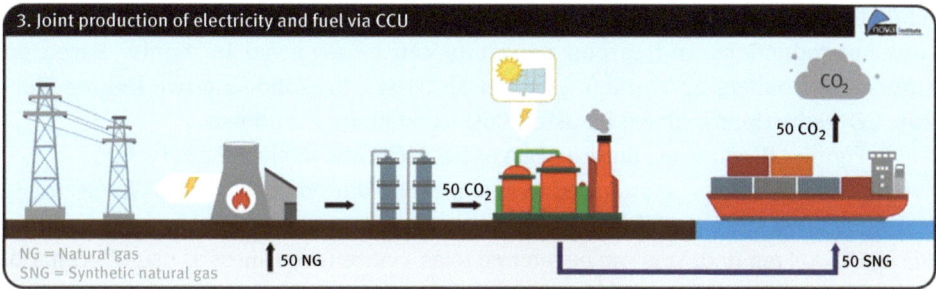

Figure 2.5: Joint production of electricity and fuel via CCU.

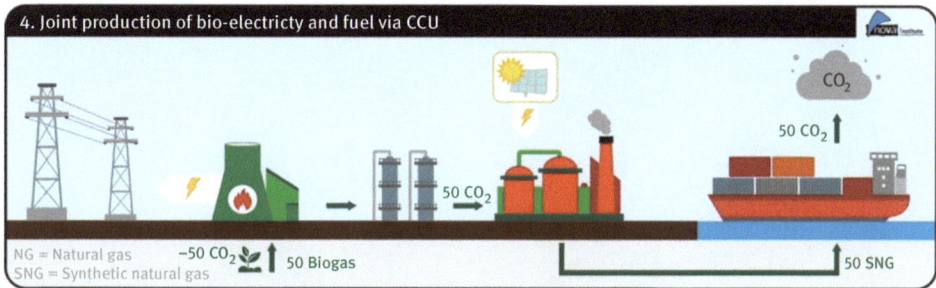

Figure 2.6: Joint production of bio-electricity and fuel via CCU.

Figure 2.7 finally replaces the renewable biomass with a direct air capture process to provide the CO$_2$ for the CCU process. Utilising renewable energy, the CCU process can produce SNG for both the electricity generation in the power plant and as a ship fuel. Providing full circularity, the emissions from the power plant are directly fed back into the CCU process, while the direct air capture covers the ship transport emissions. The resulting net balance of GHG emissions and CO$_2$ capture and utilisation is zero, indicating a 100% reduction in GHG emissions.

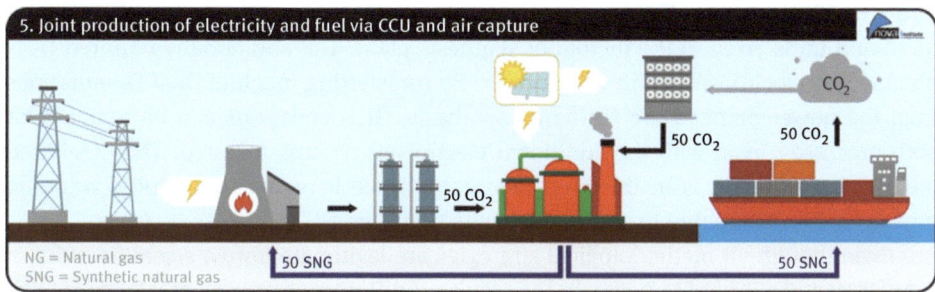

Figure 2.7: Joint production of electricity and fuel via CCU and air capture.

While the above figures model idealised conditions, it is necessary to verify that the emission reductions and carbon neutrality can be achieved in reality. Based on a number of papers by Van der Assen et al. (2014), the following two figures illustrate CO_2 reductions realised by using CCU for methanol synthesis.

In Figure 2.8, actual results of an LCA of electricity and methanol production through two separate processes is visualised. A coal power plant produces 1,273 kWh electricity and emits 1,090 kg CO_2, while for the synthesis of 1,000 kg of methanol from NG, another 745 kg CO_2 are emitted. This can be referred to as a current "business as usual" scenario, with methanol synthesis and electricity generation as two independent activities.

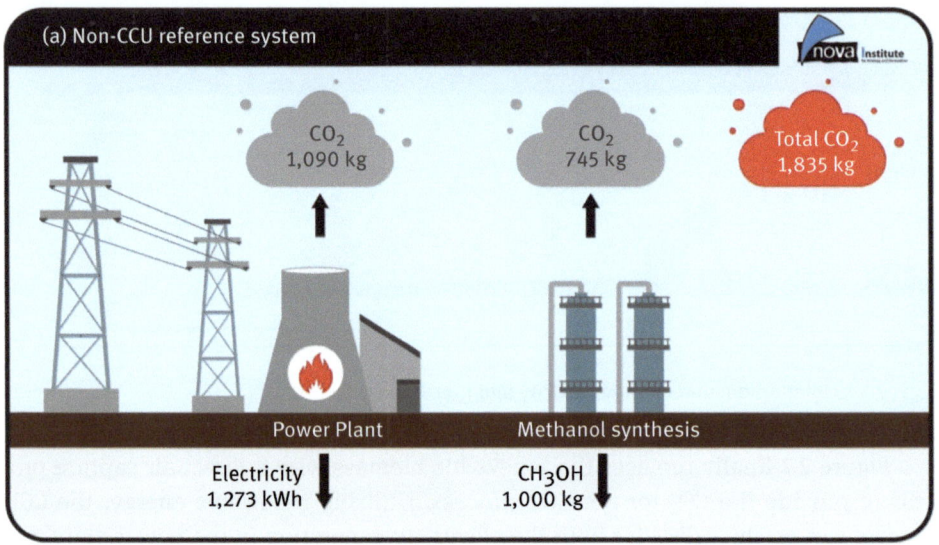

Figure 2.8: CO_2 emissions of a reference system producing electricity and methanol without using CCU technologies.

Figure 2.9 shows the greenhouse gas emissions results of an LCA for a system that combines the electricity generation of the power plant with the methanol synthesis through a CCU process. The gas containing emission stream of the power plant is captured and transferred to the methanol synthesis plant. The additionally required H_2 is provided by electrolysis using wind power. By transferring much of the CO_2 emissions from the power plant to the methanol synthesis, GHG emissions can be reduced for both processes, even with the additional electrolysis process required. The CO_2-based methanol production is in this case favourable since it reduces the global warming impact by 59% compared to methanol production from NG.

Even though all methodological strategies are legitimate from a scientific point of view, they make it hard to compare the results of different environmental footprint calculations. As a consequence, the percentage point savings listed in Table 2.3 may not be directly compared with one another.

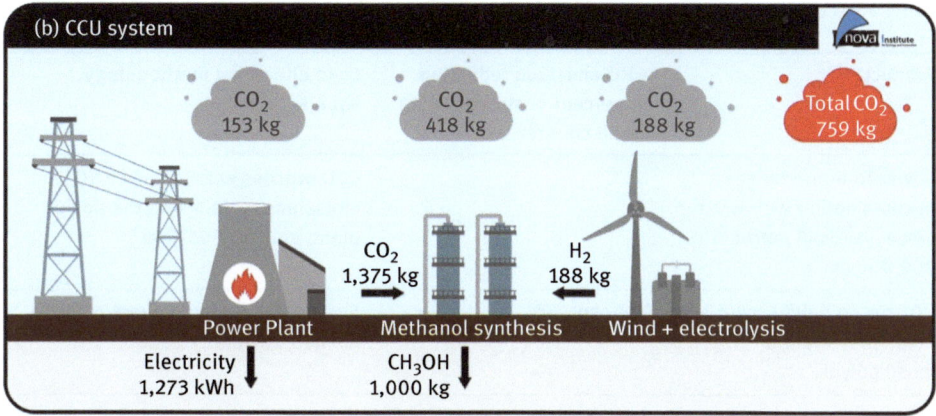

Figure 2.9: CO$_2$ emissions of the CCU system producing the same amount of electricity and methanol.

Table 2.3: Percentage point GHG reductions for different products on the basis of CO$_2$ in combination with renewable energies and on the basis of different methods and sources.

Product	GHG emission reduction in percent compared to its fossil counterpart	Load allocation methodology, source
Methanol derived from industrial CO$_2$ in combination with geothermal energy (Iceland)	90 (this value was never matched by any bio-fuel when the same calculation method was used)	Allocation: energy, methodology according to the EU Renewable Energy Directive (RED), CO$_2$ load-free, as it is an emission/waste, source: ISCC 2016
Kerosene via Fischer-Tropsch or methanol route	ca. 90	Allocation: energy, methodology according to the EU Renewable Energy Directive (RED), CO$_2$ load-free, as it is an emission/waste, source: DBFZ (German Biomass Research Centre) 2015
Power-to-liquid with air capture CO$_2$ vs fossil diesel	40–95% (depending on the power and heat source)	Allocation: energy, methodology according to the EU Renewable Energy Directive (RED), CO$_2$ load-free, as it is an emission/waste, but: all loads for the capturing/provision allocated to CO$_2$, source: Stuttgart University 2015

Table 2.3 (continued)

Product	GHG emission reduction in percent compared to its fossil counterpart	Load allocation methodology, source
Power-to-liquid biogas CO_2 in combination with wind power vs fossil petrol, after 200,000 km	85%	CCU process gets load-free CO_2, emissions remain with the power plant, source: Audi 2015
CO_2-based polyols with a 20 resp. 30% share of CO_2 vs fossil polyol	20 resp. 30%	System expansion, Covestro/RWTH Aachen University 2016
Polyproylene carbonate (PPC) vs PE and PET	50% (vs PE) and 60% (vs PET)	?, SK Innovation 2015 [7]
Methanol (CCU based on flue gas vs fossil)	58%	System expansion, Van der Assen et al. 2014 [9]

2.5 CO_2 utilisation and the circular economy

The utilisation of CO_2 is an integral component of an extended circular economy "where the value of products, materials and resources is maintained in the economy for as long as possible, and the generation of waste minimised, [...in order to...] develop a sustainable, low carbon, resource efficient and competitive economy" [8]. While the energy and traffic sector become more and more carbon-independent ("decarbonisation") and change over to electric power (or the direct use of solar heat), a few mobility applications will continue to depend on carbon-containing fuels (in particular the aviation industry with its kerosene-powered planes) as well as the entire organic chemicals and plastics industry, which will permanently depend on a renewable carbon source. This is why the circular use of carbon, for example, in the guise of CO_2, takes centre stage for a sustainable production of raw materials.

In a circular economy, carbon will be pulled from the atmosphere or from industrial emissions to be converted and utilised. At the end of the utilisation phase, it will be recycled and finally, at the end of the entire life cycle, it will be released into the atmosphere again. From there it may be re-used as a raw material, that is, it will be recycled.

Figure 2.10 shows a rendering of a comprehensive circular economy, which fully integrates various raw material, process, product and recycling paths as well as metals, minerals, fossil energy sources, biomass and also CO_2 as a raw material.

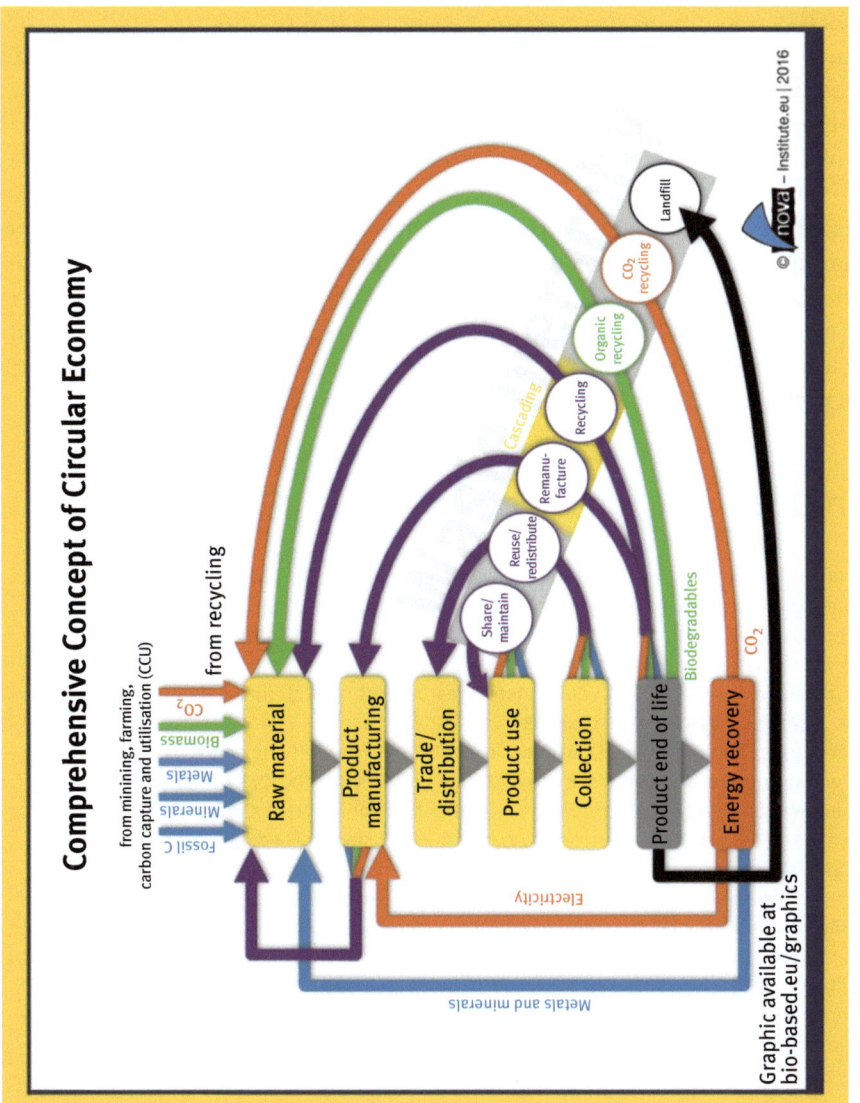

Figure 2.10: Incorporation of CO₂ within a circular economy.

From this resource perspective, we have basically three different kind of resources: (1) non-renewable resources such as minerals, metals and fossil C (2) renewable resources such as biomass and (3) CO_2 from air capture or off-gases. These resources are the building blocks of all products that we produce and consume. To efficiently utilise the resources, they should be kept as long as possible in the use. The waste hierarchy within the circular economy framework gives a clear hierarchy to keep resources in the economy with as little effort as possible. This can be obtained with sharing and reuse, aiming to replace products by services. If this is not feasible, material recycling or energy recycling can be a viable option. While energy recycling usually leads to a complete loss of resources, CCU can here complete the circular use by implementing a new short carbon cycle and is therefore a central element in the circular economy turning wastes into resources.

References

[1]   Original title in German: „Sylvicultura Oeconomica oder Anweisung zur wilden Baum-Zucht".
[2]   Translated from the German Edition, Carlowitz, Reprint from 1713, S. 150.
[3]   http://www.environmentandsociety.org/tools/keywords/hans-carl-von-carlowitz-and-sustainability (date accessed 28 November 2017)
[4]   Bund, Misereor (Hrsg) (1996): Zukunftsfähiges Deutschland, Berlin.
[5]   New Scientist, 14 October 2017; p. 41.
[6]   Searchinger, T.D., Beringer, T., Strong, A. 2017: Does the world have low-carbon bioenergy potential from the dedicated use of land? Energy Policy 110 (2017) 434–446.
[7]   SK Innovation (2012): Presentation „CO_2 Embedded Polyalkylene Carbonate GreenpolTM" by Myung-Ahn Ok at the Conference on CO_2: Carbon Dioxide as Feedstock for Chemistry and Polymers.
[8]   European commission 2015: Communication from the commission to the European parliament, the council, the European economic and social committee and the committee of the regions. Closing the loop – An EU action plan for the Circular Economy. COM/2015/0614 final, Brussels, 02.12.2015.
[9]   Von der Assen, N., Voll, P., Peters, M. and André Bardow, A.: Life cycle assessment of CO2 capture and utilization: a tutorial review. Chem. Soc. Rev. 2014 01 20.

Katy Armstrong

3 Communication regarding CO_2 utilisation

3.1 Introduction

The growth in the CO_2 utilisation technologies is primarily driven by industry and investors looking for new renewable feedstocks, in conjunction with searching for methods to reduce emissions. Many of the observed actors in the field are small and medium-sized enterprises (SMEs), often spun out of universities (Carbon8, Liquid Light, Novomer (now part of Aramco)) or large companies that have invested resources in the field to look for sustainable products (Bayer/Covestro, Audi, BASF). CO_2 utilisation technologies span a wide range of TRLs [1] with mineral carbonation, poly-urethanes and methanol amongst the most advanced technologies that have reached the market. A number of processes have reached small-scale commercial or demonstration phases in the last few years including synthetic diesel, dimethyl ether (DME), new CO_2-derived fertilisers and ethanol. The products that have reached commercialisation in some cases are exploiting specific, favourable market factors. For example, Carbon Recycling International use the abundant geothermal power in Iceland to manufacture methanol, and Carbon8 Aggregates utilise waste fly ash and air pollution control residues that attract gate fees for disposal.

The route to developing these new technologies to market involves a range of external stakeholders. Kant [2] identified seven different groups that can play a role: investors, employees, partners, competitors, customers, governments and society. Each of these groups brings different opportunities and barriers, and reflection on how to engage with them to achieve desired outcomes is advantageous. It cannot be assumed that the same strategy should be used on all groups, and therefore careful thought to each group's motivation should be given.

Once market ready, the success of a new technology or product is dependent on the desire of consumers to accept and purchase the product [3]. As such, the acceptance and awareness of products produced from non-conventional carbon sources such as CO_2 may be a key driver to commercial deployment. A few studies have begun to explore this field, for example, Jones et al. [4–6] and van Heek et al. [7];

Note: The work discussed in this chapter was part of the Horizon 2020 CarbonNext project, SPIRE5; GA no: 723678.

K Armstrong, A Bazzanella, H. Bolscher, S-K Hau, J van der Laan, D. Kramer, P. Sanderson, P. Styring and Elska Veenstra, www.carbonnext.eu (2018).

Katy Armstrong, UK Centre for Carbon Dioxide Utilisation, Chemical & Biological Engineering, The University of Sheffield, Sheffield S1 3JD, United Kingdom

https://doi.org/10.1515/9783110563191-003

however, much is still needed to be explored. There is a greater depth of research into the related areas of carbon capture and storage (CCS) technologies, renewable energy and the chemical industry in general, as these technologies/areas are more established. Hence, some lessons can be learnt from these fields. The importance of gauging public opinion should never be underestimated, nor the necessity to engage the public at an early stage to give better outcomes [4].

As CO_2 utilisation is an industrial process, research on the awareness and acceptance of the chemical industry can provide some insight on how the new technology area may be perceived. A number of studies have been conducted into the public perception of the chemical industry, for example, [8–10]. These studies highlight the attitude towards the industry in general and can help frame broad discussions. The lack of understanding around chemistry and chemicals can hinder discussing products and processes. The Royal Society of Chemistry [9, 11] found that 60% of the public agree that "everything is made of chemicals" and 19% of the public think that "all chemicals are dangerous and harmful". This may stem from the fact that the word "chemical" has multiple meanings, one of which is the association with dangerous or hazardous substances as well as in its true sense: the interactions of a substance as studied in chemistry. While the above views relate to the chemical industry as a whole, since CO_2 utilisation will be used to make products within this sector, such findings should be considered, especially as it is known that public awareness of CO_2 utilisation is low.

3.2 Key stakeholders

A variety of external stakeholders are interested in CO_2 utilisation. Key stakeholders can include policy makers, non-governmental organisations (NGOs), large and small companies, investors and the general public. It is vital to understand the background and motivation for a specific stakeholder's engagement with CO_2 utilisation, as this can help frame discussions and provide and understanding for their motivation. Key motivations can include reducing CO_2 emissions to the atmosphere, interest in buying greener products or new business opportunities that are perceived as being greener.

The motivation for policy makers, NGOs and large companies is often observed to be due to the use of CO_2 as feedstock and climate change mitigation. From a policy maker's perspective, CO_2 utilisation can often be compared and/or linked to CCS; however, this is often unhelpful from a stakeholder engagement perspective. This is because these are two distinct pathways, with differing potentials and purposes albeit both involving captured CO_2 [12]. Small companies and SME are observed to be often addressing a specific market opportunity or technology area such as the production of a specific chemical or integrating waste remediation and CO_2 utilisation. Their motivation can be based on exploiting research from academia to create a new successful business. Currently, engagement with the general public often takes the form of the media reporting CO_2 utilisation opportunities or through academic research to understand public perception of the technologies. Because of the range of differing motivations for those interested in CO_2 utilisation, it is advised that an assessment of interested stakeholder's motivation is conducted before engaging with communication strategies, so that tailored communications can be produced.

3.3 Awareness of CO_2 utilisation

As only a small number of products made from CO_2 are beginning to emerge on to the market, research into the public acceptance of these products is limited. Initial studies from the University of Sheffield [4, 6] focused on investigating perceptions across the whole field of CO_2 utilisation. The studies looked to understand views on perceived risks, benefits and the preference of specific technologies. These studies were helpful in understanding how people perceived the area of CO_2 utilisation as a whole (general positively) but raised concerns about long-term benefits and whether CO_2 utilisation just shifted the problem, allowing more CO_2 to be created just so it could be utilised. The studies highlighted the problems with discussing the whole field of CO_2 utilisation due to the diverse, complex nature of the field. In general, as was expected, the studies found that participants had limited knowledge of CO_2 utilisation and therefore explanations of the technology were necessary. This further highlighted that the topic is complex to explain and often left the participants asking for further detailed information or feeling that the information was biased towards positive views. The results gave insight on how "green" beliefs affected participants' perceptions of CO_2 utilisation when looking at the technology as a whole. This gave rise to the thought that presenting a single product made from CO_2 may have different acceptance outcomes as opposed to communicating the entire field. These initial studies found that people believed that CO_2 utilisation would have economic benefits but were wary of the long-term environmental benefits, generally thinking it was just shifting the problem. When asked to compare CO_2 utilisation technologies between methanol, cement production, plastic production, fuel production, enhanced oil recovery and a base-case of CCS, it was found that the participants preferred methanol production followed by cement production, with CCS as the least preferred option.

Van Heek et al. [7] conducted the first study into the perception of plastic products manufactured from CO_2. This study was one of the first to look at a single specific end product (CO_2 utilisation for mattresses; referred to as CCU plastics in the study) rather than to examine a range of CO_2 utilisation technologies. Therefore, insights into specific product acceptance can be derived. Their study was divided into two parts: the first looked at a number of acceptance factors and the second included perceived health complaints relating to the product. It was found, once the factor of health complaints was included, this had a significant negative impact on the acceptance of the product. Overall, respondents were quite positive to the idea of CCU mattresses. The key factor in acceptance was disposal conditions, followed by saving resources. If the product was perceived to have worse disposal conditions, that is, more emissions on disposal, the product was rejected. Therefore, the product was found to require at least as good disposal conditions to give acceptance. Saving resources, that is, carbon avoided by reducing fossil inputs, was found to be more influential than the amount of CO_2 contained in the mattress. In common with other research into CO_2 utilisation acceptance and awareness, it was found that perceived and real knowledge of CO_2 utilisation was low amongst participants, emphasising the need to communicate clearly and in a way that is seen as unbiased and avoiding "green washing". Van Heek et al. [7] conclude that:

> Considering the potential of CO_2-utilization for reduction of CO_2 emissions and fossil resource use, it will be vitally important to consult future users in CCU product development processes and to inform them about CCU in order to reach broader acceptance of CCU products. It should be considered that purely delivering information about CCU is not sufficient, because future users have to feel well informed and have to be able to rely on sources of information.

3.4 Key issues

There are a number of key issues that need to be considered when discussing the use of CO_2 as an alternative feedstock for the process industry. These key issues can affect the understanding of stakeholders regarding how CO_2 utilisation technologies work, their effects on the environment, resource efficiency and other aspects. By identifying common key issues and their possible effect (risk) on the acceptance of the sector, approaches for communication strategies to combat them can be suggested. In this section, nine key issues are highlighted and discussed and then a summary of the risk, the stakeholders involved and possible strategies to mitigate the risk are presented. At the end of the section, a simple communication guide for CO_2 utilisation products is included. This communication guide does not cover all aspects of CO_2 utilisation but aims to provide a basic starting point for developing clear communication with stakeholders.

Key issue 1: Lack of public knowledge on how CO_2 could be used

Primarily, it should not be underestimated that the public may not understand the basic principle that CO_2 is being used as a carbon source to create a new molecule and that the gas itself is not trapped inside the product waiting to leak out [13]. The public does in general understand the role of CO_2 as a "bad gas" and its contribution towards climate change, but does not generally understand how it could be used to make a product. This needs to be explained simply and clearly, that is, once CO_2 has reacted/used to make the product it is not CO_2 any more, it is a new molecule. CO_2 cannot just escape back out from the product without the product being combusted. Similarities could be drawn to photosynthesis as most people have knowledge that plants use CO_2 to grow. Using the example of a tree that uses CO_2 to grow, then it is released only when burnt to provide heat may help. This lack of knowledge of the conversion of CO_2 to make the product has led to perceived health concerns [7]. Combating this lack of knowledge that CO_2 is a being used as carbon source is the key in order to increasing understanding and acceptance.

Risk	CO_2 products are not accepted as the use of the "C" molecule is not understood
Suggested strategy	Explain that the produce is made from carbon from "waste CO_2" rather than using carbon from fossil fuels. Explain that the product directly replaces the conventional product it is just manufactured differently
Possible communicating stakeholders	Companies, researchers, NGOs
Possible addressees	General public, investors

Key issue 2: Comparing CCS and CO_2 utilisation

Comparisons between CCS and CO_2 utilisation are common. However, these should be considered as two separate but linked technologies [12]. Commonly, CCS and CO_2 utilisation are compared as they both involve capturing CO_2 (Figure 3.1). CCS is an emission-reduction technology to capture CO_2 from emitters and sequester it in geological formations for long time periods. CCS has the potential to sequester large quantities of CO_2, significantly contributing to emissions reduction targets. CO_2 utilisation, whilst also contributing to a reduction in emissions, is primarily seeking to find alternative sources of carbon as a feedstock for industry. In doing so, it is creating new sustainable and economic production pathways that avoid the use of fossil fuels. Both CCS and CO_2 utilisation have an important place in our low carbon future.

Currently there is a move towards combining the two technologies under the banner of CCUS: carbon capture, utilisation and storage. This approach has been adopted by the American Institute of Chemical Engineers, US Department of Energy, European Commission and Mission Innovation. Whilst the carbon capture stage of both technologies may be common, it is still important to identify the differences in approach that can be taken due to the final destination of the CO_2 (storage or use).

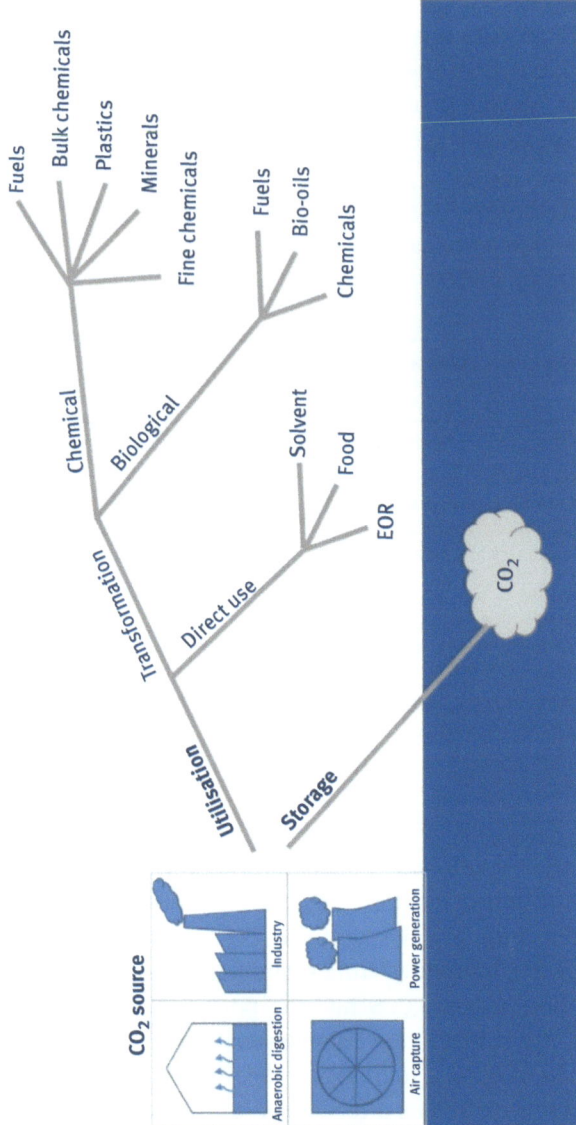

Figure 3.1: Example routes for CO_2 storage and utilisation. Reproduced with permissions from CO2Chem Media and Publishing.

Risk	CO_2 utilisation benefits are not fully understood. CO_2 utilisation is only compared with mitigation potential of CCS
Suggested strategy	Understand the motivation of the stakeholder you are communicating with. Explain the advantages of the CO_2 utilisation technology. Explain all benefits including emissions reduction, resource efficiency, circular economy and so on
Possible communicating stakeholders	Companies, researchers, NGOs
Possible addressees	Policy makers, NGOs, investors, general public

Key issue 3: Issues with presenting CO_2 utilisation as "the solution" for climate change

CO_2 utilisation is not "the" solution for climate change as no other single technology is. Nevertheless, it is one of a suite of technologies that can be employed to mitigate emissions. The IEA ETP[1] scenarios give many different pathways to decrease emissions and CO_2 utilisation is one option that could contribute by reducing the emissions from various chemical processes by switching to using CO_2 as feedstock. Best projections give the total amount of CO_2 that could be utilised as 1–7 GT per annum, but this reflects the amount of CO_2 used not the total avoided, which will be much less once the full life-cycle assessment (LCA) is taken into account (see Issue 5).

Risk	Overstating the potential of CO_2 utilisation technologies
Suggested strategy	CO_2 utilisation should be presented as part of the solution to mitigating CO_2 emissions. Ideally comprehensive LCA should be conducted to clarify emissions reduction before communication of potentials; otherwise, limitations of the claims should be presented
Possible communicating stakeholders	Companies, researchers, NGOs, policy makers
Possible addressees	Policy makers, researchers, NGOs, investors, general public, companies

---

1 https://www.iea.org/etp2017/summary/

Key issue 4: Danger of "green washing" – over-estimating the benefits of a product

The lack of transparency and comprehensive guidelines for CO_2 utilisation LCAs can lead to claims of "green washing". LCA will provide details of the environmental impacts of each technology, but inconsistency in the boundaries applied in each case can lead to varying results. A holistic approach that takes into consideration all inputs and energy sources is required. Information from reputable external organisations regarding health and environmental impact assessments may be beneficial to alleviate fears of "green washing" by the manufacturer by providing an independent opinion.

Risk	Lack of confidence in the potential of CO_2 utilisation due to perceived biased information
Suggested strategy	Conduct transparent, peer-reviewed life-cycle studies before claiming benefits of a product. Clearly state the limitations of the study and include all relevant environmental indicators not just global-warming impact. Using an external organisation to carry out the study can add credibility. All studies should be peer-reviewed and reviewers stated
Possible communicating stakeholders	Companies, researchers, NGOs
Possible addressees	Policy makers, NGOs, investors, general public, companies

Key issue 5: Confusion between CO_2 avoided, CO_2 used and re-release of CO_2

Different CO_2-based products have different lifetimes. These can span from hours, to days or even years depending on the use of the product. For mineralisation products, CO_2 can be thought of as permanently sequestered as the products end of life does not necessitate the re-release of CO_2.

For many other products, chemicals, fuels, polymers and so on, the end of life for the product often involves combustion or decomposition, which results in the release of CO_2. Therefore, the question arises "does creating the product from CO_2 actually make a difference to CO_2 emissions?" The amount of CO_2 used in the process is not the same as the amount reduced. All processes require energy and other inputs of one type or another and these have related emissions (even renewable energy has a carbon footprint). To calculate the emissions from the process, a complete LCA must be undertaken.

This can then be compared with the traditional production of the same material or the material the CO_2-based product is replacing to assess the amount of CO_2 that has been avoided. Figure 3.2 shows an example of how a CO_2-derived fuel could avoid emissions. Usually, the end-of-life process for a CO_2-based and traditionally produced product would be the same. It is just the production method that differs. Therefore, it can be concluded that if the CO_2-based product has a smaller carbon

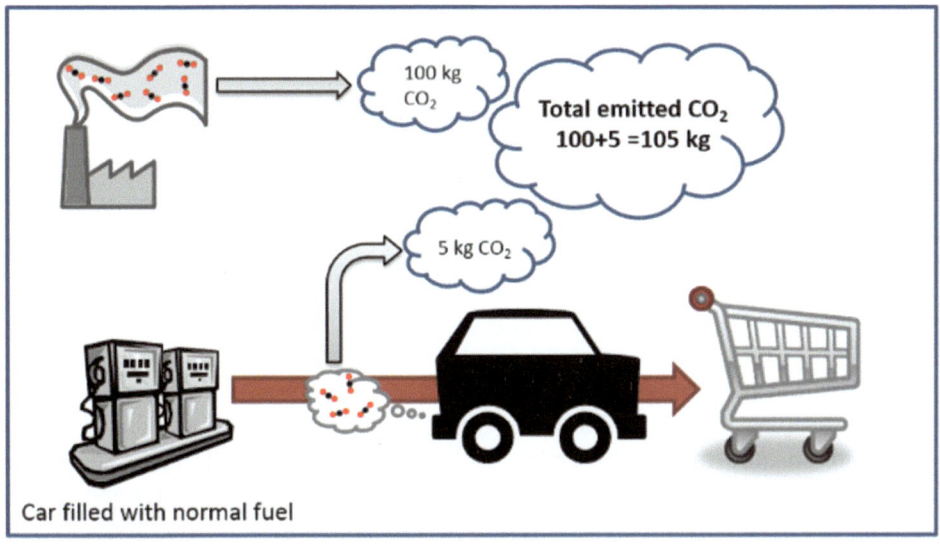

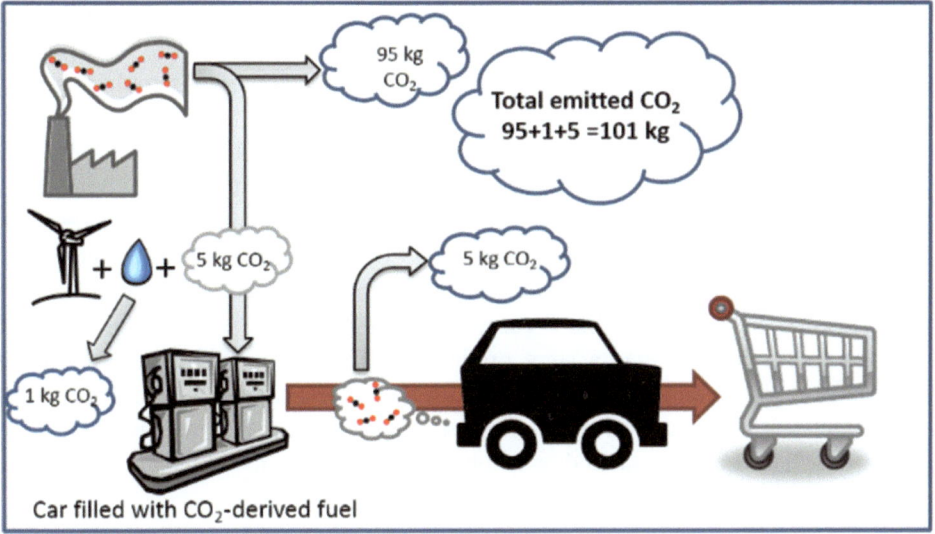

Figure 3.2: Simplified representation of avoided CO_2 in fuel use. This representation does not include the small amount of emissions resulting from carbon capture [14].

footprint to produce than the traditionally based product, it does help. An example of this is the Covestro Polyol Dream Process, which has a 15% lower carbon footprint.[2]

Risk	Misunderstanding over the CO_2 mitigation potential of a product
Suggested strategy	Conduct transparent, peer-reviewed life-cycle studies before claiming benefits of a product; using an external organisation for this can provide more credibility. Be careful with terminology such as "carbon negative". Be clear about what happens to the product at its end of life
Possible communicating stakeholders	Companies, researchers, NGOs
Possible addressees	Policy makers, NGOs, investors, general public

Key issue 6: Perception of using CO_2 utilisation as an excuse not to discontinue fossil fuels use

CO_2 should not be produced solely to be used in CO_2 utilisation, as this is contrary to the principles of sustainability and green chemistry [1]. We should be taking significant steps to reduce CO_2 emissions to meet and exceed the Paris Agreement targets (https://unfccc.int/process/the-paris-agreement/nationally-determined-contri butions/ndc-registry) from COP 21. However, we will always have to produce some CO_2, for example, in fermentation and chemical processes and thus CO_2 utilisation can be useful to reduce these emissions whilst making useful products. Technologies to capture CO_2 directly from the air before it is used to make products could also help in reducing atmospheric CO_2 levels.

The first research into the public acceptance of CO_2 utilisation conducted by Jones et al. [4] highlighted that the public feared CO_2 utilisation could be seen as preventing societal change to reduce CO_2 emissions. Participants in this research felt that if there was a recognised use for the CO_2 this would encourage people to continue with a wasteful and polluting lifestyle. It addressed the symptom not the root cause of the problem. This argument can be understood. It can be reasoned that the research participants were predominantly thinking about CO_2 produced from fossil fuels for energy uses for which there are low-carbon alternatives (wind, solar, nuclear), as the wide variety of emission sources is often not discussed. CO_2 utilisation should not be pitched as a solution to the continuous use of fossil fuels.

---

2 https://www.covestro.com/en/cardyon/overview

Switching from fossil fuels to renewable, low-carbon energy must be the primary aim. However, we need many carbon-based products, and utilising the CO_2 released from hard to decarbonise industrial processes to produce these is an option.

Risk	CO_2 utilisation is not accepted due to the perception that it inhibits creation of new renewable energy supplies and gives an excuse for producing CO_2
Suggested strategy	Explanation that there are many CO_2 sources, many of which especially industrial processes are hard to decarbonise and here CO_2 utilisation may be of benefit. Introduce discussion of direct air capture if necessary. Reference to IEA emission reduction diagrams may be useful to show where emissions originate and that there will still be emissions
Possible communicating stakeholders	Companies, researchers, NGOs
Possible addressees	Policy makers, NGOs, investors, general public

Key issue 7: Risk of underestimating the cost and thus speed of commercial uptake

CO_2 utilisation is a relatively new, emerging technology area, and as with any new technologies, initial costs are generally high and uptake is slow. Subsequently, an awareness regarding the realistic timescales of deployment and potential of the technologies is useful. In general, CO_2 utilisation products will directly replace products conventionally produced from fossil resources. These processes have been optimised over many years to maximise efficiency and revenue. New CO_2 products must be able to compete with them in the market. However, because of the interest in technologies that can help reduce CO_2 emissions and contribute to developing a sustainable, circular economy, there has been an observed interest from governments to understand the potential of these technologies. The Global CO_2 Initiative (GCI) [15] has assessed that at full deployment, five key products have the potential to utilise 7 gigatons of CO_2 per year, creating a market in excess of US $800 billion by 2030 (Figure 3.3). However, they state this would only be achievable in optimal conditions, which have been created by strategic actions in policy, research and markets. Without these strategic actions, GCI estimates that 1 gigaton of CO_2 can be utilised. Thus, it can be concluded that the speed of commercial uptake will be dependent on external factors providing mechanisms, which make CO_2 utilisation costs lower and hence increase commercial uptake.

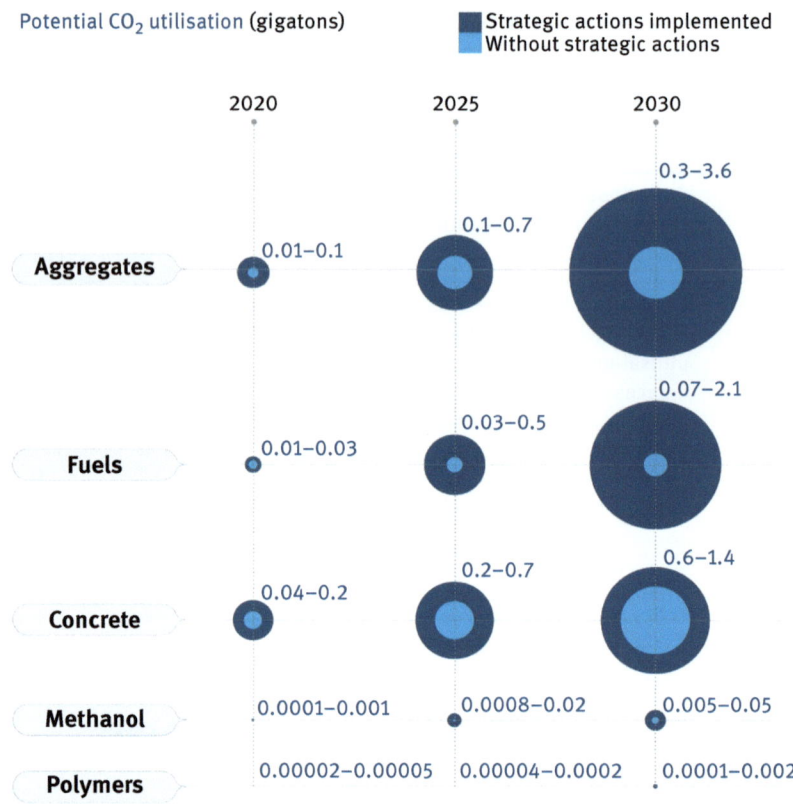

Figure 3.3: Potential increase in market size due to implementation of strategic actions. Global CO_2 Initiative [15].

Risk	CO_2 utilisation costs and the time needed for market penetration are underestimated. The speed of impact of technologies is exaggerated
Suggested strategy	Understand the motivation of the stakeholder you are communicating with. Techno-economic assessments (TEA) should be used to predict costs and future deployment scenarios. Use references to other new start-up green technologies and their speed of deployment. References to deployment scenarios for CO_2 utilisation should take into account all assumptions, market conditions and mechanisms employed to create the scenario
Possible communicating stakeholders	Companies, researchers, NGOs
Possible addressees	Policy makers, NGOs, investors, general public

Key issue 8: Lack of understanding regarding how the CO$_2$ product properties compare with normal product

There can be confusion about the differences between the CO$_2$-based product and the product it replaces. All similarities and differences should be clearly explained. If it is a direct like for like replacement it should be clearly stated, for example, methanol or DME. For polymer products, both Covestro (cardyon™) and Saudi Aramco (Converge™) report some properties of their products to be superior. This enables the consumer to make informed choices especially if the CO$_2$-based product has a higher price point. Here it is also important to explain any end-of-life disposal differences as highlighted by van Heek et al. [7], as it has been found to be a major contributor to acceptance of the product.

Risk	CO$_2$ utilisation products are not accepted and consumers not willing to purchase them
Suggested strategy	Clearly explain how the product is produced using carbon from "waste" CO$_2$ and that it directly replaces the conventional product. If it is chemically identical state so. State any superior properties to the conventional product. Use independent verification if necessary
Possible communicating stakeholders	Companies, researchers, NGOs
Possible addressees	Policy makers, NGOs, investors, general public

Key issue 9: Awareness regarding the integration with low-carbon, renewable energy

All CO$_2$ utilisation products should have a lower carbon footprint than the product that they are replacing, so they avoid CO$_2$ emissions. As we move towards a decarbonised electricity supply, most of these processes will need to utilise low-carbon or renewable energy and this potential dependency should be understood. Furthermore, the ability to seasonably store or utilise excess renewable energy at times of oversupply will be critical as the share of low-carbon renewable energy increases in the energy mix and the creation of CO$_2$ fuels or chemicals could provide one such method.

Using renewable energy for CO$_2$ utilisation processes needs careful consideration, and questions arise as to the best use of the available renewable energy. Although "guarantees of origin" can be provided to prove that renewable energy has been used, this may just cause renewable energy to be diverted from one application to another. Hence, these may not actually provide an emissions reduction when the whole system is taken into account. To ensure this does not happen, it may be necessary to create new sources of renewable energy or utilise "excess or surplus" renewable energy supply, which is produced in excess of consumer demand.

Risk	CO_2 utilisation is perceived as being in conflict with renewable energy deployment
Suggested strategy	Explain how CO_2 products could provide solutions for seasonal energy storage. Clearly state the energy sources used in production. If possible, LCA and TEA studies should be performed to show renewable energy impacts
Possible communicating stakeholders	Policy makers, companies, researchers, NGOs
Possible addressees	Policy makers, NGOs, investors, general public, companies

3.5 Simple guidance for communication of CO_2 utilisation

- Know your audience. Different audiences have different backgrounds and motivations for their interest in CO_2 utilisation.
- The motivation for CO_2 utilisation is not the same for each product. It is simpler to focus on specific products rather than the whole field of CO_2 utilisation.
- Explain that the product is made from carbon from "waste CO_2" rather than using carbon from fossil fuels.
- Explain how the product replaces the conventional product, but it is manufactured differently from CO_2 instead of fossil fuels. If it is chemically the same, explain this.
- Explain that in the manufacturing process, the carbon in CO_2 is used and is no longer CO_2.
- Explain the carbon footprint – this product has X% lower carbon emissions than the usual product across its whole life span. If necessary, include the end of life process (recycling, burning of fuels), how could CO_2 be re-released and whether this differs from the conventionally made product. It can be helpful to use a diagram such as Figure 3.4.
- Explain product properties – this product has the same/improved properties than the conventionally made product or alternative.

- CO_2 utilisation is not a replacement for CCS. Both can contribute to reducing CO_2 emissions.
- CO_2 utilisation focuses on adding value to CO_2 by using it as a carbon source to create new products.

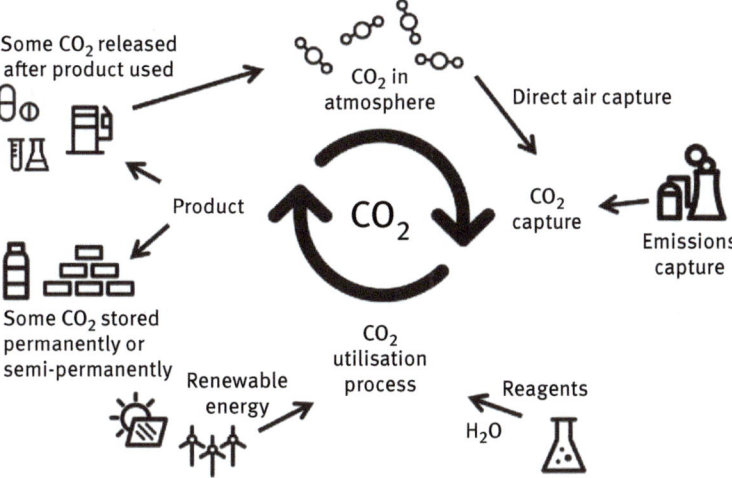

Some CO_2 released after product used

CO_2 in atmosphere

Direct air capture

Product

CO_2

CO_2 capture

Emissions capture

Some CO_2 stored permanently or semi-permanently

CO_2 utilisation process

Renewable energy

Reagents

H_2O

Figure 3.4: CO_2 utilisation cycle. Reproduced with permission from CO_2Chem Media and Publishing.

3.6 Conclusion

CO_2 utilisation is not a single simple process. It is a whole suite of technologies that utilise CO_2 as a resource to make new products. CO_2-derived products are diverse, as are the processes to make them. Therefore, creating an effective communication strategy is important and can present unforeseen issues. Research into understanding the acceptance and awareness of CO_2 utilisation process is still in its infancy, with only a few published studies available. As more studies are published and more CO_2-derived products are established in the market, both the understanding of how to develop effective communication strategies along with general awareness of the field will evolve.

It is recommended that when communicating about new CO_2 utilisation products, the interests and motivations of the stakeholders are carefully considered from the start. Furthermore, there are a range of considerations and misconceptions that should be taken into account when deciding on communication strategies for CO_2 utilisation. In general, it is simpler to convey a single product or product group rather than discuss the whole range of CO_2 utilisation technologies. Careful consideration should be given to explain that the product is made from carbon from CO_2, its properties and the amount of CO_2 emissions that are avoided by manufacturing the product from CO_2. By taking into account these considerations, it is hoped that commonly observed pitfalls can be avoided.

References

[1] Anastas, P.T. and Warner, J.C., 1998. Principles of green chemistry. Green chemistry: Theory and practice

[2] G. Wilson, Y. Travaly, T. Brun, H. Knippels, K. Armstrong, P. Styring, D. Krämer, G. Saussez, and H. Bolscher, *A Vision for Smart CO2 Transformation in Europe: Using CO2 as a resource.* 2015.

[3] M. Kant, "Overcoming Barriers to Successfully Commercializing Carbon Dioxide Utilization," *Front. Energy Res.*, vol. 5, no. September, p. 22, 2017.

[4] S. Ram and J. N. Sheth, "Consumer resistance to innovations: the marketing problem and its solutions," *J. Consum. Mark.*, vol. 6, no. 2, pp. 5–14, 1989.

[5] C. R. Jones, R. L. Radford, K. Armstrong, and P. Styring, "What a waste! Assessing public perceptions of Carbon Dioxide Utilisation technology," *J. CO2 Util.*, vol. 7, pp. 51–54, 2014.

[6] C. R. Jones, B. Olfe-Kräutlein, and D. Kaklamanou, "Lay perceptions of Carbon Dioxide Utilisation technologies in the United Kingdom and Germany: An exploratory qualitative interview study," *Energy Res. Soc. Sci.*, vol. 34, 2017.

[7] C. R. Jones, D. Kaklamanou, W. M. Stuttard, R. L. Radford, and J. Burley, "FDCDU15 - Investigating public perceptions of Carbon Dioxide Utilisation (CDU) technology: a mixed methods study," *Faraday Discuss.*, vol. 183, pp. 327–347, 2015.

[8] J. van Heek, K. Arning, and M. Ziefle, "Reduce, reuse, recycle: Acceptance of CO2-utilization for plastic products," *Energy Policy*, vol. 105, no. October 2016, pp. 53–66, 2017.

[9] S. Castell and M. Clemence, "Public Attitudes to Science 2014," 2014.

[10] TNS BMRB, "Public attitudes to chemistry," 2015.

[11] A. Emily and H. Ceng, "Public Attitudes towards the UK Oil, Gas and Chemical Industries IMechE Oil, Gas and Chemical Committee," no. September, 2016.

[12] TNS BMRB, "Public attitudes to chemistry Communication ToolKit," 2015.

[13] T. Bruhn, H. Naims, and B. Olfe-Krautlein, "Separating the debate on CO2 utilisation from carbon capture and storage," *Environ. Sci. Policy*, vol. 60, pp. 38–43, 2016.

[14] A. Zimmerman and M. Kant, *CO2 Utilisation Today*, no. September. 2017.

[15] K. Armstrong and P. Styring, "Assessing the Potential of Utilization and Storage Strategies for Post-Combustion CO2 Emissions Reduction," *Front. Energy Res.*, vol. 3, no. March, pp. 1–9, 2015.

[16] Global CO2 Initiative, "A Roadmap for the Global Implementation of Carbon Utilization Technologies," 2016.

Katy Armstrong, Peter Sanderson and Peter Styring

4 Promising CO_2 point sources for utilisation

4.1 Introduction

All CO_2 utilisation technologies require a source of CO_2 as a feedstock. Sources of CO_2 differ in purity, concentration and volume. Therefore, identifying potential sources and matching them with specific CO_2 utilisation technologies can be beneficial. CO_2 is formed from one atom of carbon covalently double-bonded to two atoms of oxygen and is naturally occurring in our atmosphere. CO_2 is produced in numerous ways including respiration, combustion of organic materials (including fossil fuels) and fermentation. CO_2 is a necessary part of the carbon cycle where plants use CO_2, light and water to create carbohydrate energy and oxygen; however, excess CO_2 contributes to global warming.

The carbon molecule in CO_2 can be used as a feedstock to create new valuable carbon-based products. As we move into an increasingly carbon constrained environment, the ability to re-use carbon molecules multiple times could become a key component in the drive to reduce carbon emissions and ensure the sustainability of the chemical industry. The identification of the most promising sources of these carbon emissions enables new and existing industries to identify symbiotic opportunities, which could enhance deployment.

Carbon capture aims to capture CO_2 from point sources or from the air using physical or chemical processes so that it can be stored or used. Different sources have different properties, which lead to differing ease and cost of capture. CO_2 capture is recognised as a key-enabling technology to reduce CO_2 emissions and hence there is a significant depth of research in the field. However, the majority of the research is focused on carbon capture for storage (CCS) not utilisation. In this regard, the volume of CO_2 that can be captured is the key driver, so that these emissions can be significantly sequestered. Hence, large-scale emitters such as power stations that individually can emit in excess of 20 Mt CO_2/yr have been a key research focus. CO_2 utilisation, however, has different priorities. The quantity of CO_2 that can be utilised varies between different applications. Therefore, a variety of sources can be used depending on the specific application. In general, processes that produce

Notes: The work contained in this chapter was part of the Horizon 2020 CarbonNext project, SPIRE5; GA no: 723678.

D. Kramer, K Armstrong, H. Bolscher, P. Sanderson, P. Styring, A Bazzanella, S-K Hau and Elska Veenstra, www.carbonnext.eu (2018).

Katy Armstrong, Peter Sanderson, Peter Styring, UK Centre for Carbon Dioxide Utilisation, Chemical & Biological Engineering, The University of Sheffield, Sheffield, United Kingdom

https://doi.org/10.1515/9783110563191-004

fuels not only require most CO_2 but also have the largest energy demand, whereas processes to produce pharmaceuticals and fine chemicals have a lower CO_2 demand. Matching supply and demand is key to ensure the economic viability of the process. Predicting the amount of CO_2 that can be utilised globally is difficult. Several studies have been conducted that give a range between 300 Mt/y in 2016 [1] and 7 Gt/y by 2030 [2]. However, the commonly accepted view is a range of 1.5–2 Gt/yr for future consumption [3–5]. As Europe's share of GDP is about 23% and its share of chemical production is 29%, it has been estimated that around 25% of global CO_2 utilisation could take place in Europe, that is, up to 500 Mt/yr [6].

A key target for future CO_2 utilisation processes is sourcing CO_2 from the atmosphere, known as direct air capture (DAC). DAC would provide a closed-loop cycle, where CO_2 would be used, then either sequestered in a long-term product or re-emitted after a product such as urea or methanol are used. This re-emitted CO_2 could then be captured again, providing an essentially closed loop. The significant advantage of DAC is that the capture unit can be sited at any location, and adjacent to the utilisation facility, negating the need for transport of the CO_2 feedstock. However, DAC can be an expensive capture mechanism as the concentration of CO_2 in the atmosphere is approximately 410 ppm (0.041%) and therefore large amounts of air must be processed to achieve the required amount of CO_2, and the sorbent material or process must be highly selective towards CO_2 over other gases. The majority of DAC technologies are currently in low- to mid-level technology readiness, with a few reaching small-scale deployment. However, costs will need to be significantly reduced before widespread deployment is observed.

4.2 CO_2 emissions

If current trends in greenhouse gas emissions continue, it is predicted that global temperatures will rise by between 3.7 °C and 4.8 °C above pre-industrial levels by 2100 [7]. There is a general agreement with IPCC views that we should be aiming to limit warming to a maximum of 1.5 °C. To reach this objective, annual CO_2 emissions need to reduce by 45% from 2010 levels by 2030 and achieve net zero emissions by 2050 [8]. There are several mechanisms needed to achieve this goal. Scenario modelling by the International Energy Agency gives a number of mitigation options, which are combined to reach the necessary targets. These include increasing renewable energy capacity, efficiency measures, expansion of nuclear energy generation and fitting carbon capture and storage units to existing emitters and bio-CCS. These must be deployed in increasing capacity to curb emissions. Another approach is to significantly curtail the use of fossil fuels, rapidly switching energy production to low-carbon sources. McGlade and Ekins [9] state that to give at least a 50% chance of a lower than 2 °C rise, over 80% of global current coal reserves, 50% of gas reserves and 33% of oil reserves must not be used. Either of these approaches to reduce greenhouse gas emissions necessitate a step-chance in technology and policy commitment to achieve them.

There are numerous point sources of CO_2 emissions, and targeting appropriate sources for CO_2 utilisation can help to reduce capture costs and influence location choices of CO_2 utilisation technologies. The global energy sector emits the most CO_2 (Figure 4.1). However, the energy industry may not be the most appropriate source of CO_2 when it is to be used as a feedstock for CO_2 utilisation processes due to the large quantity and varying composition of the emissions and relative lower concentration of CO_2 when compared with other sources. Most CO_2 utilisation processes would need only a small percentage of the total CO_2 emitted from the energy provider, which could be taken via a slipstream. However, this is not likely be the most economically viable route to sourcing CO_2 unless CCS and CCU technologies are integrated (i.e., CCS is deployed to decarbonise the energy provider and some captured CO_2 is diverted for use instead of storage) as other higher purity, lower volume sources are more likely to match the requirements for CCU.

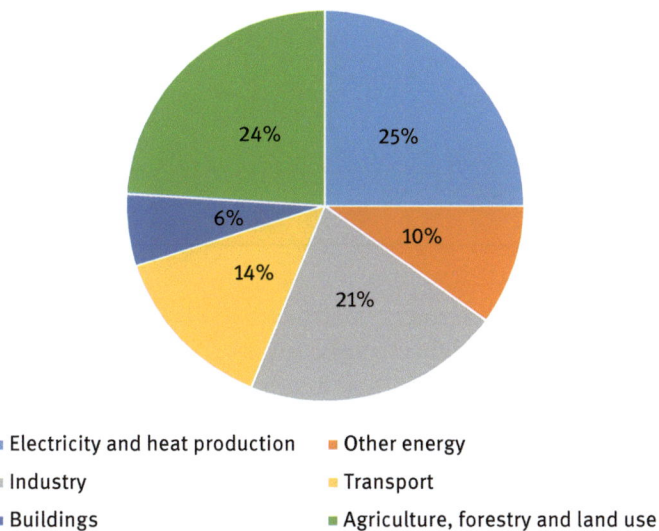

Figure 4.1: Global GHG Emissions by sector. IPCC 2014 [10].

Other large emission categories are the transport and agriculture, forestry and land use sectors. Many of these emissions originate from non-stationary sources: non-point sources can only be captured using technologies that take CO_2 directly from the air (called DAC) once the CO_2 has entered the atmosphere. Industrial processes accounted for 21% of emissions in 2014. In order to limit global warming to 1.5 °C, industrial emissions will need to decrease by 65–90% by 2050 [8]. Although many industrial processes strive for emission reductions through efficiency measures, the

IPCC has stated that efficiency will not be sufficient to meet required reductions [8]. Therefore, pathways to reduced emissions via alternative or new technologies are necessary such as electrification and CCUS. For a number of industries, emission reduction technologies such as CCU that also could have an economic benefit or lower economic penalty are being preferentially explored.

Industrial emissions come from a range of different processes. Table 4.1 lists the emissions from industrial sectors in Europe. The European Pollution Release and Transfer Register (E-PRTR) is a compulsory Europe-wide register of pollutants arising from industrial facilities. It contains data reported annually by more than 30,000 facilities within nine industrial sectors across Europe. A facility must report data annually to the E-PRTR if it exceeds certain thresholds – the threshold for CO_2 is 0.1 Mt/yr.

Table 4.1: Sources of CO_2 emission in the EU categorised by sector. Adapted from E-PRTR [12].

Sector	CO_2 emissions in Europe in 2014 [M tonnes]
Chemicals	245.1
Construction (including manufacture of cement)	144.1
Food and agriculture	5.9
Metal (including iron and steel industry)	166.0
Mining	7.1
Paper	77.1
Waste	55.5
Other	13.1

The three largest emitters of CO_2 are the chemical, construction and metals industries. These industries are identified as more difficult to decarbonise and the chemicals industry in particular is interested in the use of CO_2 as a carbon feedstock [11].

4.3 Identifying optimal sources of CO_2 for utilisation

Sources of CO_2 for CO_2 utilisation have been studied by Naims [13] and Von der Assen et al. [6]. The two works use different methodologies to select the most promising sources. The Naims' assessment was based on economics and the Von der Assen et al.'s study was based on an environmental merit order. Both studies used extensive literature searches to ascertain benchmarked data for best-practice scenarios for CO_2 emitters. Naims compared the cost of CO_2 captured and CO_2 avoided to create a merit order, whilst Von der Assen et al. created environmental merit order curves based on environmental impacts as defined by comparative life-cycle analysis (LCA) studies. The two assessments do overlap, although both do not cover exactly the same CO$_2$ sources. For example, the Naims study does not analyse retrofit post-combustion onto power generation, only pre-combustion, and neither does it consider fermentation processes within Europe.

Both studies conclude that the purest CO$_2$ sources should be targeted first:
- Hydrogen production
- Gas processing
- Ethylene oxide manufacture
- Ammonia production
- Bio-ethanol fermentation (assessed by Naims only and based on North America and Brazil)

Followed by subsequent targets of lower purity:
- Paper and pulp industry
- Integrated gasification combined cycle (IGCC)
- Iron and steel
- Cement

These conclusions are unsurprising since the major inhibiting factor in CO$_2$ capture is the energy required for the capture and separation processes. The energy needed will affect both the cost and environmental implications of the process. Therefore, targeting higher purity streams of CO$_2$ will keep energy requirements to a minimum as smaller volumes of emitted gas will need to be processed to result in the same volume of purified CO$_2$ when compared with a more dilute source. Naims concludes that for near-term scenarios, high-purity CO$_2$, which can be captured for low cost of approximately €33/tonne, should be sufficient. Von der Assen notes that increases in ethanol plants and biogas fermentation will lead to new relatively pure CO$_2$ sources, which will be environmentally beneficial from a capture perspective. Because of the lower CO$_2$ concentration in the emissions from the power sector, most of the largest emitters are not included in the primary target sources. The power sector, due to its large emission portfolio, may also be more suited to carbon capture and storage technologies, whereby larger volumes of CO$_2$ could be sequestered than could be dealt with in CO$_2$ utilisation processes.

4.4 Description of the most promising CO$_2$ sources

4.4.1 Steam Methane Reforming to produce Hydrogen

Steam methane reforming (SMR) is used to produce most of the current hydrogen supply. Methane (CH$_4$) is reacted with high-temperature steam at 700–1,000 °C under pressure using a catalyst to promote the reaction $CH_4 + H_2O \rightarrow CO + 3H_2$.

Subsequently, the water–gas shift reaction takes place to convert the produced carbon monoxide and remaining water to hydrogen and CO_2. The process results in high-purity hydrogen and CO_2 streams. SRM facilitates generally emit between 0.1–0.8 Mt of CO_2 per annum.

4.4.2 Natural gas processing

Natural gas does not have the purity needed for further processing when it is extracted. CO_2 and other acid gases such as H_2S must be removed from natural gas before it can be used. Typically, amine adsorption processes are used to remove the CO_2 leading to a high-purity CO_2 stream, which could be utilised to produce CO_2-derived products. Global emissions of CO_2 from natural gas processing that could be utilised are estimated to be around 50 Mt/year [14] and typical plant emissions range from 0.1 to 1 Mt/y.

4.4.3 Ethylene oxide production

Ethylene oxide is produced by the oxidation of ethylene and requires a silver catalyst to promote the reaction. Ethylene oxide is a used as an intermediate to produce many industrial chemicals including polymers and ethylene glycols. High-purity CO_2 is produced during the production of ethylene oxide that must be removed.

4.4.4 Ammonia production

Ammonia produced via the Haber process is a bulk chemical and predominantly used as a fertiliser, often further processing into urea. Production combines hydrogen (predominately from natural gas, CH_4) with nitrogen to produce ammonia, NH_3. CO_2 is produced during the production of the hydrogen. Ammonia production contributes around 1% of global greenhouse gas (GHG) emissions [14]. The production of urea from ammonia and CO_2 utilises more than 100 Mt/CO_2 per year [1]; however, there are still significant CO_2 emissions that are not utilised.

4.4.5 Paper pulp industry

The paper pulping industry is highly energy and raw material intensive and has high CO_2 emissions [15]. A JRC report [15] found that CCS would not be a cost-effective technology to deploy in the paper industry in Europe; however, bio-CCS may be an option. Subsequently the use of the CO_2 to generate income via CCU may be a future option. The locations of the paper pulp facilities may inhibit the use of the CO$_2$ in the process sector as they can be in areas close to raw materials (forests) but not to existing infrastructure for chemical plants. Therefore, cost of building the necessary infrastructure or transporting the CO$_2$ to the required location for use may be prohibitive.

4.4.6 Integrated gasification combined cycle

Integrated gasification combined cycle (IGCC) power plants use a gasifier to turn a carbon feedstock (usually coal) into synthesis gas (CO and H$_2$), which is then used in gas and steam turbines to produce electricity. IGCC can be combined with carbon capture technologies as the higher concentrations of CO$_2$ in the exhaust streams make capture easier than in traditional power plants where the CO$_2$ is more dilute.

4.4.7 Iron and steel production

The production of iron and steel is highly energy intensive and hence emission levels are high. Steel production occurs in two stages: first iron production and then steel making. The iron production process produces the most emissions (70–80% of the total emissions), as iron ore is reduced to metallic iron, usually with coke. The sector is continuously looking to reduce emissions and therefore the utilisation of CO$_2$ is seen as a promising pathway. Many facilities are located near to chemical parks and hence the required infrastructure and expertise for utilising the CO$_2$ in the process industry is available.

4.4.8 Cement industry

Cement production contributes to 5–7% of global GHG emissions, with every ton of cement producing around 900kg CO$_2$ [16]. The production of cement is highly energy intensive and CO$_2$ emissions occur in predominantly two areas of the cement-making process. Emissions arise from limestone (calcium carbonate) being calcined to produce calcium oxide and the kilns require heating necessitating the burning of fossil fuels. The cement industry is deploying efficiency measures to reduce emissions, but will need carbon capture and utilisation/storage to decarbonised completely.

4.5 Identifying promising CO$_2$ sources in Europe

We will consider CO$_2$ utilisation technologies of considerable interest in Europe both from an emissions reduction and circular economy perspective [5, 17–19].

Hence, Europe is used here as an example of how promising opportunities for CO_2 utilisation can be identified. The use of CO_2 as a feedstock in the circular economy necessitates identifying the amount of CO_2 available, the locations of emitters and the surrounding infrastructure. CO_2 emitters in Europe must publish data on their emissions if they emit more than 0.1 Mt/yr; this data is gathered in the E-PRTR.[1] By utilising this data, mapping can be produced showing the location of emitters to chemical parks and other industry and hence identifying possible symbiotic opportunities. Figure 4.2 shows the location of emitters over 0.1 Mt/yr across Europe and clusters of large emitters in northern Europe. Although this shows that there are numerous point sources, mapping and identifying the key sources of CO_2 identified above is more beneficial to gain insight to target locations for CCU technologies.

The utility of this data can be increased by limiting it to only those industries identified as key CO_2 sources for utilisation (Figure 4.3 and Table 4.2). These key sources total emissions of over 350 Mt CO_2/yr, which if utilised would represent a significant reduction in emissions. To enable industrial symbiosis, CO_2 sources located close to current process industries are highly desirable as this reduces the costs of transport pipelines and associated infrastructure. Therefore, Figure 4.3 also shows the proximity of these key emitters to key chemical parks. It can be observed from the mapping that the most prevalent source of CO_2 is the cement industry, as although the steel industry is a greater contributor to emissions there are fewer point sources. However, it is predicted that CO_2 capture from the cement industry will be more costly than other options; therefore, although less numerous, higher purity and cheaper capture cost sources should be targeted first.

4.6 Which sources will be available in the long term

It is expected that the main change in CO_2 availability will arise from reductions in emissions arising from coal power generation, which is expected to decrease significantly by 2100. Over the medium term, "clean coal" technologies such as IGCC or pressurised fluidised bed will improve combustion efficiencies and in the longer term there is expected to be a move away from coal altogether. The relatively ambitious IEA 2 °C scenario of the ETP2015 model foresees a reduction of coal as fuel input for electricity and heat generation from 33.8 EJ (1 EJ = 10^{18} J) in 2012 to 5.1 EJ in 2050, corresponding to a 85% reduction [20]. With around 46% of the global emissions arising from fossil fuel combustion currently coming from coal [21], reductions in coal use will impact CO_2 availability. However, the impact upon CO_2 utilisation may be limited, as CO_2 from this source is generally of low

---

1 https://prtr.eea.europa.eu/#/home.

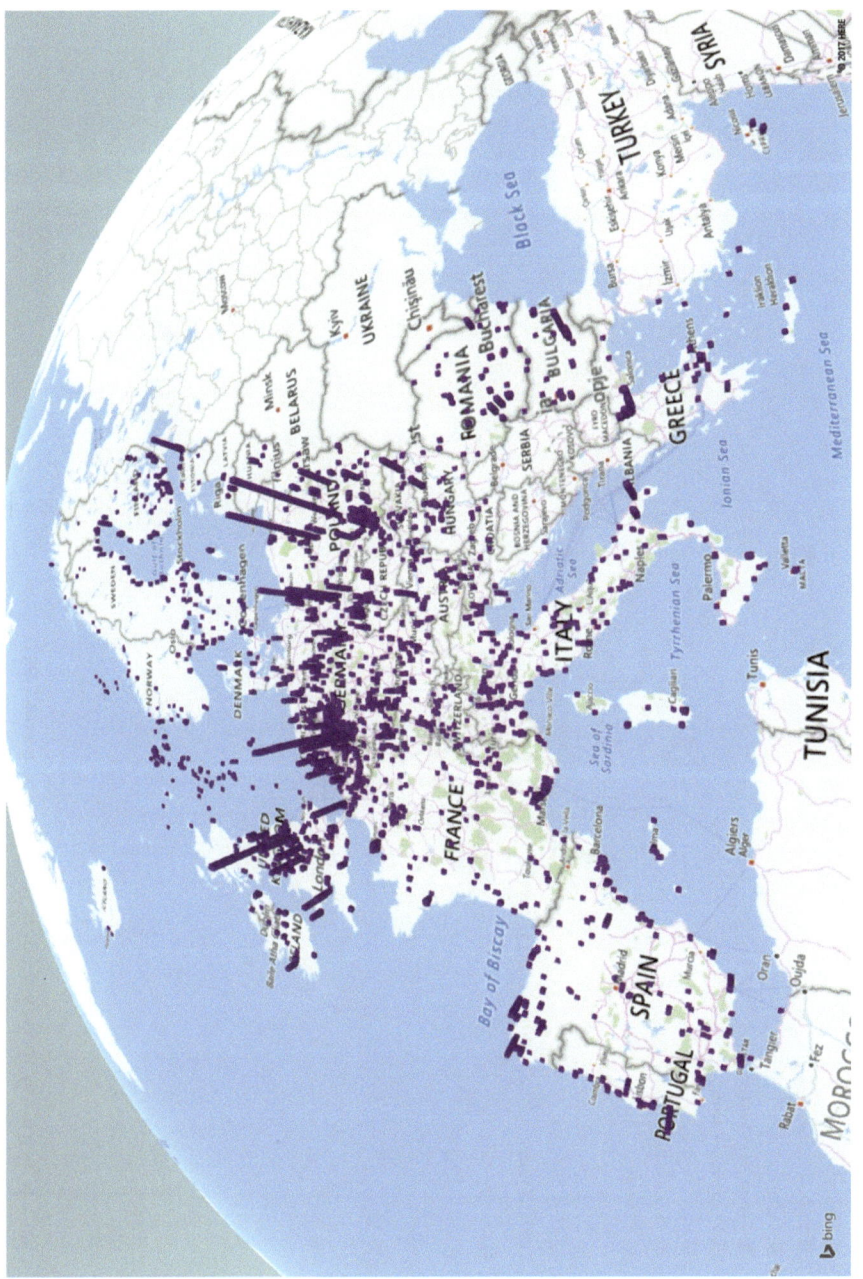

Figure 4.2: Map of location of sources of CO₂ emission in the EU. Adapted from E-PRTR.

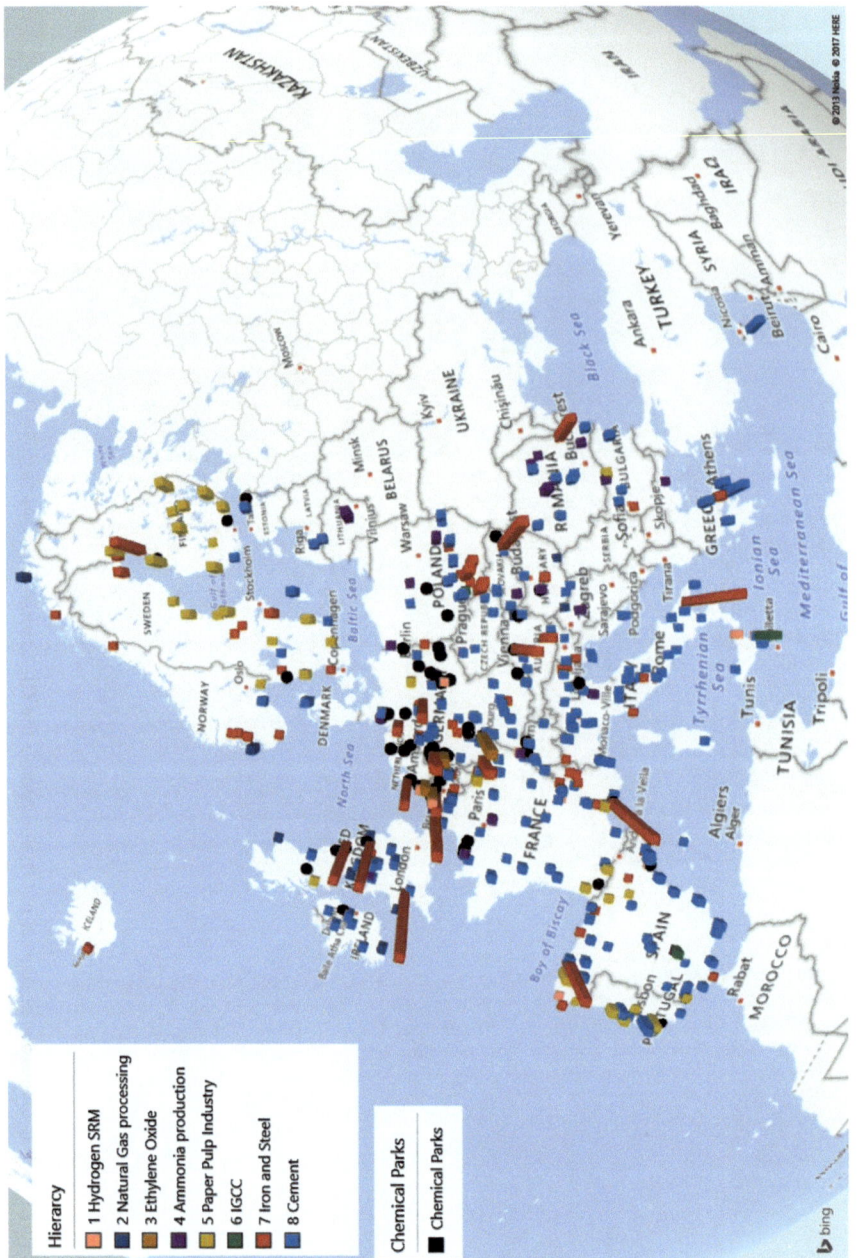

Figure 4.3: Map of key sources of CO$_2$ in Europe. Adapted from E-PRTR.

Table 4.2: Key sources of CO$_2$ in Europe. Adapted from E-PRTR and Naims, 2016.

CO$_2$ source	CO$_2$ concentration (%)	Emission per year (Mt CO$_2$/yr)	Cost (€/t CO$_2$)	Number of point sources emissions over 0.1 Mt/yr
Hydrogen production	70–100	5.3	30	15
Natural gas production	5–70	5.0	30	10
Ethylene oxide production	100	17.7	30	6
Ammonia production	100	22.6	33	27
Paper pulp industry	7–20	31.4	58	35
Coal to power (IGCC)	3–15	3.7	34	3
Iron and steel	17–35	151.3	40	93
Cement	14–33	119.4	68	212
Total		356.4		

concentration at 12–14% [14] and can be contaminated with sulphur and heavy metals such as mercury, making capture and purification (clean-up) more expensive. Consequently, it is expected that CO$_2$ arising from purer sources will be preferentially utilised as described previously. As a result of the 2018 Special Report from IPCC [8], the revised target of less than 1.5 °C temperature rise will require even more effort in terms of emissions reduction.

The report of the Energy Technology Transitions for Industry: Strategies for the next industrial revolution, published by the IEA in 2009 [22], looks at five industrial sectors: iron and steel; cement; chemicals and petrochemicals; pulp and paper and aluminium. It concludes that in order to reach a global emissions reduction of 50% by 2050, industry would need to reduce emissions by 21%, which assumes a near complete decarbonisation of the power sector. However, because of strong growth in demand, such reductions are not expected to be achieved by efficiencies and technology improvements alone. It is projected to be achieved only by including CO$_2$ capture within the adopted strategies, giving rise to possible opportunities for use of the CO$_2$ as a feedstock.

Technology changes will determine future industrial emissions and some industrial sectors will be able to reduce emissions more significantly than others. Around 96% of global H$_2$ production is currently from steam reforming of methane, oil-based or coal gasification [23], which result in CO$_2$ emissions that will only increase if projected increases in H$_2$ usage transpire. However, a switch to electrolysis of water using renewable energy will mean that CO$_2$ availability from this source will decrease significantly. Other low-carbon technologies, such as photocatalytic water-splitting or biohydrogen/fermentative production, are further from commercial reality. Currently H$_2$ usage is split roughly 50:50 between hydro-treating/hydro-cracking by refineries and ammonia/nitrogen-based fertiliser production by the chemical industries.

One major CO_2 source, natural gas processing, is expected to increase in the medium-term as power generation shifts away from coal and natural gas is used to balance intermittent renewable generation. Projections suggest that natural gas use will increase by 85% between 2007 and 2050 [24], and so CO_2 emissions arising from processing/cleaning the gas prior to its eventual combustion will rise.

4.7 Key challenges for CO_2 utilisation

There are a number of key challenges regarding the use of CO_2 as a feedstock for the process industry. The largest is creating economically viable processes that simultaneously have a positive environmental impact when compared to traditional production methods. Theoretically, CO_2 can be used as a carbon source in many process; however, whether the theory can be industrially deployed is a complex question. Comprehensive techno-economic analysis is required combined with robust LCA to ensure process viability. A key cost is the capture of the CO_2. As described above there are plentiful sources of CO_2, but the capture of CO_2 from these sources may not be economically viable using currently available technologies. Other key challenges include:
- matching volumes of CO_2 from emitters to technology solutions,
- reducing transportation distance,
- decreasing costs of air capture so CO_2 utilisation technology locations can be decoupled from CO_2 sources,
- decreasing the energy penalty of capture technologies,
- CO_2 storage for utilisation,
- standardisation of LCA to allow comparison between technologies for environmental impacts and CO_2 reduction,
- integration with mechanisms such as EU ETS and carbon taxes,
- if CCS deployment to the power industry takes place, putting mechanisms in place to utilise a proportion of the captured CO_2,
- emitters with no significant process industry in close proximity, should CO_2 be transported or new industry be encouraged to locate close to the emitter?
- the need to decarbonise carbon intensive industry may provide a push for deployment of CO_2 utilisation technologies.

4.8 How green is my carbon?

One aspect of CO_2 utilisation that needs to be addressed is the environmental credentials of the carbon in recycled CO_2. It has been argued that any product using captured CO_2 from a process that uses fossil-based fuels is in fact using fossil-based CO_2. This is an interesting argument where one needs to consider the fate of the

CO$_2$. If we burn gas, one molecule of methane becomes one molecule of CO$_2$. Both contain the same carbon atom that was derived from a fossil resource. We could capture that CO$_2$ molecule and convert it into methane again through hydrogenation (Figure 4.4). The likelihood at this time is that the hydrogen also came from a fossil-oil resource through SMR. Thus, is the new methane molecule derived from fossil oil? Indirectly yes, it is. Now let us consider what happens if the CO$_2$ molecule is released to the atmosphere. It could persist or it could react in a photosynthetic process to produce carbohydrates and oxygen. The carbohydrates produced by the plant contains the same carbon atom that was emitted from the combustion of fossil methane. Using the analogy of the synthetic methane, the carbohydrate produced, while now biomass, should also be considered to be derived from a fossil resource (Figure 4.5).

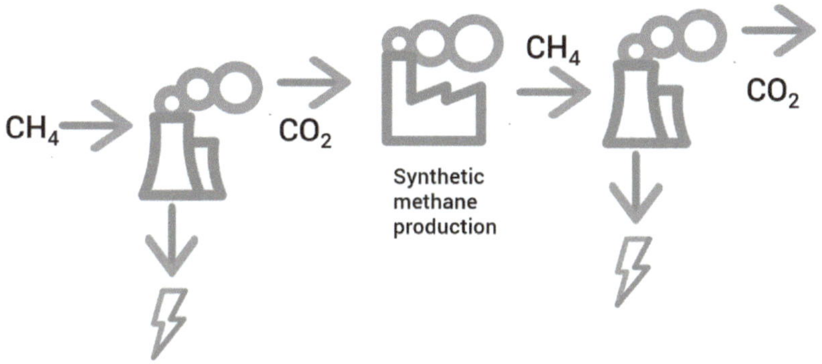

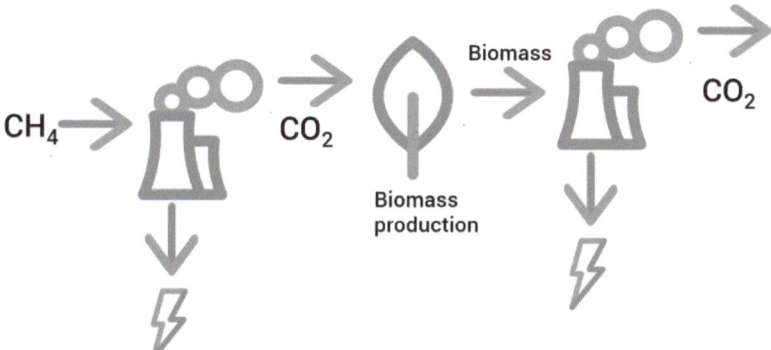

Figure 4.4: Following the carbon atom in producing synthetic methane from fossil methane.

Figure 4.5: Following the carbon atom in producing biomass from fossil methane.

Both the synthetic methane and the carbohydrate contain a carbon that was initially present in a fossil resource. However, both are now second-life products. When these are consumed by combustion, the carbon is again released as CO$_2$. Should this be considered to be fossil carbon? The answer lies in the carbon cycle. Fossil oil and gas was initially plant and animal life that over millions of years reacted to become hydrocarbons. There is always an interchange between the biosphere and the "hydrocarbonsphere". The difference is that while natural fossilisation occurred quite literally over geological timescales, the chemical and catalytic reduction of CO_2 to hydrocarbons takes place over minutes or hours. Science is being used to accelerate the carbon cycle. As the Lansink waste hierarchy proposes, if we cannot avoid waste, we should recycle and reuse the waste, in this case CO_2, to produce next-generation products that do not rely on primary fossil-oil resources. Sourcing CO_2 emissions to use in product manufacture is therefore essential.

4.9 Conclusions: Future outlook and potential impact

There are plentiful sources of CO_2 that could be used as a carbon feedstock. Primary targets for sourcing CO_2 should focus on those sources with the highest concentration of CO_2 (hydrogen production, natural gas processing, ethylene oxide manufacture and ammonia production) as the higher concentration of CO_2 reduces the cost of capture. However, larger volumes of CO_2 are available from the iron and steel industry and cement industries, albeit at lower CO_2 concentration. As industries look to decarbonise (particularly the iron and steel and cement sectors), there is an observed market pull to deploy CO_2 utilisation technologies to provide an economically beneficial method of reducing CO_2 emissions. As next-generation carbon capture technologies reach the market, other sources of CO_2 may become increasingly economically viable.

This chapter has been adapted in part from the CarbonNext Project www.carbonnext.eu.

References

[1] M. Aresta, A. Dibenedetto, and A. Angelini, "The changing paradigm in CO2 utilization," *J. CO2 Util.*, vol. 3–4, pp. 65–73, Sep. 2013.
[2] Global CO_2 Initiative, "A Roadmap for the Global Implementation of Carbon Utilization Technologies," 2016.
[3] G. Centi and S. Perathoner, "CO2-based energy vectors for the storage of solar energy," *Greenh. Gas Sci. Technol.*, vol. 1, pp. 21–35, 2011.
[4] K. Armstrong and P. Styring, "Assessing the Potential of Utilization and Storage Strategies for Post-Combustion CO2 Emissions Reduction," *Front. Energy Res.*, vol. 3, no. March, pp. 1–9, 2015.
[5] VCI and DECHEMA, "Position Paper: Utilisation and Storage of CO2," 2009.

[6]   N. Von Der Assen, L. J. Muller, A. Steingrube, P. Voll, and A. Bardow, "Selecting CO2 Sources for CO2 Utilization by Environmental-Merit-Order Curves," *Environ. Sci. Technol.*, 2016.

[7]   Intergovernmental Panel on Climate Change, "Climate Change 2014 Synthesis Report Summary Chapter for Policymakers," *Ipcc*, 2014.

[8]   "IPCC – Intergovernmental Panel on Climate Change." [Online]. Available: https://report. ipcc.ch/sr15/index.html. [Accessed: 25-Jan-2019].

[9]   C. Mcglade and P. Ekins, "The geographical distribution of fossil fuels unused when limiting global warming to 2 °C," *Nature*, vol. 517, no. 7533, pp. 187–190, 2014.

[10]   T. F. Stocker, D. Qin, G. K. Plattner, M. M. B. Tignor, S. K. Allen, J. Boschung, A. Nauels, Y. Xia, V. Bex, and P. M. Midgley, *Climate change 2013 the physical science basis: Working Group I contribution to the fifth assessment report of the intergovernmental panel on climate change*, vol. 9781107057, no. 7. 2013.

[11]   The Royal Society, "The potential and limitations of using carbon dioxide," 2017.

[12]   "E-PRTR." [Online]. Available: https://prtr.eea.europa.eu/#/home. [Accessed: 20-Nov-2018].

[13]   H. Naims, "Economics of carbon dioxide capture and utilization—a supply and demand perspective," *Environ. Sci. Pollut. Res.*, vol. 23, no. 22, pp. 22226–22241, 2016.

[14]   L. Metz, B., Davidson, O., de Coninck, H., Loos, M., Meyer, *IPCC Special Report on Carbon dioxide capture and Storage*, no. October. 2005.

[15]   J. A. Moya and C. C. Pavel, *Energy efficiency and GHG emissions: Prospective scenarios for the pulp and paper industry*. 2018.

[16]   E. Benhelal, G. Zahedi, E. Shamsaei, and A. Bahadori, "Global strategies and potentials to curb CO2emissions in cement industry," *J. Clean. Prod.*, vol. 51, pp. 142–161, 2013.

[17]   G. Wilson, Y. Travaly, T. Brun, H. Knippels, K. Armstrong, P. Styring, D. Krämer, G. Saussez, and H. Bolscher, "A Vision for Smart CO2 Transformation in Europe: Using CO2 as a resource," 2015.

[18]   A. M. Bazzanella and F. Ausfelder, "Low carbon energy and feedstock for the European chemical industry."

[19]   A. Zimmerman and M. Kant, *CO2 Utilisation Today*, no. September. 2017.

[20]   IEA, "Energy Technology Perspectives (Executive Summary)," p. 14, 2015.

[21]   J. A. H. W. P. Jos G.J. Olivier, Greet Janssens-Maenhout, "Trends in global co2 emissions 2016," p. 40, 2016.

[22]   IEA, *Energy Technology Transitions for Industry*. 2009.

[23]   IEA (International Energy Agency), *ETP – Energy Technology Perspectives 2012*. 2012.

[24]   International Energy Agency, *Energy Technology Perspectives: Scenarios & Strategies to 2050*. 2010.

Katy Armstrong, Arno Zimmermann, Leonard Müller,
Johannes Wunderlich, Georg Bucher, Annika Marxen,
Stavros Michailos, Peter Sanderson, Stephen McCord,
Henriette Naims, André Bardow, Peter Styring
and Reinhard Schomäcker

5 Techno-economic assessment and life cycle assessment for CO_2 utilisation

This chapter is mainly based on the *Techno-Economic Assessment and Life Cycle Assessment Guidelines for CO₂ Utilisation* [1] written by the authors. This chapter provides a brief introduction to techno-economic assessment (TEA) and life cycle assessment (LCA) for CO_2 utilisation, and all topics are explained in further detail in the Guidelines mentioned above.

5.1 Introduction

Research into CO_2 utilisation has been growing in momentum as industry, academia and policy makers seek solutions to reduce their carbon emissions, create new circular economy business opportunities and look to new feedstocks for carbon-based materials. These increased levels of research and new opportunities lead to the necessity to evaluate proposed new technologies to assess their commercial and environmental viability. Informed decision-making raises questions such as how sustainable is the process, what are the profit margins, how can renewable energy be used and what are the environmental impacts? (Figure 5.1). To enable such decision making, a systematic evaluation is required and hence TEA and LCA have a key role to play in the development, deployment and commercialisation of CO_2 utilisation technologies.

TEA is a methodology framework to analyse the technical and economic performance of a process, product or service. LCA is a methodology to account for the environmental impacts of a product or service throughout its entire life cycle. For CO_2 utilisation, it is common that technological, economic and environmental factors are all driving influences in developing and deploying a new process. A CO_2-derived product can be desirable due to their ability to reduce environmental impacts, but simultaneously that product must be economically viable if it is to find a place on the market. Hence, it is common that both TEA and LCA are required and the results from one study can impact the other.

Methods to conduct a TEA and LCA for CO_2 utilisation can vary. LCAs are governed by International Standards ISO 14040 and ISO 14044; however, there is still flexibility as to how the LCA is conducted based on the goal of the assessment and

https://doi.org/10.1515/9783110563191-005

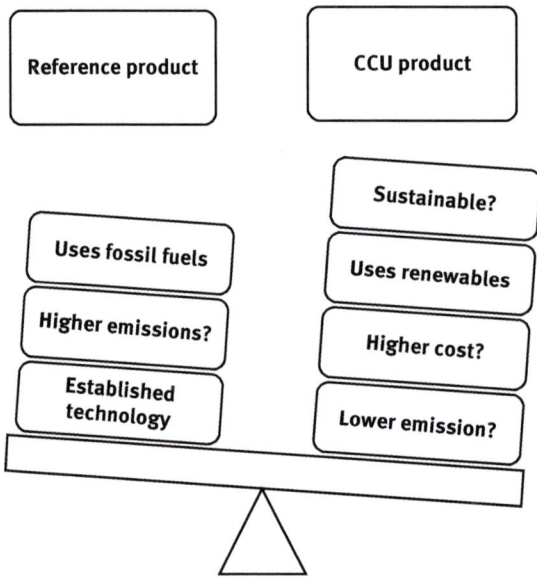

Figure 5.1: Possible factors in choosing CCU product or reference product.

the assessment indicators used. TEAs are not subject to international standards, but follow engineering conventions to assess the economic and technical viability of a proposed plant. The variation in methods applied can lead to issues with the comparability of studies and differing results for what appears to be similar products or even the same product. Therefore, it is essential that both TEA and LCA are conducted in a transparent manner to allow readers to understand the methodological choices that have been made and the scenarios that have been employed. An example of transparency in TEA could be reporting key performance indicators such as the energy inputs (kWh of electricity, GJ of natural gas) alongside the estimated utility cost (which will utilise sensitive data such as a current price in a specified location). This enables readers to draw conclusions based on the complete picture and compare results from different assessments.

CO_2 utilisation can produce a wide range of products, from fuels to construction materials, bulk agricultural chemicals to pharmaceuticals. Such products may be at varying levels of technology development and have different applications, but all have the similarity of using CO_2 as a feedstock. Because of the desire to reduce greenhouse gas emissions and find alternative carbon sources for the process industry, there is an observed increase in the interest in results from TEA and LCA of CO_2 utilisation processes. Policy makers, investors and funding bodies are seeking clarity on the benefits of implementing such technologies and require methods to compare them. TEA and LCA can provide data for such comparisons; however, care should be taken to ensure assessments are truly comparative, comparing like to like.

As a consequence, the CO_2 utilisation community have expressed a desire for guidelines to ensure consistent assessment approaches, to enhance comparability between studies and to clarify methodological choices. This chapter discusses such an approach, highlighting common pitfalls and suggesting approaches to combat them.

5.2 Techno-economic Assessment

TEA is used to determine the technological and economic performance of a system. TEA is a commonly used engineering method to evaluate a new product system and highlight any potential hotspots and pinch points. An example of a hotspot may be the cost of hydrogen production, which contributes to a significantly large proportion of the operating cost. An example pinch-point may be the recycling of a catalyst that results in a bottleneck in the process. A TEA will calculate a number of indicators that will enable decision makers to analyse a process system and determine recommendations. Common indicators include capital expenditure (CAPEX), operational expenditure (OPEX) and profitability indicators such as net present value and payback time. Indicators can be normalised to eliminate units of measurement so that multiple indicators can be compared, and their relationships explored.

TEA studies can be carried out using theoretical or real data. TEAs often feedback into process design and are therefore conducted in parallel to aid process developments. A TEA is conducted to aid and support decision making and therefore results are interpreted, but further decisions such as technology development or identifying a business case are carried out separately.

To conduct an accurate TEA, specific assumptions are required based on the context of the system. Therefore, it is usually necessary to assign the system a location and time horizon to enable data collection. TEAs can be carried out in a generalised context, but the result will only have a generic application as there is wide variations on input prices based on system location; for example, electricity prices vary greatly between countries.

5.3 Life cycle assessment

LCA is used to establish the environmental impacts of a product throughout its life cycle, which would typically extend from the extraction of raw materials to the disposal/end of life of the product. During this life cycle, reuse and recycling of the product can be accounted for.

The concept of LCA first started in the 1960s as methods of analysing resources and energy efficiency. The field was developed and standardised by the International

Standard Organisation (ISO) in 1996 and then further updated in 2016. The current standards are ISO 14040 [2] and 14044 [3].

The ISO standards divide an LCA into four main phases, as presented in Figure 5.2. The phases are interdependent of each other and are all based on the goal and scope set in the first phase. LCA is carried out iteratively and, hence, phases are revisited as data become available.

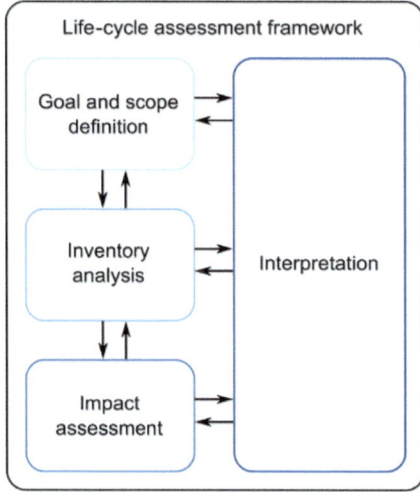

Figure 5.2: Phases of an LCA [2].

1) **Goal and scope definition.** Here, the aims of the study are stated along with the intended application and the target audience. The system is defined in the scope, including the system specifications, boundaries and any assumptions. The scope of the study should address the goal and any requirements for comparative assessment.

2) **Inventory analysis.** Here, data for the analysis are gathered, and the process is modelled. Flowcharts should be used, showing all flows into and out of the system.

3) **Impact assessment.** Here, the potential environmental impacts of the system are calculated using the data collected in the inventory. This is often done using computer software.

4) **Interpretation and presentation of results.** Here, the results of the impact assessment are evaluated and interpreted to reach conclusions based on the goal defined at the start of the assessment. This stage is iterative and sensitivity analysis can be applied to determine the significance of impacts on unit processes.

5.4 A common approach to TEA and LCA

As previously stated, TEA and LCA are both commonly used to assess the viability of CO_2 utilisation processes. The order in which TEA and LCA is carried out can change depending upon the requirements of the commissioner of the assessment. It can be desirable to conduct an early stage TEA to assess the commercial viability of a product; for other cases ensuring that environmental impacts, such as CO_2 emissions, are lower than the conventional route may be the starting point, as this can impact technology choices. TEA and LCA can also be carried out simultaneously and hotspots from the individual assessments used to optimise the process to the desired goal. In all cases, the quality and amount of data available for the process is a defining factor in the accuracy and depth of the analysis.

As both TEA and LCA are needed to be able to fully assess a CO_2 utilisation technology, it follows that a linked approach to assessment make sense [1]. The approach to conducting an LCA is defined by international standards ISO 14040 and 14044, which are further elucidated in many handbooks and methodologies, such as, ILCD Handbook [4] and ReCiPe [5]. TEA does not have such international standards but is well documented in chemical-engineering practice. By transferring some basic approaches from LCA methodology to TEA for CCU, a more cohesive approach can be created, which aids both the practitioner and the commissioner of the study; for instance, choosing a functional unit, setting system boundaries and using technology readiness levels (TRLs) in the same way in TEA and LCA. This approach further helps if results from either assessment are required to be integrated; for example, the cost of CO_2 abated can be calculated as a combined economic and environmental indicator. This approach is more comprehensive than that of life cycle costing as a wider variety of indicators can be considered.

5.5 Key aspects in defining goals and scopes for CCU

Common to both types of assessment, certain parameters for the study should be identified from the start. Studies require large amounts of data that can either be primary (data collected from the plant) or secondary (data from literature or simulations). In both cases, the scope and goal of the study determines the data required. Data collection is often the most time-consuming part of conducting a TEA or an LCA. Setting a clear goal and scope is covered in the *ILCD Handbook* [4] and further CO_2 utilisation specific guidance is given in the *Techno-Economic Assessment and Life Cycle Assessment Guidelines for CO_2 Utilisation* [1], but key aspects are described below.

5.5.1 Identifying system boundaries

The scope of the assessment study must be identified before data can be collected. Studies can be carried out covering different aspects of the product system. Complete boundaries for whole product system can be described as from "cradle" (raw material extraction) to "grave" (disposal/end of product life). However, not all studies are carried out from "cradle to grave", some stop or start at the "gate" (the factory gate) and only consider production processes inside the factory (Figure 5.3). In general, TEA studies are usually conducted "gate to gate", taking into account the costs of the raw materials but not the technical aspects of producing them. The boundaries of an LCA study should start at the cradle, so as to assess all the environmental impacts of the raw materials; however, the end boundary can vary depending on the use and composition of the product. If the CO_2 utilisation product has an identical chemical structure and composition to the conventional product it is being compared against, the study can be conducted cradle to gate as the usage and end-of-life impacts will be the same for both the conventional reference and CO_2 utilisation product. If the composition and structure is different, a cradle to grave study should be conducted. If a combined LCA and TEA is to be undertaken, the boundaries for each study should be identical to allow combined indicators to be calculated.

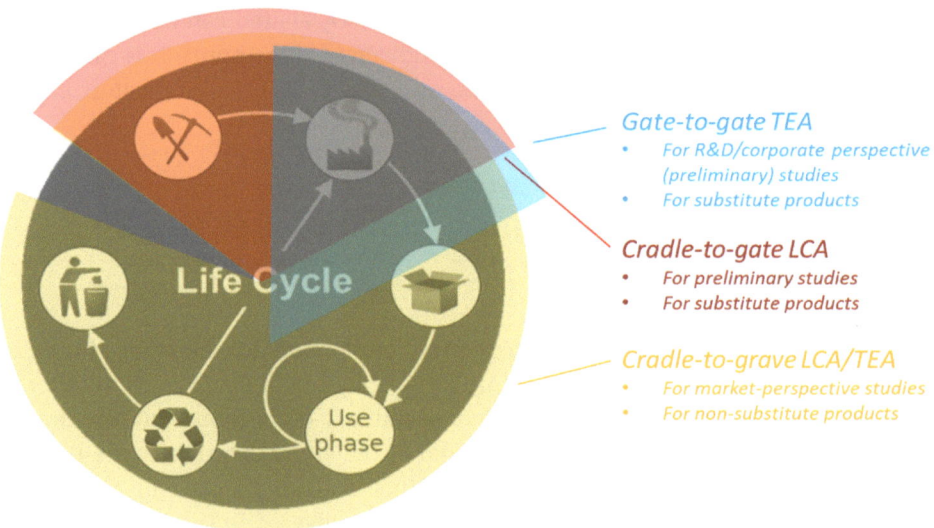

Figure 5.3: Boundaries of a system [1].

5.5.2 Identifying technology readiness

The level of detail to which a TEA or an LCA can be carried out is a factor of the accuracy and availability of data, which can be dependent on the maturity of the technology. All technologies progress in maturity from basic research through development to commercialisation. TRLs are a commonly used methodology to assess the maturity of a new technology and give a quick indication of a technology's standing on the development path to commercialisation. TRLs work on a scale from 1 to 9, through from basic research (TRL 1–3) to full commercial deployment (TRL 9), as seen in Table 5.1. There are a number of commonly used TRL concepts including those from NASA [6] and the European Commission H2020 [7]. Therefore, it is helpful to state which set of TRL definitions have been used in an LCA or a TEA as the concepts can vary slightly. However, in general, TRL 1–3 corresponds to the research phase, TRL 4–6 to the development phase and TRL 7–9 to the deployment phase.

Table 5.1: Technology readiness levels (TRLs) as defined by the European Commission Horizon 2020 funding program.

TRL	EC H2020
TRL 1	Basic principles observed
TRL 2	Technology concept formulated
TRL 3	Experimental proof of concept
TRL 4	Technology validated in lab
TRL 5	Technology validated in relevant environment (industrially relevant environment in the case of key enabling technologies)
TRL 6	Technology demonstrated in relevant environment (industrially relevant environment in the case of key enabling technologies)
TRL 7	System prototype demonstration in operational environment
TRL 8	System complete and qualified
TRL 9	Actual system proven in operational environment (competitive manufacturing in the case of key enabling technologies, or in space)

TEA and LCA can be carried out at any level during technologies development and hence identifying the TRL can be a helpful first step as this can indicate the types of data that can be obtained and the complexity/depth of analysis that can be undertaken, as well as the degree of certainty in the result. As the TRL increases so does the certainty in the data and therefore the resulting assessment (Figure 5.4).

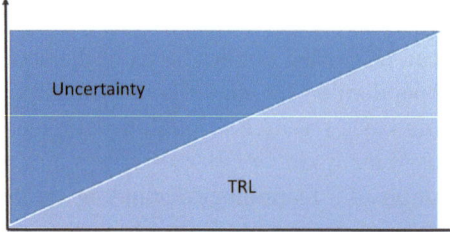

Figure 5.4: As TRL increases, uncertainty decreases.

Most CO_2 utilisation processes contain a number of unit processes that are joined together to create the entire utilisation process system. For example, a CO_2 methanol system can consist of carbon capture, hydrogen production and finally methanol production. Each of these unit process can be at a different maturity or TRL; therefore, it is useful to assess the maturity not only of the whole system but also of the unit process. When identifying the TRL for the whole system, the lowest TRL value for a unit process should be applied. In Figure 5.5, three unit processes make up the overall product system. As the lowest TRL (the CCU process) is TRL 5, the whole system should be described as being at this level of maturity or less.

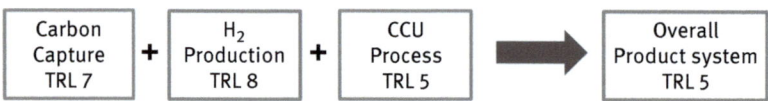

Figure 5.5: System maturities for a CO_2 utilisation product system.

For low TRL processes, it is most useful to use a TEA or an LCA as a hotspot analysis. The data used to perform the study at a low TRL are most likely to be generic (from a database) or from a process simulation. Therefore, a high degree of uncertainty may be present. Using the results of the TEA or LCA to identify hotspots in the process that are having a significant influence on the technical, economic or environmental performance of the system can guide research and aid future process design.

5.5.3 Defining a reference case

In the majority of cases, the assessment will be used to compare a CO_2 utilisation technology to another process making the same or a similar product. This is called the reference case. The reference case will need to be defined along with the CO_2

utilisation system. The reference case may be the current industry standard method of producing the product or an alternative route such as bio-based technologies. More than one comparison can also be carried out, that is, the standard route, a bio-based and CO_2 utilisation route can be compared.

The reference case and CCU system should have the same functions (see multifunctionality below), that is, if the CCU system is making product A and product B, the reference case should also make the same products. This is particularly important if the reference case should be sited in a geographically similar area and the assessment carried out using the same scenarios.

When comparing the reference case and the CO_2 utilisation system, it is necessary to take into account the TRLs of each system. Most reference cases will be of high maturity and been optimised over long time periods when compared to CO_2 utilisation systems that are at earlier stages of development and deployment. Comparing low-maturity systems to higher-maturity systems can lead to over- or under-estimation of some aspects, for example, energy consumption in low TRL systems, and therefore impacts can be over- or under-estimated. The practitioner should keep this in mind in the interpretation phase of the studies and make this clear in the assessment report.

5.5.4 Choosing functional units for CCU

The functional unit is the basis of comparison for the system and qualitatively and quantitatively describes the system. It is important that the correct functional unit is chosen for the assessment to ensure sound comparisons. For example, *decorative paint coating for a 1 m^2 wall to an opacity of 90%* or *1 tonne of methanol used as a chemical feedstock* is a more accurate description than *1 litre of paint* or *1 tonne of methanol*, as the former clearly describes how the product will be used.

CO_2 utilisation products can differ in chemical structure and composition but perform the same function as the reference product. Therefore, care must be taken to ensure a correct functional unit is derived. Figure 5.6 shows a decision tree for determining the functional unit for CO_2 utilisation process systems. Deriving the function unit depends on the application of the CO_2 utilisation product and the same product can have differing applications. For example, regarding CO_2-derived fuels, if the fuel has an identical chemical structure and composition, that is, fossil DME is replaced with CO_2-derived DME, the energy content (LHV) can be used as the functional unit. However, if CO_2-derived DME is being compared as a replacement to fossil diesel, the functional unit must be the energy service, that is, the energy needed to drive a specific car 1 km.

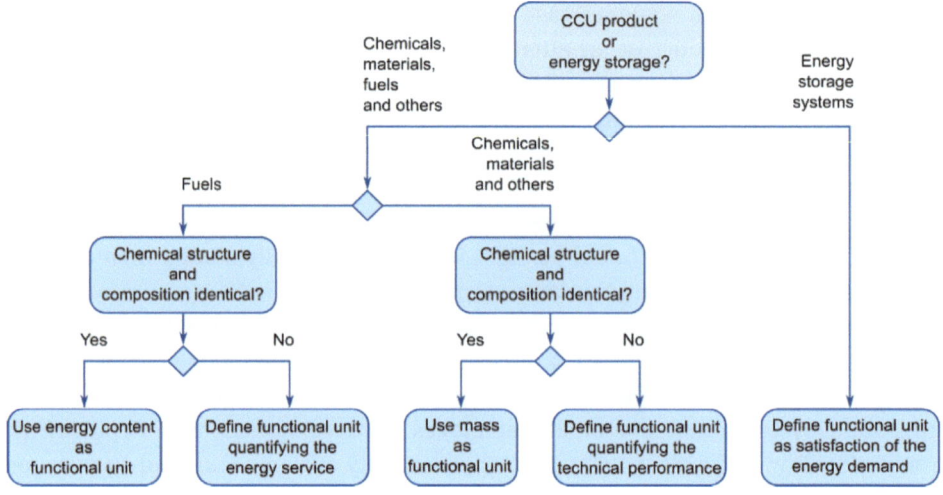

Figure 5.6: Decision tree for the selection of a suitable functional unit for CCU [1].

5.6 Pitfalls common to both TEA and LCA

There are a number of common pitfalls that have been identified in CO_2 utilisation TEA and LCA studies [1, 8]. Some are common to both types of assessment and others are usually TEA or LCA specific. A number of common pitfalls such as identifying TRLs, defining product systems, boundaries and functional units have already been described. Further common pitfalls are discussed below.

5.6.1 Energy sources and scenarios

Many CO_2 utilisation systems have considerable energy inputs. To decrease environmental impacts, renewable electricity sources such as wind or solar power are often used to provide the required energy. Consideration should be given to the quantity of renewable energy that is required for the system and as to whether it can be realistically provided at the system location. Similarly, the impact of the intermittency of renewable energy on operation of the CO_2 utilisation system should be considered and methods to account for intermittency reported. Future energy scenarios can also be used in the assessment to assess any differences in the impacts. The IEA in the Energy Perspectives Report [https://www.iea.org/etp/] provides electricity mix scenarios for the future and these can be used as the basis for future scenario models. Ideally, energy sources and scenarios should be specific to the location of the proposed facility, taking into account local energy provision and future predictions. However, this is not always possible and more generic scenarios can be used.

5.6.2 Data transparency

There is often the necessity to compare results between TEAs and LCAs and therefore understanding how the study has been conducted and where the data have come from (simulations, pilot plant, demonstration plant, etc.) is essential. However, often the data in published studies are limited and only the results are presented. To enhance transparency and comparability, it is recommended that the data sources, calculation methods and assumptions are presented in the report so that studies can be reproduced if necessary. If confidential data are used, relevant parts can be redacted to avoid confidentiality issues, but this should also be clearly documented.

5.6.3 Uncertainty and sensitivity analysis

It is important to state in both TEA and LCA the sources and certainty of the data used for the inventory and assessment. The quality of the data impacts the robustness, reliability and credibility of the results. Uncertainty in data decreases with rising TRL, as knowledge of system performance increases and access to real process data becomes possible. Uncertainty analysis allows the practitioner to assess the uncertainty of both the data, models, scenarios and context used in the assessment. Uncertainty analysis can quantify the practitioner's confidence in the assessment outcomes and therefore can be beneficial to decision making.

Sensitivity analysis allows the identification of the most influential variables upon the assessment results. By conducting a sensitivity analysis, the effect of varying specific parameters on calculated indicators or impacts can be observed. For example, if the cost of electricity decreases, the resultant impact on the price of the product can be analysed.

5.7 TEA specific pitfalls

5.7.1 Selecting CO_2 prices

All CO_2 utilisation processes require CO_2 as an input. It is essential to carefully estimate the cost of the CO_2 as this can be a major input to the process. Common pitfalls in selecting a CO_2 price are as follows:

- Assuming zero cost without specifying why
- Assuming an emissions trading price or tax as the cost
- Assuming the cost of CO_2 avoided is the same as CO_2 captured

CO_2 will always have a cost associated with its capture, processing and delivery to the point of use, although this may be minimal if direct flue gas can be used without further processing. The CO_2 price is determined by the product system and the system elements included inside the boundaries of the system. If the capture process is outside the system boundaries, a suitable market price for CO_2 can be assumed. If the capture process is inside the boundaries, the cost should be calculated from the capital and operational (CAPEX and OPEX) costs associated with the capture, transportation, purification and compression of the CO_2.

5.7.2 Incentive mechanisms

There are a number of incentive mechanisms that can apply to CO_2 utilisation processes, which can increase the profitability of the process. Such mechanisms can include carbon taxes, renewable fuels production incentives and waste disposal avoidance schemes. The use of such incentives should be clearly reported and used with caution. For example, in the case of a waste mineralisation process, a gate fee can apply for use of hazardous ashes from incineration, which would otherwise be disposed at a waste processing facility. These gate fees are a long-standing mechanism and are unlikely to change; therefore, they can be incorporated in the assessment. Other incentives such as a carbon tax or renewable fuel credit should only be applied at the current incentive level and if there is proof that the CO_2 utilisation product will qualify for such a scheme. Applying such incentives mechanism for future scenarios is complex as incentive schemes can alter based on political decisions. If an incentive mechanism is to be applied, it is always recommended that it is applied as a scenario, rather than in base case calculations for the CO_2 utilisation system and all assumptions about the application are clearly documented.

5.8 LCA specific pitfalls

5.8.1 Dealing with multi-functionality

Many CO_2 product systems have more than one function and are described as multi-functional. This is often due to the CO_2 source producing a main product, that is, electricity or ammonia and CO_2 as a by-product simultaneously. In many cases, CO_2 is an unwanted product of this system; however, when it is then used as a feedstock in a CO_2 utilisation process the impacts and products of both processes need to be taken into account. The CO_2 utilisation system will then fulfil two functions as it produces two products. It is therefore multi-functional. To be comparable, the reference system should include the same functions as the CO_2 utilisation system; otherwise, the two systems are not comparable (Figures 5.7 and 5.8). Multi-functionality is not specific to CCU but does occur frequently in CCU processes. The ILCD Handbook gives a hierarchy of methods to solve multi-functional issues such as systems expansion and sub-division that should be followed.

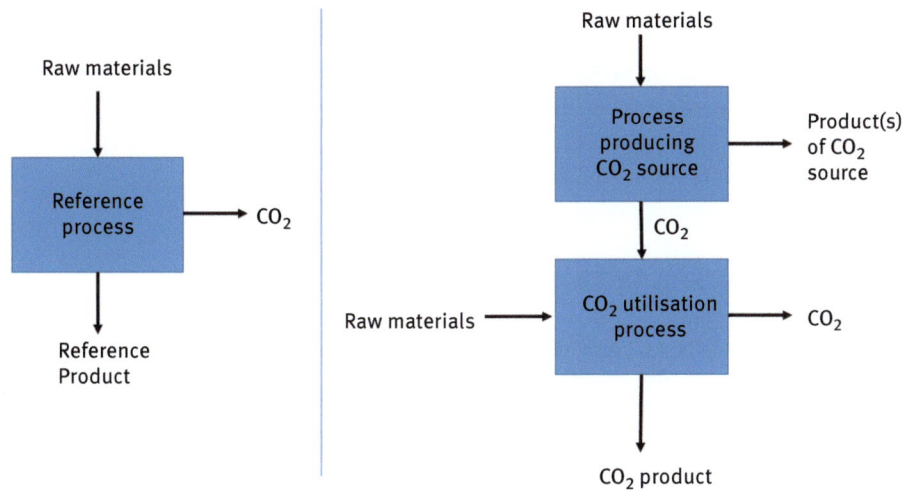

Figure 5.7: Systems with non-comparable functions.

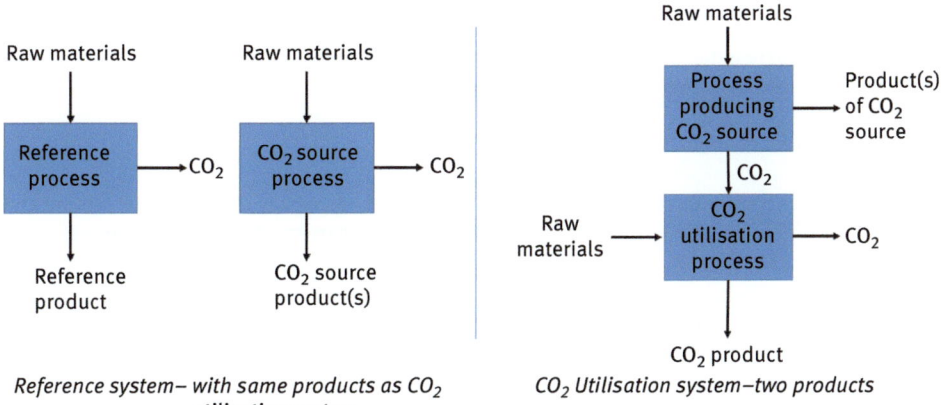

Figure 5.8: Systems with comparable functions.

5.8.2 Reporting only CO_2 impacts

It can be tempting, when evaluating CO_2 utilisation technologies, to only report global warming (GW) environmental impacts as these are directly related to CO_2 emissions. This is understandable as often the reduction of CO_2 emissions is a significant driver in the development of the CO_2 utilisation technology. However, this significantly reduces the transparency of the study. All environmental impacts of the process should be included as they may be greater than the reference case. For example, the CO_2 utilisation system may have reduced GW impacts but have higher human-toxicity and eutrophication potential impacts. In such a case, a decision would need to be made as to whether the higher impacts are acceptable or could be mitigated. Therefore, although an initial assessment may be looking to ensure CO_2 emissions are reduced, it is advised that all environmental impacts are reported.

5.8.3 Carbon avoided and negative CO_2 emissions

In some studies, it has been presumed that as CO_2 utilisation processes are using CO_2 as a feedstock, they will automatically have negative or zero CO_2 emissions. This is not necessarily true. A CO_2 utilisation process will have an input of CO_2 but also emit CO_2 due to the impacts of using other feedstocks and energy inputs. Additionally, depending upon the lifetime of the product, CO_2 can be emitted when the product reaches its end of life. However, there are cases where a CO_2 utilisation system can be carbon neutral or even carbon negative, meaning the system uses and stores more CO_2 than it releases.

A system can be carbon neutral (zero overall emissions) if:

> All emissions over the life cycle of the product are zero *and* the CO_2 has been captured from the atmosphere either biogenically *or* via air capture before being released at the end of life *or* the CO_2 comes from a point source and is permanently sequestered in the product.

A system can have negative emissions if:

> The CO_2 has been captured from the atmosphere either biogenically or through air capture and the CO_2 is permanently sequestered in the product and the overall CO_2 emissions are less than the amount sequestered.

Carbon avoided is different from carbon neutral or carbon negative. Carbon avoided refers to the amount of CO_2 that is not emitted by switching from the current product technology to the CO_2 utilisation technology. For example, if methanol produced by steam reforming of methane produces 50 kg CO_2/t and CO_2-based methanol produces 35 kg CO_2/t, it can be concluded that 15 kg CO_2/t are avoided. In many CO_2 utilisation processes, the amount of CO_2 avoided can be significant when compared to the amount utilised in producing the CO_2-derived product.

5.9 Conclusions

TEA and LCA are important tools in assessing the technological, economic and environmental potential of CO_2 utilisation technologies. The need for such assessment is growing as CO_2 utilisation systems increase in technology readiness from basic research to full commercial deployment. In particular, LCA is being increasingly used by decision-makers to assess opportunities and guide investment decisions and decarbonisation strategies. Existing methodologies and ISO standards should be used to conduct such assessments, but there are a number of common pitfalls that occur during the assessment of CO_2 utilisation technologies resulting in lack of transparency and comparability between results. Careful consideration of the CO_2 utilisation system ensuring correct boundaries, TRL, functional units and assessment scenarios can avoid many of these pitfalls and lead to transparent and comparable reports in which decision makers can have greater confidence [1, 8].

References

[1]   A. W. Zimmermann, J. Wunderlich, G. A. Buchner, L. Müller, K. Armstrong, S. Michailos, A. Marxen, H. Naims, A. Bardow, R. Shomäcker and P. Styring, *Techno-Economic Assessment & Life Cycle Assessment Guidelines for CO2 Utilization*. 2018.
[2]   International Organization for Standardization, "ISO 14040-Environmental management – Life Cycle Assessment – Principles and Framework," *Int. Organ. Stand.*, vol. 3, p. 20, 2006.
[3]   The International Standards Organisation, "Environmental management – Life cycle assessment – Requirements and guidelines – ISO 14044," *Int. J. Life Cycle Assess.*, vol. 2006, no. 7, pp. 652–668, 2006.
[4]   European Commission – Joint Research Centre – Institute for Environment and Sustainability, *International Reference Life Cycle Data System (ILCD) Handbook – General guide for Life Cycle Assessment – Detailed guidance*. 2010.
[5]   M. Goedkoop, R. Heijungs, M. Huijbregts, A. De Schryver, J. Struijs, and R. Van Zelm, "ReCiPe 2008," *Potentials*, pp. 1–44, 2009 .
[6]   T. Mai, "Technology Readiness Level," 2015 https://www.nasa.gov/directorates/heo/scan/engineering/technology/txt_accordion1.html Accessed 21/05/2019.
[7]   Horizon 2020, "Technology readiness levels (TRL)," *Horiz. 2020 – Work Program. 2014–2015*, no. 2014, p. 4995, 2015.
[8]   N. von der Assen, J. Jung, and A. Bardow, "Life-cycle assessment of carbon dioxide capture and utilization: avoiding the pitfalls," *Energy Environ. Sci.*, vol. 6, no. 9, p. 2721, 2013.

Farnaz Sotoodeh, Tjerk J. de Vries and Geert F. Woerlee

6 CO_2 as a solvent

6.1 Introduction

To most people today, the name "carbon dioxide" (CO_2) refers to one of the main greenhouse gases produced and released largely by oil and gas, coal and hydrocarbon plants. Although CO_2 is known also for its vital role in plant life and survival through photosynthesis process, environmental safety concerns and the general perception of CO_2 being regarded as a hazardous substance that pollutes the atmosphere have greatly overshadowed its unique aspects.

It was more than a century ago when the solvent properties of liquid and compressed CO_2 were first discovered to dissolve various liquids and solids. Different outcomes during the time shortly led to further experiments in the near-critical or supercritical regions. Phenomenal results were obtained showing that certain solids could dissolve and consequently precipitate in CO_2 by varying the pressure. These discoveries initiated a wave of activities afterwards to explore the CO_2 solvent power of various substances and were particularly stimulated by developing technologies in the field of process engineering. The reports in this regard are numerous and CO_2 properties have been exploited ever since in many different applications.

To grasp the nature of CO_2 as a solvent, let us refer to the CO_2 phase diagram depicted in Figure 6.1. As shown in this figure, the critical point of CO_2 is at 73.8 bar and 31.1 °C [1]. Below this temperature, CO_2 can exist both as a liquid and as a gas. In this region, CO_2 can be transformed from a gas into a liquid by increasing the pressure so that the molecular attraction forces increase and overcome the kinetic energy as a result of decrease in the distance between the CO_2 molecules. Above the critical temperature, however, the average kinetic energy of the CO_2 molecules exceeds the potential energy between them regardless of the intermolecular distance, and hence CO_2 cannot condense to a liquid anymore. It is in this region that CO_2 density can vary continuously from gaseous-like to liquid-like without phase boundaries by relatively small changes in temperature or pressure. This is key to note since the solubility of a substance in supercritical CO_2 is a strong function of CO_2 density.

The largest change in the density of CO_2 occurs in the vicinity of the critical point. In the supercritical region, CO_2 acts as a solvent due to its liquid-like density and is completely miscible with many polar, non-polar, aprotic and protic organic solvents.

Farnaz Sotoodeh, Tjerk J. de Vries, Geert F. Woerlee, FeyeCon Development and Implementation, Weesp The Netherlands

https://doi.org/10.1515/9783110563191-006

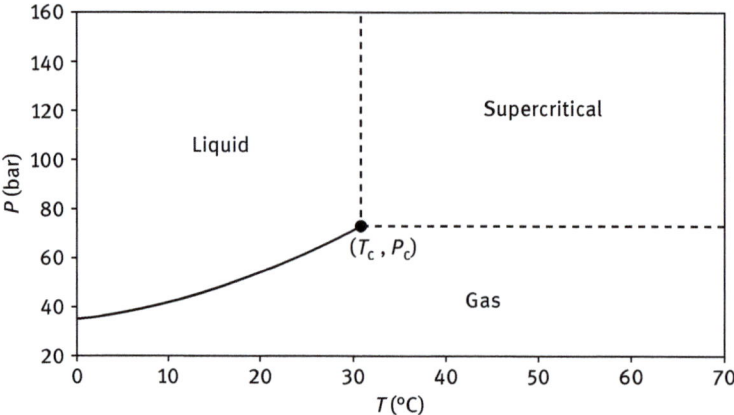

Figure 6.1: Phase diagram of carbon dioxide with critical point at $Pc = 73.8$ bar, $Tc = 31.1$ °C.

Other than the solubility power of CO_2, physical properties such as viscosity and diffusivity are related directly to the density and show steep variations near the critical point. Figure 6.2 illustrates changes in CO_2 viscosity as a function of density. It should be noted that even at high CO_2 densities, the viscosity of CO_2 is still much lower than that of conventional liquid solvents.

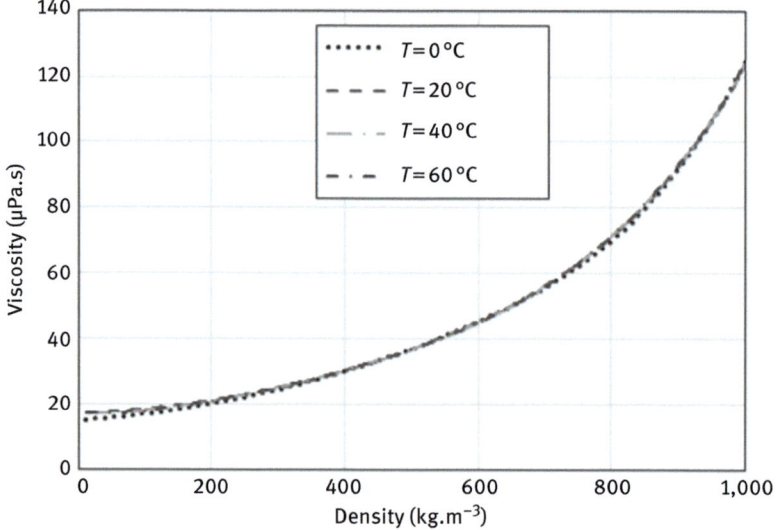

Figure 6.2: Viscosity correlation of pure CO_2 to density. Data obtained from NIST chemistry database [2].

Diffusion coefficients have close correlation to the viscosity and hence to the variations in density (see Figure 6.3 as an example). This is particularly pronounced in a separation process where diffusivities contribute significantly to the overall mass transfer in the system [3].

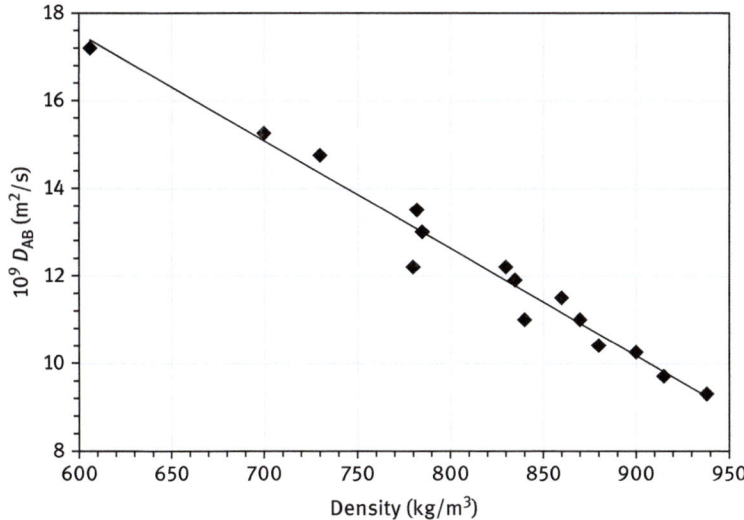

Figure 6.3: Diffusion coefficient of toluene in supercritical CO$_2$ as a function of CO$_2$ density. Diffusivity is inversely proportional to the density. Reproduced from data reported by J.J. Suárez et al. [4].

In fact, in supercritical region we have the solubility power of a liquid combined with the mass transport properties of a gas. This means that CO$_2$ can increase the speed of a process with an order of magnitude or more. By intensifying the process, the size of the equipment can be decreased or a reduction of time for the operation can be achieved. This translates commercially into an operational cost reduction and when combined with the environmental advantages, more favourable legislation or safety measures, results in a potentially favourable business case.

6.2 Existing industrial processes

The past few decades have seen the advent of rapidly expanding applications of CO$_2$ in industry. It was the successful extraction of caffeine from coffee 40 years ago that first drew attention to commercial usage of supercritical CO$_2$. Nowadays, CO$_2$ has found applications in dry cleaning, extraction of spices, dyeing of textiles, particle formation (for food, cosmetics, agriculture, health and pharmaceuticals), biomedical scaffolding, chemically reactive systems, polymer synthesis, surface coating, material processing in optics and electronics, as well as bio-separation processes. All these developments involve supercritical CO_2 technologies and offer a broad perspective for further utilisation of CO_2 as a carrier, processing and production medium. Some of these processes have been commercialised although many still hold great promise for industrial growth and scale up.

Although supercritical CO_2 and its unusual properties are known for many years, its industrial implementation, as for any new technology, has been impeded by obstacles related predominantly to financial risks associated with infrastructure change and market uncertainties in manufacturing new products. Tackling these barriers in adoption of the technology and maintaining a prolonged existence in the market have been primarily possible where there has been a distinct advantage offered either with respect to the product quality, or where CO_2 technology has been capable to create profit margin or increased production volume.

CO_2 processes normally require the use of high-pressure equipment. Most industrial CO_2 processes are operated over the pressure range of 50–300 bar and under mild temperatures. These mild temperatures are a definite benefit for food applications or thermally unstable materials. There are certain CO_2 processes that require higher temperature and pressure conditions, for example, polymer foaming, or extraction of materials that are difficult to dissolve in CO_2 in which pressure can reach as high as 500–1,000 bar. Operating at such conditions poses additional cost and feasibility constraints. From an economic point of view, CO_2 processes are best performed at lowest pressure and using as little CO_2 as possible.

In this respect, focus will be given to reviewing processes where CO_2 technology has made a successful business case and has been implemented on a commercial scale. This section is not exhaustive and mainly aims to highlight the most widespread usage of CO_2 in industry.

6.2.1 Extraction

Extraction using liquid or supercritical CO_2 is probably the most common process employing pressurised CO_2. CO_2 extraction of natural solid materials and active ingredients from herbs, hops or other plants that can be used in foods and pharmaceuticals, or CO_2 removal of residual impurities for producing high-grade products are now common practice in industry and are operated at large scale in several countries.

The basic set-up for an industrial CO_2 extraction unit is shown in Figure 6.4 (left). CO_2 is circulated continuously by switching between the extraction and separation vessels, and the extract is drained periodically from the separator. High diffusivity and adjustable solubility power of supercritical CO_2 allow for penetration of CO_2 into various complex structures and mixtures and selective extraction of the target compounds. Depending on the composition profile of the feed, more than one separation vessel may be required. In this case, the separation vessels are kept

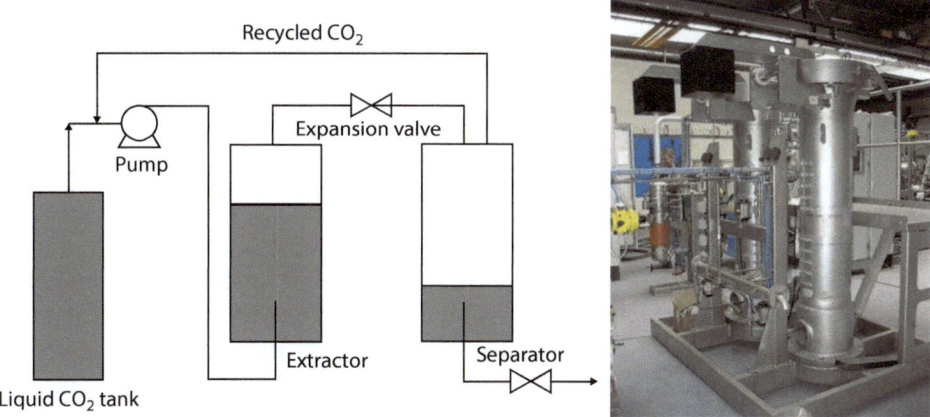

Figure 6.4: (Left) Basic schematic of industrial set-up for CO$_2$ extraction; (right) photo of industrial CO$_2$ extraction unit; installation by FeyeCon Development & Implementation (D&I) in Pakistan.

at different pressures to collect separately the "heavy" components (e.g. waxes and fats) and the "light" components (such as essential oils, flavours and aromas).

Purification of food and pharmaceutical products is done commercially on a relatively small scale. Extraction of free fatty acids and rancid fractions from fats and oils, as well as removal of unwanted components such as monomers, solvents or byproducts from pharmaceuticals are few practical examples. An industrial-scale CO$_2$ extraction unit is illustrated in Figure 6.4 (right).

Although the capital cost associated with CO$_2$ extraction equipment is relatively high, most of the industrial success of this process is owed to the straightforward single-step isolation of the extract, the absence of organic solvents and most importantly the exceptional quality of final product. Another attractive aspect in employing such process is the low operating costs. Since the extraction can be operated in continuous mode for each loading batch, costs associated with labour and personnel are reduced significantly. These factors have made the CO$_2$ extraction process very competitive and, in most cases, a successful business case compared to conventional extraction methods.

In the coming sections, application of the CO$_2$ extraction process in different industries for making various products is briefly reviewed.

6.2.1.1 Essential oils, flavours and fragrances

Industrial application of liquid and supercritical CO$_2$ in the extraction of essential oils, flavours and fragrances from plants, herbs and flowers has grown rapidly since the 1980s. CO$_2$ extracts have received increasing demand in perfumery and cosmetics, health-related products, food and beverages. The characteristic smell and flavour of a plant or herb is due to the presence of a large variety of compounds (many among which being volatile, thermally instable or prone to hydrolysis) and their complex interactions. Synthetic flavours and fragrances normally lack the character of the original. In the case of essential oils, reduced bioavailability of active ingredients such as antioxidants in synthetic oils is a major drawback. On the other hand, natural extracts obtained via traditional extraction methods in which degradative temperatures (in case of steam distillation) or toxic solvents are applied do not generally retain the natural freshness due to the loss of volatiles, degradation of labile constituents and formation of off-flavours in the extract [5, 6]. For this reason, a separation method that allows for efficient separation under mild operating conditions without employing aggressive solvents would seem very advantageous.

Extraction under supercritical CO_2 is considered a mild process since the temperature is kept at maximum 50–60 °C for such applications. In addition, separation in CO_2 is normally based on the volatility of compounds. Considering essential oils' and aromas' main constituents to be terpene hydrocarbons with varying boiling points, extraction and further isolation of certain fractions using the supercritical CO_2 method is usually found to be very effective. By adjusting CO_2 density in a flexible fashion, exceptional selectivity towards compounds of interests can be obtained. As a result, extracts obtained via CO_2 method normally maintain the true composition of the starting material and possess the natural sensory quality.

There are technical and operating challenges, however, involved in CO_2 separation processes, which often relate to a high solvent to feed ratio required due to limitations in the solubility of target compounds in supercritical CO_2. Using large amounts of solvent leads to a reduction in mass transfer rate and hence the separation efficiency. As a result, the concomitant processing costs will increase. One of the major goals in the design and operation of a supercritical CO_2 separation process is to reduce the required amount of solvent by adjusting the process conditions so that the right density for maximum solubility is reached.

Commercial-scale capacities used for CO_2 extraction of active ingredients, flavours and fragrances are normally in the range of 40–50 tonnes/year based on intake. Nowadays, such industrial processes are actively operated by many companies across the world. Ecomaat Bulgaria [7] (a partner of FeyeCon D&I) utilises supercritical CO_2 technology at large scale for extraction of essential oils from Bulgarian roses and lavender flowers. The obtained natural extracts are used in perfumery as well as health and aromatherapy applications. Other examples include Flavex Naturextrakte GmbH (Germany), Evonik Industries (Germany), Hops Extract Corporation of America (USA) and Fuji Flavor Co. (Japan), to name a few.

6.2.1.2 Edible oils

Extraction of edible oils derived from plant, fish or microbial sources such as micro-algae biomass and their purification and fractionation can be carried out effectively under a supercritical CO$_2$ environment. CO$_2$ is a non-polar compound and in super-critical state acts like a lipophilic solvent with its density being flexibly adjusted by varying the temperature and pressure. This feature allows CO$_2$ to selectively extract different fractions of oil under relatively low pressures (120 bar) and near ambient temperatures (ca. 40 °C). Mild operating conditions are necessary for preserving the oil quality in particular when handling oils containing omega-3 polyunsaturated fatty acids (Ω-3 PUFAs) such as fish or microalgae oil. Ω-3 PUFAs (most notably ei-cosapentaenoic acid and docosahexaenoic acid) are popular dietary supplements but highly susceptible to oxidation upon exposure to light, oxygen and heat, caus-ing off-aromas and poor taste. A suitable commercial process for extraction and concentration of PUFAs has therefore been of growing interest. "Cold processing" of edible oils under supercritical CO$_2$ not only offers distinct advantage with respect to the product quality due to the low process temperature, but also provides excep-tional selectivity (>80%) for separation and concentration of desired oil fractions, in particular PUFAs.

6.2.1.3 Cannabinoids from cannabis

Cannabidiol (CBD), a non-psychotropic cannabinoid, and Δ^9-tetrahydrocannabinol (THC), a psychoactive component, are two very well-known constituents existing in cannabis plants. Although the psychoactive properties of THC have been known for several decades, only recently, a broad scope of pharmacological activities of CBD has been discovered. Ever since, growing interest has been given to CBD-based pharmaceutical products. Conventional extraction methods employing organic sol-vents such as methanol, pentane, hexane and heptane are more and more restricted concerning the toxicity of solvent residues remained in the extract after the extrac-tion process. This is, in particular, of significant concern if the product is targeted for food, pharmaceutical or cosmetic applications. In addition, due to the heat sen-sitivity of cannabinoids (decarboxylation mechanism is activated at elevated tem-peratures), processes that operate at mild conditions are favoured for offering a higher quality product.

Without using any chemicals or toxic solvents, supercritical CO$_2$ extraction pro-cess operating at temperatures as low as 40 °C has the advantage to isolate CBD with high purity and process yield. Recently, many companies in the United States have started to employ supercritical CO$_2$ to extract CBD from cannabis.

6.2.1.4 Fibres

Natural cellulosic fibres derived from plant and vegetables (e.g. cotton, jute, hemp and flax) or animals (e.g. silk and wool) as well as synthetic ones (such as polyester used in textile) can be treated with CO_2 to produce nanocellulose particles [8] or remove impurities including mostly organic solvents, fats and waxes. Supercritical CO_2 extraction at low temperatures (<40 °C) is applied as a benign technology if the extraction involves thermolabile compounds, delicate texture or when the presence of solvent residues in the fibre affects the product quality. This holds particularly for cases when the extracted substance is supplied as an ingredient for food or pharmaceutical applications. Cocoa butter present in cocoa shell solid waste and waxy lipids such as ceramides extracted from natural wool are some examples [9].

6.2.2 Dyeing with CO_2

Started as a grant project, installations made by DyeCoo Textile Systems BV [10] (a spin-off company from FeyeCon D&I and the world's first supplier of industrial CO_2 dyeing equipment) for dyeing of polyester in supercritical CO_2 are now in operation in several countries such as Taiwan and Thailand. An example of a plant is shown in Figure 6.5.

Conventional dyeing of polyester fabric uses enormous quantities of water and some additives to suspend the dye in water. Clean water is a serious problem in many countries where textiles are dyed. The water quality needed for conventional dyeing is even higher than the quality of tap water in most of these countries. Dyeing of polyester in supercritical CO_2 does not require water or additives. Many of the conventional disperse dyes used for polyester dyeing have high enough solubility in supercritical CO_2 and are used in pure form. This along with the high diffusivity of supercritical CO_2 allows for fast and homogenous impregnation and dyeing of the textile. By passing the supercritical CO_2 used in the dyeing process over a separator, CO_2 is reused in every run except for a small portion that cannot be reclaimed economically; the last bars in the dye vessels are vented to the atmosphere [11, 12].

Next to the technical and economic advantages of this process, because of the much lower energy requirements (~30% lower) in supercritical CO_2 dyeing, the carbon footprint is reduced significantly compared to the conventional dyeing methods. Dyeing with supercritical CO_2 has been a successful business case, and currently other materials such as various synthetic fabrics, cotton, leather and paper are also being investigated for dyeing using this technique.

Figure 6.5: Photo of six quick closures of dyeing vessels (2,500 L each) for commercial beam dyeing of polyester fabric; installation by DyeCoo Textile Systems BV in Thailand.

6.2.3 Tissue cleaning

Tispa Medical BV (a spin-off company from FeyeCon D&I) developed a machine that uses supercritical CO$_2$ and ethanol to extract water and fats from human tissues obtained from biopsy [13, 14]. The Tispa Medical process consists of exchange/extraction with ethanol and exchange/extraction with CO$_2$ in order to dehydrate the tissue. Subsequently, the treated tissue is impregnated with paraffin to enable further analysis. This process is by several factors faster compared to the conventional techniques. Several hospitals in the Netherlands, Finland and Germany are now using this machine to replace the conventional method, which employs toxic organic solvents such as xylene during dehydration and alcohol/solvent exchange processes. Illustrated in Figure 6.6 are images of fatty tissues processed by Tispa

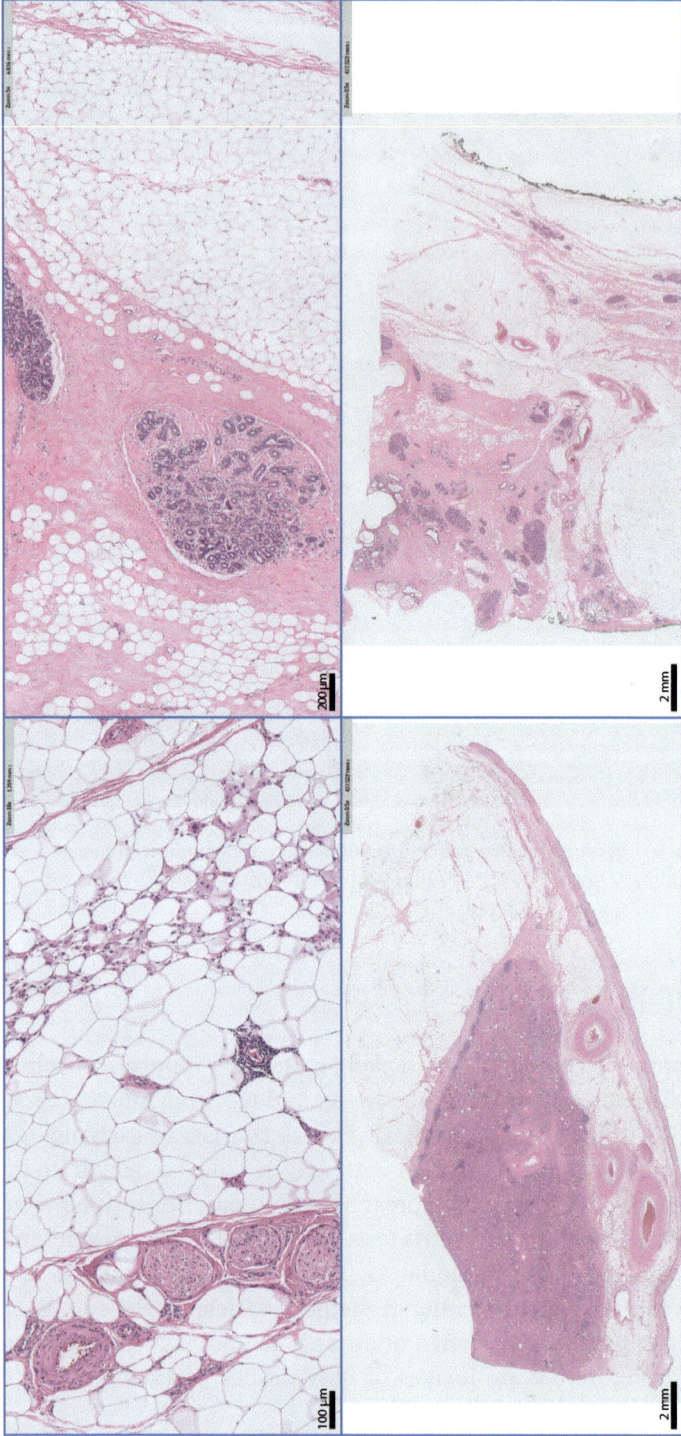

Figure 6.6: Scanning electron microscopy (SEM) images of fatty tissues processed by Tispa Medical CO_2 equipment. No pre-treatment with aggressive solvents such as acetone is required, and better quality is achieved in shorter processing time compared to conventional processing methods.

Medical CO$_2$ equipment. The structure is mostly preserved while no pre-treatment with chemical solvents such as acetone is required.

6.2.4. Cork closure

At industrial-scale, innovative company DIAM [15] uses a patented supercritical CO$_2$ process called "Diamant" to remove volatile components and mainly 2,4,6-trichloroanisole (TCA) from cork stoppers. The latter is known to be primarily responsible for "cork taint" in cork-bottled wine [16], an unpleasant off-flavour that affects the wine quality and has caused serious economic issues for wine industries for many years. The occurrence of TCA in cork could originate from the cork chlorine-bleaching process or the interactions between microorganisms and chlorinated compounds that exist naturally in cork. Supercritical CO$_2$ process is used successfully by DIAM to selectively extract TCA from cork stoppers and provide a significantly improved sensorial quality to the cork stoppers.

6.2.5 Methanol synthesis

The world's first commercial plant for production of methanol from renewable sources has been constructed and in operation by Carbon Recycling International (CRI) [17] in Grindavik, Iceland. The production capacity is 4,000 tonne methanol/year. In this process, CO$_2$ is captured from industrial waste and reacts with hydrogen (produced from renewable energy sources such as wind, water, geothermal and solar) in the presence of a catalyst. Methanol is produced as a result of this catalytic hydrogenation reaction. The significance of this process is the direct utilisation of captured waste CO$_2$ and transforming it into an added-value product.

6.2.6 Particle formation

Several techniques exist for creating particles in supercritical CO$_2$. A solute can be dissolved in CO$_2$ and sprayed to atmosphere, creating homogeneous nano-sized materials. Unfortunately, this process requires an immense amount of CO$_2$ as all solutes of interest have limited solubility in supercritical CO$_2$ (typically >1,000 kg CO$_2$/kg product is needed). A far more economical option in this case is the spraying of an emulsion, solution or suspension in CO$_2$ (via a single-fluid nozzle) or with CO$_2$ (via a two-fluid nozzle). CO$_2$ and organic solvent can both be recycled. The cheapest particle generation technique, however, is spraying a melt of fat/wax/oligomers or polymers saturated with CO_2 into the atmosphere. CO_2 helps to greatly reduce the viscosity of the melt. See, for example, Figure 6.7 illustrates the viscosity of an anhydrous milk fat (AMF) sample as a function of saturation with CO_2.

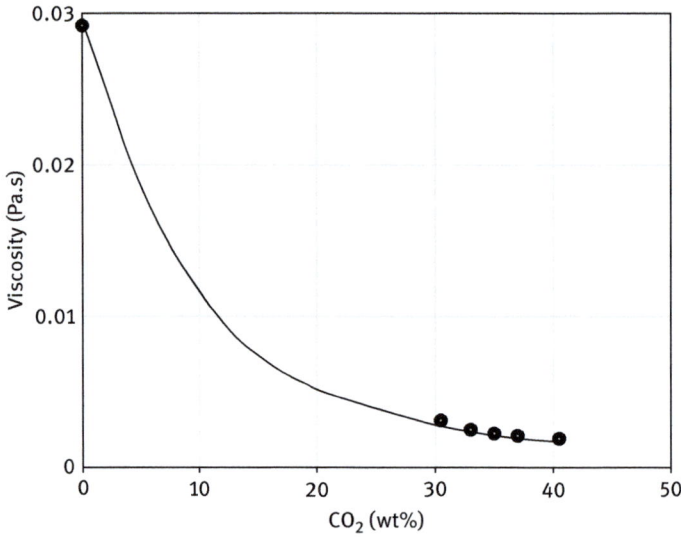

Figure 6.7: Viscosity of anhydrous milk fat (AMF) saturated with CO_2 (data points) and interpolation of data points (solid line) using $\eta_{mix} = \eta_{CO2}{}^{\omega CO2}\, \eta_{AMF}{}^{\omega AMF}\, \exp(G_{CO2\text{-}AMF}\omega_{CO2}\omega_{AMF})$, with binary parameter $G_{CO2\text{-}AMF} = -3.24$. η_{mix} is the mixture viscosity, η_{CO2} and η_{AMF} are CO_2 and AMF viscosities, respectively, and ω is the component mass fraction.

Depending on the process conditions, solid or sponge-like particles can be generated. In this respect, FeyeCon D&I has developed a process for producing structuring agent used in edible dispersions such as margarines or spreads [18]. Unilever uses this process on a commercial scale to create fat particles for margarine production. The fat particles (crystals) are used to create the structure in margarine. This process has been a successful industrial business case due to the low operating cost and uniqueness of the product. Finer (sponge-like) crystalline fat particles are created using this technique compared to the conventional chilling or crystallisation of fat in an oil/water mixture, thus reducing the saturated fat content in margarine.

6.3 Most promising and ongoing processes

As described above, most industrial implementations of supercritical CO_2 technology to date have been centred on natural CO_2-processed products that have shown proven market values mainly in cosmetics, pharmaceuticals and food applications.

New promising developments, nevertheless, have been emerging recently by both academia and industry, which reveal a broader horizon in utilisation of CO$_2$.

This section reviews processes in which application of CO$_2$ offers a specific advantage. These processes are either under development or ongoing on a rather small scale and limited only to certain niche. It is important to bear in mind that for any new development to sustain, CO$_2$ technology not being an exemption, it must prove its unambiguous benefit over the existing ones. In business, this mainly means profit and product quality.

6.3.1 Chemical synthesis

CO$_2$ is a non-toxic, non-flammable and non-halogenated compound; characteristics that combined with being inexpensive and readily accessible make its utilisation in chemical synthesis very advantageous. CO$_2$, as discussed previously, has unique properties in the supercritical state. What makes supercritical CO$_2$ particularly suitable for chemical synthesis is mainly related to its low viscosity, relative chemical inertness, exceptional wetting characteristics, miscibility, high diffusivity, no surface tension specially in interacting with solid surfaces, and above all the tunable density and solvent properties that facilitate its separation from the reaction mixture [1, 19, 20].

These benefits have led to very exciting innovations reported on CO$_2$, captured from industrial waste, being directly transformed into other chemical structures via catalytic fixation and cycloaddition mechanisms [21–26] for synthesis of, for example, bio-based polymers that can be used in packaging applications. These contributions provide a valuable perspective on utilisation of CO$_2$ both as a reactant and a processing solvent for making either new products or replacing existing ones. However, such processes have not yet been fully scalable due to technical constraints with regard to the separation of (by)products, catalytic selectivity and product performance. In particular, in case of a new product, the commercial success has most often and dominantly been challenged by the lack of market-driven incentives, not to mention the technical adaptation required for the process. In this respect, a quick overview of a few applications is given in the following sections.

6.3.1.1 Molecules and polymers

Many chemical reactions in supercritical CO$_2$ have been described in academia. Homogeneous and heterogeneous catalytic and enzymatic reactions for the conversion of small molecules have been widely investigated. Most of these studies focus on increased selectivity and reactivity [20]. For industry, most promising reactions are based on the replacement of halogenated solvents (e.g. synthesis of fluorinated compounds) and reactions involving gasses, that is, hydrogen, carbon monoxide and oxygen. The latter is due to the fact that these gasses can form a single phase with the other reactants in supercritical CO_2, thus eliminating mass transfer issues. Another advantage, in the case of oxygen and hydrogen system, is the inertness of CO_2 towards oxygen as well as the expansion of the non-explosive range in H_2/O_2 mixtures in the gas phase [27].

In the field of polymerisation, mechanisms including radical, cationic, anionic, transition metal catalysed and step growth polymerisations as well as polymer–fluid interactions have been extensively investigated in supercritical CO_2 [28–30]. Almost all polymerisation processes are based on precipitation or suspension polymerisation. Only some perfluorinated [31] or special polymers can remain in solution in supercritical CO_2 [27]. Certain fluorinated polymers synthesised in supercritical CO_2 have shown to offer superior performance compared to those made conventionally. In this respect, Dupont commercialised a process for polymerisation of perfluorinated monomers [32] and constructed a small pilot plant in 1999, followed by a semi-production-scale plant in 2002 to develop the process further. Unfortunately, DuPont stopped its development and the present position is not clear.

Catalytic carbonation processes for the synthesis of alkylene carbonates [33], to replace the conventional and toxic synthesis route via phosgene, copolymerisation [34], solvent-free coupling [35], epoxidation [36], hydrogenation [37], dehydrogenation [38] and cyclisation reactions [39] using CO_2 either directly as a reactant or as a synthesis medium, are among the continually expanding applications of (supercritical) CO_2 in chemical synthesis.

6.3.1.2 Catalysts

The drive for utilisation of supercritical CO_2 in catalyst synthesis stems from its unusual and exceptional characteristics, which allow to tune the reaction medium properties with respect to the solubility and transport ability. Most industrial chemical processes are catalysed heterogeneously, where the reactants adsorb onto an active solid surface rather than reacting in a homogeneous gas or liquid phase. The solid surface of the catalyst acts like a playground. It provides active sites on which the reactants are activated, interact and convert into products. The number of products produced within a certain reaction time is therefore directly linked to the number of active sites that are accessible to the reactants. To maximise the availability of these active sites, commercial catalysts are mostly prepared so that the active sites are dispersed in the form of small particles, normally in the order of one to several nanometres, throughout a porous support material. The porous structure of the support allows for the reactants to diffuse from the outer surface of the catalyst towards the active sites on which the reaction takes place. The smaller the active particles are, the larger would be the total surface area per unit mass of the catalyst and hence, the higher the number of available sites per unit mass. Typical examples of such supported catalysts are transition metals such as Ni or Co, their oxides or sulphides, distributed on a high surface area carrier such as SiO$_2$ or Al$_2$O$_3$. These supported catalysts are then manufactured into various shapes, such as pellets, for large-scale industrial use.

For a reactant to convert into a product at a considerable rate, it must reach the active sites by diffusing through the reaction medium and inside the catalyst porous network. Once adsorbed on the active site, it interacts with other species on the surface and transforms into a product. The product desorbs from the surface, diffuses out of the catalyst pores and into the reaction medium, where it can be measured in the reaction mixture via analytical tools. These processes are related at multiple scales (see Figure 6.8), and depending on the reaction conditions as well as the applied geometry can become slow and rate-limiting, thus controlling the overall performance of the system. In chemical processes that are structure sensitive or limited by diffusion of reactive species on the surface, the arrangement of active sites can critically influence the activity, selectivity and stability of a catalyst. Inhomogeneity in particle sizes and their distribution on the support can

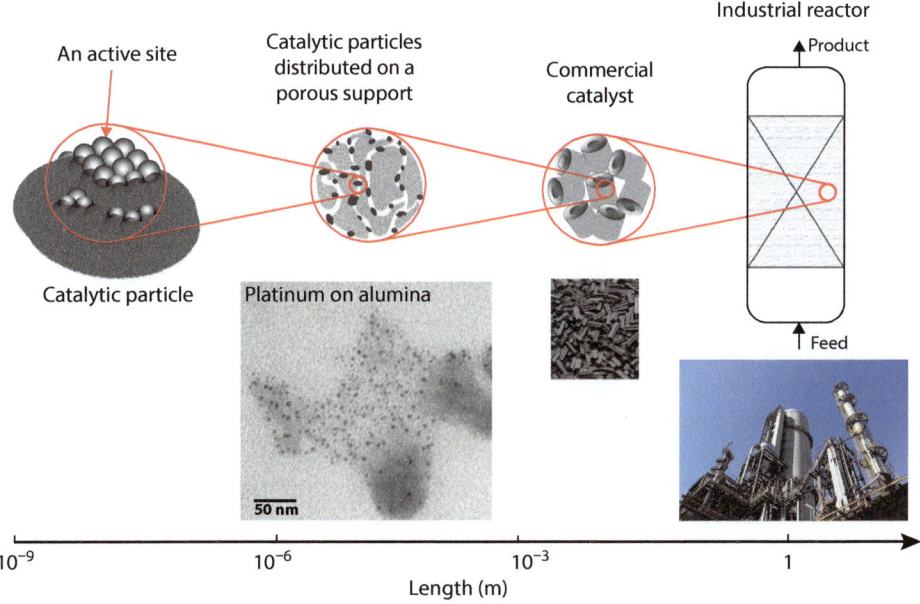

Figure 6.8: Multiscale nature of heterogeneous catalysis from nano- to macroscopic level. Shown in the middle is the transmission electron microscopy (TEM) image of an alumina support impregnated with a platinum metal fluoride precursor in supercritical CO$_2$. Pt crystallites are in the range of 1–5 nm; ©FeyeCon D&I, the Netherlands.

change the selectivity or cause catalyst deactivation by triggering undesired side reactions [40–42]. Exothermic side reactions produce temperature gradients on the surface resulting in deformation and blockage of the catalyst porous nanoscale network. Thus, poor turnover rates and catalytic activities are observed on the macroscopic scale. Achieving defined and uniform site distribution in catalyst synthesis is hence a primary goal and is complicated due to the complex and multiscale nature of heterogeneous catalysis.

In academia, computational techniques and mechanistic models [43–49] are being continuously improved to bridge between these various scales to be able to guide experiments. From the real-world catalysis point of view, direct translation of such models into actual catalytic performance in chemical processes can be invaluable in terms of industrial design, scale up and economical aspects. This, however, is extremely challenging because of the complex surface morphology of catalysts typically used in commercial scale. In this respect, efficient design and synthesis methods that can produce stable, reactive and selective surfaces are of great significance.

Heterogeneous catalysts used most commonly in industry are traditionally made using various techniques, including incipient wetness, wet, dry or capillary impregnation, chemical or physical vapour deposition, precipitation and sol–gel. In the case of making supported catalysts, the way a precursor is added to the support transforms the synthesis endeavour into an art since this step has a crucial influence on the distribution of active sites through the support and defines the final properties of the catalyst. Fast or inhomogeneous addition of the precursor leads to agglomeration of the active sites, inadequate site/support interaction and hence reducing the structural stability and catalytic activity under reaction conditions. The addition of the precursor becomes even more complicated in achieving high dispersion when more than one active phase is needed.

The above-mentioned challenges suggest the intriguing prospect of using supercritical CO_2 in the synthesis process, where support impregnation with catalytic precursor is carried out in a homogeneous and controlled fashion. Structures containing aromatic ligands, organometallic complexes and inorganic salts such as $PtCl_2$ are insoluble in CO_2. In order to improve the solubility of catalytic precursors in CO_2 strategies such as addition of a co-solvent, CO_2-philic ligands or fluorine atoms have been investigated [20]. A clean and controlled way to produce catalyst nanoparticles of narrow size distribution is via using supercritical CO_2 as a benign and nontoxic volatile acid. CO_2 dissolves in water at mild pressures <100 bar and readily forms carbonic acid, H_2CO_3. Under this condition, a number of catalyst and promoter precursor salts in aqueous solution can be dissolved in CO_2. The choice of metal precursor in this case is important. Quick variations in the pressure would allow for fast and uniform change in the pH throughout the mixture, resulting in uniform precipitation of active material and thus enabling precise control of the particle size. Supercritical CO_2 technology is used at FeyeCon D&I to synthesise nanoporous catalysts with high dispersion (see Figure 6.8 for an example of Pt metal dispersed on a porous alumina support). Techniques such as rapid expansion of supercritical solutions, supercritical antisolvent [50, 51] and supercritical fluid chemical deposition carried out in supercritical CO$_2$ are among the most exciting and industrially emerging catalyst synthesis methods for achieving highly dispersed and homogeneously sized active nanoparticles.

Supercritical CO$_2$ can also be used for the synthesis of highly porous support materials with large surface areas. Aerogels, or the so-called frozen smoke, are exceptionally light solids [52], which make them one of the most attractive materials in the world's manufacturing and production. Due to their extremely high porosity, implementation of aerogels in catalysis improves the mass transfer and reduces the risk of formation of hot spots on the surface. Wet-gel drying using supercritical CO$_2$ during the synthesis process yields very well-structured nanoporous materials by circumventing the capillary stress in drying via evaporation, which causes considerable structural damage. Moreover, it reduces the processing costs enormously due to the significantly faster drying procedure compared to the conventional drying techniques. Porous supports produced with this method can be impregnated with suitable CO$_2$-soluble precursors for the synthesis of active catalysts.

6.3.2 Material processing

CO$_2$ has several chemical advantages, which make it exceptionally attractive in material processing. It has zero surface tension in the supercritical state, which allows complete wetting of the surface and penetration of CO$_2$ into complex structures during impregnation or extraction processes. CO$_2$ cannot be oxidised and is therefore an excellent solvent for performing oxidation reactions. Moreover, due to its low viscosity, heat transfer rate in a CO$_2$ mixture is normally high.

To act as a solvent, a sufficient "liquid-like" density is required. CO$_2$ has a density of 447 kg/m^3 at 50 °C and 110 bar. To get the same density for methane at that temperature, a pressure of >5,000 bar is needed! It is important to realise that density is the parameter that mainly defines the properties of CO$_2$ as a solvent, not pressure or temperature. It is true that at higher temperatures usually more material dissolves in CO$_2$. This is, however, due to lower intermolecular forces between the solute molecules such as, for example, hydrogen bonding, which becomes much weaker at higher temperatures.

CO$_2$ provides benefits not only chemically, but also environmentally. Through its perceived benign properties, CO$_2$ contributes to the sustainability of a process by minimising the environmental and ecological impacts. Unlike conventionally employed toxic and organic reagents and solvents, CO$_2$ does not leave residue in the material after processing and hence does not require the typical post-processing solvent recovery procedures.

6.3.2.1 CO_2 for microencapsulation

Microencapsulation refers to a process in which a continuous film (or shell) is formed around a dispersed liquid or solid, enclosing it in the form of a capsule [53, 54]. The coating acts as a barrier to protect and separate the core from its surrounding environment or to help with the storage. Microencapsulation is particularly important for preservation of active ingredients that are susceptible to degradation upon exposure to moisture, light, oxygen, heat, friction or organic solvents. In this respect, a release mechanism is typically devised in order to utilise the coated material after encapsulation. Depending on the application, microencapsulates can vary in morphology and particle size ranging from 0.5 to 200 μm.

Diverse microencapsulation techniques have been employed in various industries, including simple and complex coacervation [55, 56], extrusion coating, centrifugal and rotational suspension and phase separation, gelation, interface polymerisation [57, 58], electrostatic or mechanical encapsulation [59] and spray drying [60–63]. Conventional spray drying has been the most commonly used technique since it offers several technical and economical advantages over other methods. Major shortfalls, however, in forming large agglomerates with low surface area and degradation of core material associated with typical usage of hot air (170–200 °C) as a drying medium, severely limit the application of this process particularly for the encapsulation of heat-sensitive ingredients used in food, agriculture and pharmaceutical industries [63, 64].

Recent advances have overcome these drawbacks by utilising supercritical CO_2 [65–70] as a nontoxic, selective and environmentally benign solvent. In this respect, a commercially interesting technique is spraying of an aqueous solutions or emulsions containing both coating material and active ingredients with CO_2 (via a two-fluid nozzle for mixing and atomisation). Compared to hot air, CO_2 spray drying is done under mild and eco-friendly operating conditions (typically 40 °C and 100–150 bar), reducing thermal degradation of the product. Unique transport and thermophysical properties of supercritical CO_2 (low viscosity, low surface tension, high diffusivity relative to liquids and high density compared to gases) allow for fast and effective removal of the liquid medium and creating particles with high surface area and porous structure. This enables faster dissolution of the particles in cold water compared to hot air-dried particles. The operating conditions used in spray drying process including temperature, pressure, CO_2 flow rate and the degree of atomisation are chosen to ensure high enough CO_2 density and solvent power for complete removal of the liquid medium and hence formation of dried particles with desired size. These parameters are therefore key to the final properties and performance of the obtained microencapsulates.

Figure 6.9 illustrates an example of microencapsulated particles formed under supercritical CO_2 environment using biodegradable wall material. The choice of wall material greatly influences the efficiency of the process and depends on the

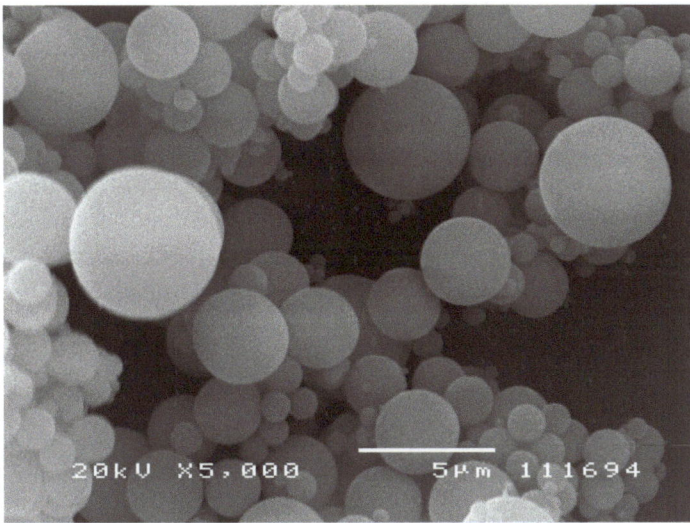

Figure 6.9: Scanning electron microscopy (SEM) image of microencapsulated spheres made using biodegradable wall material using supercritical CO$_2$ technology. ©FeyeCon D&I, the Netherlands.

physical and chemical properties of the final particles. The coating is normally selected based on parameters such as the hydro- or lipophilicity, pH sensitivity, the encapsulation ability under supercritical CO$_2$ environment, compatibility with the active ingredient, as well as cost and safety specifications. Biodegradable carbohydrate-based polymers, proteins, lipids and waxes are among the popular choices for wall material.

Although this technique is more expensive than conventional spray drying with hot air, it can compete due to its superior product quality. Nowadays, several companies active in food, cosmetic and pharmaceutical sectors are evaluating and considering using this process for producing microencapsulates for their natural formulations and products.

6.3.2.2 Drying fruit and vegetables with CO$_2$

Vegetables, herbs, fruits and even meat can be dried with supercritical CO$_2$. The process is an alternative to traditional air or freeze-drying methods, where elevated temperatures in the former result in thermal degradation of active nutrients (e.g. antioxidants, vitamins and carotenoids) and low temperatures in the latter lead to the formation of porous and brittle structures. On the contrary, the oxygen-free and mild operating conditions under supercritical CO$_2$ environment during the drying process preserves the physical quality and chemical properties of material such as texture, flavour and functionality of nutrients. Furthermore, bacterial growth is inhibited by the removal of water, and microorganisms and enzymes are deactivated due to the antimicrobial activity of supercritical CO_2. Shelf life and the safety of the food are hence significantly improved.

Above its critical temperature and pressure, CO_2 diffusion is similar to gases while its density is closer to liquids. This allows CO_2 to easily permeate inside food and vegetables and remove water. Although the solubility of water in CO_2 is limited, CO_2 can be circulated at high speed over the food material to be dried. A suitable desiccant is normally used for trapping the collected water. Basic schematic of an industrial drying process using supercritical CO_2 is illustrated in Figure 6.10 (left).

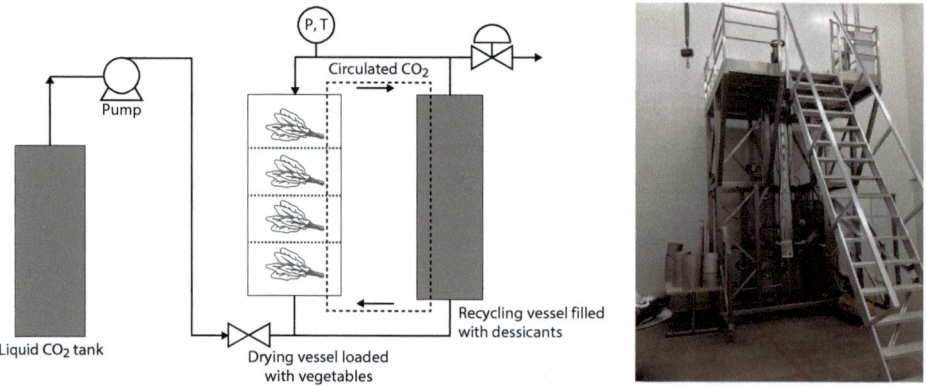

Figure 6.10: (Left) Basic schematic of the drying process using supercritical CO_2; (right) drying vessel of commercial CO_2drying equipment; installation by CO_2dry (the Netherlands).

Despite several benefits offered by CO_2 drying technology, this process has not been commonly implemented on a commercial scale. The reason has been mainly the required high investment cost. Nonetheless, currently several companies have been exploiting CO_2 drying process in food industries. CO_2dry [71] (a spin-off company of FeyeCon D&I) utilises the solvent-free CO_2 drying technology at low energy consumption for introducing high quality dried fruit and vegetables into the food market as shown in Figure 6.10 (right).

6.3.2.3 Foaming with CO_2

Compared to other benign gases, pressurised CO_2 has a high solubility in many non-polar and slightly polar polymers. This enables CO_2 to be an effective foaming agent. The techniques to foam polymers with CO_2 are generally categorised as (1) saturation of a polymer in a pressure vessel followed by expansion of the polymer in the same vessel by releasing the pressure, (2) saturation of a polymer in a pressure vessel below the foaming temperature, releasing the pressure, taking out the polymer and expansion of the polymer outside the vessel using a heat source to initiate the foaming and (3) saturation of a polymer in an extruder and foaming at the exit (die) of the extruder.

The first method is only economical with relatively expensive polymers and low expansion ratios, for example, shoe soles. Since the expansion occurs inside the pressurised vessel, capital costs are relatively high using this method. In comparison, the second method requires a much smaller vessel volume per kilogram foam produced and can therefore be much more feasible.

By using an extruder, production of foam is done continuously rather than batch. An industrial example is the production of Extruded polystyrene (XPS) foam that in response to environmental regulations switched from chlorinated blowing agents for a large part to CO$_2$, for example, Styrodur®.

6.4 Conclusions and prospects

There has been an increased public awareness in the recent years regarding the use of hazardous solvents, especially in processing food, cosmetics and pharmaceuticals. This, together with industrial regulatory restrictions increasing more than ever, has perhaps been the main drive for the recognition and commercial utilisation of CO$_2$ for obtaining a "clean" and "sustainable" product. Although research and development in the field of supercritical CO$_2$ have been going on for many years in numerous other fields as well, emergence of commercial applications has been limited due to lack of addressing the most important industrial concern: the production costs.

Because of the unique combination of gas and liquid properties, supercritical CO$_2$ can in many cases lead to new products that are currently not possible to produce with other technologies. Although extremely exciting to work on new products and processes, business building for new product and markets is much more time consuming and risky than replacing an existing product with a cheaper or better product.

Acknowledgements: FeyeCon Development & Implementation has been widely active in the field of CO$_2$ technology for over 25 years. This brief review has been an endeavour to present notable industrial applications and some of the ongoing developments in this field. In this respect, the authors would like to acknowledge the countless and illuminating discussions with all members at FeyeCon who contributed to the work and understanding of CO$_2$ processes provided in this chapter. The authors would like to particularly thank Daniela O. Trambitas for providing valuable insights. Tispa Medical B.V. for providing the microscopy images of CO$_2$-treated fatty tissues is also gratefully acknowledged.

Bibliography

[1]   R. B. Gupta and J.-J. Shim, *Solubility in Supercritical Carbon Dioxide*. CRC Press, 2006.
[2]   "NIST Chemistry WebBook." [Online]. Available: https://webbook.nist.gov/chemistry/.
[3]   G. F. Woerlee, "Hydrodynamics and mass transfer in packed columns and their applications for supercritical separations," Delft, The Netherlands : Delft University Press, 1997.
[4]   J. J. Suárez, I. Medina, and J. L. Bueno, "Diffusion coefficients in supercritical fluids: available data and graphical correlations," *Fluid Phase Equilibria*, vol. 153, no. 1, pp. 167–212, Nov. 1998.
[5]   K. Ansari and I. Goodarznia, "Optimization of supercritical carbon dioxide extraction of essential oil from spearmint (Mentha spicata L.) leaves by using Taguchi methodology," *The Journal of Supercritical Fluids*, vol. 67, pp. 123–130, Jul. 2012.
[6]   M. B. King and T. R. Bott, Eds., *Extraction of Natural Products Using Near-Critical Solvents*. Springer Netherlands, 1993.
[7]   "ECOMAAT BULGARIA," *ECOMAAT BULGARIA*. [Online]. Available: http://www.ecomaat.com
[8]   R. Ramírez, I. Garay, J. Álvarez, M. Martí, J. L. Parra, and L. Coderch, "Supercritical fluid extraction to obtain ceramides from wool fibers," *Separation and Purification Technology*, vol. 63, no. 3, pp. 552–557, Nov. 2008.
[9]   R. Menendez Gonzalez, D. O. Trambitas, S. Cantekin, I. Graveson, M. Jennekens, and S. Momin, "Supercritical CO2 Cellulose Spraydrying," WO/2017/051030, 31-Mar-2017.
[10] "DyeCoo," *Dyecoo*. [Online]. Available: http://www.dyecoo.com/
[11] M. V. Fernandez Cid *et al.*, "United States Patent: 7938865 – Method of dyeing a substrate with a reactive dyestuff in supercritical or near supercritical carbon dioxide," 7938865, 10-May-2011.
[12] Der Kraan Martijn Van, D. O. Trambitas, I. G. ION, and G. F. Woerlee, "United States Patent: 20160017539 – Process of marking a textile substrate," EP2961881 A1, 25-Feb-2014.
[13] "Tispa Medical BV." [Online]. Available: http://www.tispamedical.com/.
[14] W. G. M. Agterof, R. Bhatia, GB, and G. W. Hofland. "United States Patent: 8187655 – Dehydration method," 8187655, 29-May-2012.
[15] "Diam." [Online]. Available: https://www.diam-closures.com/manufacturer-of-wine-champagne-cork-closures.
[16] H. R. Buser, C. Zanier, and H. Tanner, "Identification of 2,4,6-trichloroanisole as a potent compound causing cork taint in wine," *J. Agric. Food Chem.*, vol. 30, no. 2, pp. 359–362, Mar. 1982.
[17] "CRI," *CRI – Carbon Recycling International*. [Online]. Available: http://carbonrecycling.is/
[18] "Dispersion structuring agent," EP2181604A1, 05-May-2010.
[19] "Supercritical Carbon Dioxide: In Polymer Reaction Engineering," *Wiley.com*. [Online]. Available: https://www.wiley.com/en-us/Supercritical+Carbon+Dioxide%3A+In+Polymer+Reaction+Engineering-p-9783527310920. [Accessed: 11-Jul-2018].
[20] P. G. Jessop, "Homogeneous catalysis using supercritical fluids: Recent trends and systems studied," *The Journal of Supercritical Fluids*, vol. 38, no. 2, pp. 211–231, Sep. 2006.
[21] M. North, P. Villuendas, and C. Young, "A Gas-Phase Flow Reactor for Ethylene Carbonate Synthesis from Waste Carbon Dioxide," *Chemistry – A European Journal*, vol. 15, no. 43, pp. 11454–11457.
[22] M. North, R. Pasquale, and C. Young, "Synthesis of cyclic carbonates from epoxides and CO2," *Green Chem.*, vol. 12, no. 9, pp. 1514–1539, Sep. 2010.
[23] T. Werner and N. Tenhumberg, "Synthesis of cyclic carbonates from epoxides and CO2 catalyzed by potassium iodide and amino alcohols," *Journal of CO2 Utilization*, vol. 7, pp. 39–45, Sep. 2014.

[24] G. Trott, P. K. Saini, and C. K. Williams, "Catalysts for CO2/epoxide ring-opening copolymerization," *Phil. Trans. R. Soc. A*, vol. 374, no. 2061, p. 20150085, Feb. 2016.

[25] R. Saada *et al.*, "Greener synthesis of dimethyl carbonate using a novel tin-zirconia/graphene nanocomposite catalyst," *Applied Catalysis B: Environmental*, vol. 226, pp. 451–462, Jan. 2018.

[26] European H2020 FP7 grant project, "Production of Cyclic Carbonates from CO2 using Renewable Feedstocks," *CyclicCO2R*, 2016-2013. [Online]. Available: http://www.cyclicco2r.eu/.

[27] E. J. Beckman, "Supercritical and near-critical CO2 in green chemical synthesis and processing," *The Journal of Supercritical Fluids*, vol. 28, no. 2, pp. 121–191, Mar. 2004.

[28] T. J. de Vries, M. F. Kemmere, and J. T. F. Keurentjes, "Characterization of Polyethylenes Produced in Supercritical Carbon Dioxide by a Late-Transition-Metal Catalyst," *Macromolecules*, vol. 37, no. 11, pp. 4241–4246, Jun. 2004.

[29] P. Marizza *et al.*, "Supercritical impregnation of polymer matrices spatially confined in microcontainers for oral drug delivery: Effect of temperature, pressure and time," *The Journal of Supercritical Fluids*, vol. 107, pp. 145–152, Jan. 2016.

[30] E. Reverchon, R. Adami, S. Cardea, and G. D. Porta, "Supercritical fluids processing of polymers for pharmaceutical and medical applications," *The Journal of Supercritical Fluids*, vol. 47, no. 3, pp. 484–492, Jan. 2009.

[31] L. Du, J. Y. Kelly, G. W. Roberts, and J. M. DeSimone, "Fluoropolymer synthesis in supercritical carbon dioxide," *The Journal of Supercritical Fluids*, vol. 47, no. 3, pp. 447–457, Jan. 2009.

[32] P. Douglas Brothers and E I du Pont de Nemours and Co, "Polymerization of fluoromonomers in carbon dioxide," US6103844A, 15-Aug-2000.

[33] W. J. Peppel, "Preparation and Properties of the Alkylene Carbonates," *Ind. Eng. Chem.*, vol. 50, no. 5, pp. 767–770, May 1958.

[34] M. Super, E. Berluche, C. Costello, and E. Beckman, "Copolymerization of 1,2-Epoxycyclohexane and Carbon Dioxide Using Carbon Dioxide as Both Reactant and Solvent," *Macromolecules*, vol. 30, no. 3, pp. 368–372, Feb. 1997.

[35] M. Rohr, C. Geyer, R. Wandeler, M. S. Schneider, E. F. Murphy, and A. Baiker, "Solvent-free ruthenium-catalysed vinylcarbamate synthesis from phenylacetylene and diethylamine in 'supercritical' carbon dioxide," *Green Chem.*, vol. 3, no. 3, pp. 123–125, Jan. 2001.

[36] F. Montilla *et al.*, "Trimethylsilyl-substituted ligands as solubilizers of metal complexes in supercritical carbon dioxide," *Dalton Trans.*, vol. 0, no. 11, pp. 2170–2176, May 2003.

[37] R. J. da Silva, A. F. Pimentel, R. S. Monteiro, and C. J. A. Mota, "Synthesis of methanol and dimethyl ether from the CO2 hydrogenation over Cu·ZnO supported on Al2O3 and Nb2O5," *Journal of CO2 Utilization*, vol. 15, pp. 83–88, Sep. 2016.

[38] J. H. Earley, R. A. Bourne, M. J. Watson, and M. Poliakoff, "Continuous catalytic upgrading of ethanol to n-butanol and >C4 products over Cu/CeO2 catalysts in supercritical CO2," *Green Chem.*, vol. 17, no. 5, pp. 3018–3025, May 2015.

[39] S. Wang and F. Kienzle, "The Syntheses of Pharmaceutical Intermediates in Supercritical Fluids," *Ind. Eng. Chem. Res.*, vol. 39, no. 12, pp. 4487–4490, Dec. 2000.

[40] F. Sotoodeh and K. J. Smith, "Structure sensitivity of dodecahydro-N-ethylcarbazole dehydrogenation over Pd catalysts," *Journal of Catalysis*, vol. 279, no. 1, pp. 36–47, Apr. 2011.

[41] P. Ghasvareh and K. J. Smith, "Effects of Co Particle Size on the Stability of Co/Al2O3 and Re–Co/Al2O3 Catalysts in a Slurry-Phase Fischer-Tropsch Reactor," *Energy Fuels*, vol. 30, no. 11, pp. 9721–9729, Nov. 2016.

[42] D. Y. Murzin, "Catalyst deactivation and structure sensitivity," *Catal. Sci. Technol.*, vol. 4, no. 9, pp. 3340–3350, Aug. 2014.

[43] K. Reuter, "Ab Initio Thermodynamics and First-Principles Microkinetics for Surface Catalysis," *Catal Lett*, vol. 146, no. 3, pp. 541–563, Mar. 2016.

[44] O. Deutschmann, *Modeling and Simulation of Heterogeneous Catalytic Reactions: From the Molecular Process to the Technical System*. John Wiley & Sons, 2013.

[45] J. Meyer and K. Reuter, "Modeling Heat Dissipation at the Nanoscale: An Embedding Approach for Chemical Reaction Dynamics on Metal Surfaces," *Angewandte Chemie International Edition*, vol. 53, no. 18, pp. 4721–4724, Mar. 2014.

[46] P. Spiering and J. Meyer, "Testing Electronic Friction Models: Vibrational De-excitation in Scattering of H2 and D2 from Cu(111)," *J. Phys. Chem. Lett.*, vol. 9, no. 7, pp. 1803–1808, Apr. 2018.

[47] T. Wang and K. Reuter, "Structure sensitivity in oxide catalysis: First-principles kinetic Monte Carlo simulations for CO oxidation at RuO2(111)," *The Journal of Chemical Physics*, vol. 143, no. 20, p. 204702, Nov. 2015.

[48] S. Matera, M. Maestri, A. Cuoci, and K. Reuter, "Predictive-Quality Surface Reaction Chemistry in Real Reactor Models: Integrating First-Principles Kinetic Monte Carlo Simulations into Computational Fluid Dynamics," *ACS Catal.*, vol. 4, no. 11, pp. 4081–4092, Nov. 2014.

[49] M. Maestri and E. Iglesia, "First-principles theoretical assessment of catalysis by confinement: NO–O2 reactions within voids of molecular dimensions in siliceous crystalline frameworks," *Phys. Chem. Chem. Phys.*, vol. 20, no. 23, pp. 15725–15735, Jun. 2018.

[50] S. A. Kondrat *et al.*, "Preparation of a highly active ternary Cu-Zn-Al oxide methanol synthesis catalyst by supercritical CO 2 anti-solvent precipitation," *Catalysis Today*, Mar. 2018.

[51] A. Baiker, "Supercritical Fluids in Heterogeneous Catalysis," *Chem. Rev.*, vol. 99, no. 2, pp. 453–474, Feb. 1999.

[52] N. Hüsing and U. Schubert, "Aerogels—Airy Materials: Chemistry, Structure, and Properties," *Angewandte Chemie International Edition*, vol. 37, no. 1–2, pp. 22–45.

[53] L. A. Luzzi, "Microencapsulation," *J Pharm Sci*, vol. 59, no. 10, pp. 1367–1376, Oct. 1970.

[54] F. E. Massoth, W. E. Hensel, and W. W. Harlowe, "Basic Studies of Encapsulation Process. Correlation of Capsule Size," *Ind. Eng. Chem. Proc. Des. Dev.*, vol. 4, no. 1, pp. 6–13, Jan. 1965.

[55] B. K. Green and L. Schleicher, "Manifold record material," US2730456A, 10-Jan-1956.

[56] B. K. Green, "Oil-containing microscopic capsules and method of making them," US2800458A, 23-Jul-1957.

[57] T. Dobashi, T. Furukawa, K. Ichikawa, and T. Narita, "Coupling of Chemical Cross-linking, Swelling, and Phase Separation in Microencapsulation," *Langmuir*, vol. 18, no. 16, pp. 6031–6034, Aug. 2002.

[58] T. M. Chang, "SEMIPERMEABLE MICROCAPSULES," *Science*, vol. 146, no. 3643, pp. 524–525, Oct. 1964.

[59] Kirk-Othmer (Editor)"Kirk-Othmer Encyclopedia of Chemical Technology, 27 Volume Set, 5th Edition," *Wiley.com*.

[60] A. Sultana, A. Miyamoto, Q. Lan Hy, Y. Tanaka, Y. Fushimi, and H. Yoshii, "Microencapsulation of flavors by spray drying using Saccharomyces cerevisiae," *Journal of Food Engineering*, vol. 199, pp. 36–41, Apr. 2017.

[61] S. M. Jafari, E. Assadpoor, Y. He, and B. Bhandari, "Encapsulation Efficiency of Food Flavours and Oils during Spray Drying," *Drying Technology*, vol. 26, no. 7, pp. 816–835, Jul. 2008.

[62] A. M. Goula and K. G. Adamopoulos, "A method for pomegranate seed application in food industries: Seed oil encapsulation," *Food and Bioproducts Processing*, vol. 90, no. 4, pp. 639–652, Oct. 2012.

[63] A. Sosnik and K. P. Seremeta, "Advantages and challenges of the spray-drying technology for the production of pure drug particles and drug-loaded polymeric carriers," *Advances in Colloid and Interface Science*, vol. 223, pp. 40–54, Sep. 2015.

[64] Anandharamakrishnan C. and Padma Ishwarya S., *Spray Drying Techniques for Food Ingredient Encapsulation*. John Wiley & Sons, Ltd, 2015.

[65] G. F. Woerlee, NL, G. W. Hofland, NL, P. S. Vermeulen, and NL, "United States Patent: 8226984 – Process for the preparation of encapsulates through precipitation," 8226984, 24-Jul-2012.

[66] A. T. Poortinga, NL, D. O. Trambitas, NL, G. W. Hofland, and NL, "United States Patent: 8637104 – Microencapsulate and process for the manufacture thereof," 8637104, 28-Jan-2014.

[67] H. Kröber and U. Teipel, "Microencapsulation of particles using supercritical carbon dioxide," *Chemical Engineering and Processing: Process Intensification*, vol. 44, no. 2, pp. 215–219, Feb. 2005.

[68] D. T. Santos, J. Q. Albarelli, M. M. Beppu, and M. A. A. Meireles, "Stabilization of anthocyanin extract from jabuticaba skins by encapsulation using supercritical CO2 as solvent," *Food Research International*, vol. 50, no. 2, pp. 617–624, Mar. 2013.

[69] W.-C. Tsai and S. S. H. Rizvi, "Simultaneous microencapsulation of hydrophilic and lipophilic bioactives in liposomes produced by an ecofriendly supercritical fluid process," *Food Research International*, vol. 99, pp. 256–262, Sep. 2017.

[70] M. K. Hrnčič, D. Cör, M. T. Verboten, and Ž. Knez, "Application of supercritical and subcritical fluids in food processing," *Food Qual Saf*, vol. 2, no. 2, pp. 59–67, Jun. 2018.

[71] "CO2Dry." [Online]. Available: http://www.co2dry.com/. [Accessed: 01-Aug-2018].

Part II: CO_2 capture and separation

Muhammad Akram

7 Current status of the CO$_2$ capture technology by absorption

7.1 Introduction

Chemical absorption technology is one of the best-known and mature technologies and has been demonstrated at commercial scale. The technology has been widely used on gas-sweetening plants for decades. The beauty of this technology is that it can be retrofitted to the existing infrastructure. Therefore, it has a very good potential for climate change mitigation at least in the short term, while more advanced and cost-effective technologies are being developed.

The CO$_2$ capture process by absorption is based on the separation of CO$_2$ by a solvent, usually based on aqueous amine solutions. The term "solvent" in this context is confusing as these are actually aqueous solution of amines. Amines react with CO$_2$ to form compounds that are soluble in water; however, the process is equilibrium limited [1]. The technology is applicable to power plants as well as to industrial sources of CO$_2$. Concentration of CO$_2$ in flue gases from power plants varies typically from 5% in gas-fired power plant to 15% in coal fired power plants. Therefore, one of the technical challenges to devise a cost-effective CO$_2$ separation technology from flue gases is lower thermodynamic driving force due to lower partial pressure of CO$_2$ [1], which makes the process economically expensive.

A huge amount of effort is being devoted to make this process economically viable. Currently, there are two commercial-scale plants, Boundary Dam and Petra Nova (details given in Section 7.3), while a number of pilot-scale research facilities are operational around the globe. Pilot-scale facilities include Technology Centre Mongstad (TCM) and SINTEF in Norway, Sulzer in Germany, Pilot Scale Advanced Capture Technology (PACT) facilities in UK amongst others. A list of pilot-scale facilities operational around the globe is listed in Table 7.1.

7.2 Principle of operation

A simplified process flow diagram of a typical plant is shown in Figure 7.1. A conventional CO$_2$ capture plant by absorption has two main units, absorber and desorber or stripper. The process works on the principle of temperature swing, that is, CO$_2$ is absorbed at a lower temperature and desorbed at a higher temperature. The gas to be treated is introduced at the bottom of the absorber where it comes in

Muhammad Akram, University of Sheffield

https://doi.org/10.1515/9783110563191-007

Table 7.1: List of worldwide absorption-based pilot-scale CO_2 capture plants.

Project	Location	CO_2 source
UKCCSRC Pilot-Scale Advanced Capture Technology (PACT)	UK	Coal, biomass, gas turbine, synthetic flue gas
The SINTEF pilot plant at Tiller	Norway	Propane burner for heating buildings
Sulzer Chemtech CCS pilot plant	Switzerland	Gas-fired burner for heating homes
Tomakomai CCS Demonstration Project	Japan	Hydrogen production
Sinopec Zhongyuan Carbon Capture Utilization and Storage Pilot Project	China	Fluidised catalytic cracker
Sinopec Shengli Oilfield Carbon Capture Utilization and Storage Pilot Project	China	Coal power plant
Plant Barry & Citronelle Integrated Project	USA	Coal power plant
KoSol Process for CO_2 Capture (KPCC) – Boryeong	South Korea	Coal power plant
CO_2 Capture Test Facility at Norcem Brevik	Norway	Cement
National Carbon Capture Center (NCCC)	USA	Coal power plant
Post-Combustion Capture (PCC) @CSIRO	Australia	Coal power plant
Shand Carbon Capture Test Facility (CCTF)	Canada	Coal power plant
Technology Centre Mongstad (TCM)	Norway	Catalytic cracker Gas fired CHP plant
TNO	Netherlands (Rotterdam)	Coal power plant

contact with a down coming liquid solvent. The solvent favourably absorbs CO_2 from the gas. The CO_2-depleted gas leaves from the top of the absorber. The gas is further processed in a wash tower to remove entrained and carried over solvent. The solvent after CO_2 absorption, termed as *rich solvent*, is pumped onto the top of the second tower where it flows down and sent to the reboiler. The solvent is heated up in the reboiler by using steam to release CO_2 from the solvent. Released CO_2 and steam produced in the reboiler due to heating up travel up in the striper and come in contact with the down coming solvent. Some of the CO_2 is released in this

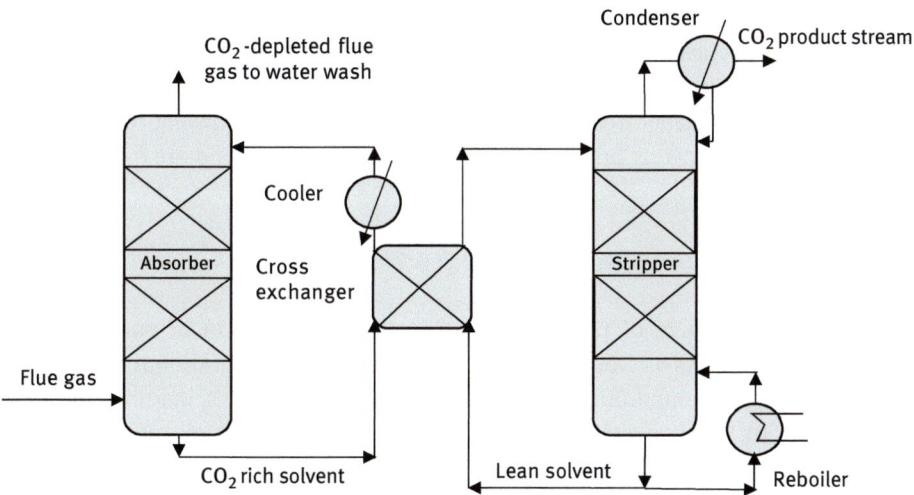

Figure 7.1: Simplified PFD of a typical CO_2 separation plant using absorption technology.

process, but majority of the CO_2 is stripped in the reboiler. The solvent after removing CO_2 is termed as *lean solvent*. The lean solvent from the reboiler is pumped back to the absorber and the cycle continues.

Total time for solvent to go through the whole cycle from absorber to desorber and back to absorber varies from plant to plant based on plant design, configuration and solvent properties. TCM has reported a calculated cycle time of approximately 45 min [2]. Similar numbers were reported for PACT pilot plant, Sheffield. Cycle time at PACT was measured by filling absorber with demineralised water (non-conductive) and desorber with amine (conductive) and measuring time for the solvent to become a homogeneous mixture [3] by measuring conductivity of the mixture at four different places in the plant.

Stripped CO_2 is passed through a cooler/condenser where steam and solvent vapours are condensed and separated from the gaseous stream and sent back to the process. The relatively dry and pure CO_2 stream can be used for different applications. It can be sent to compression process for further processing for storage purposes, process called *carbon capture and storage* (CCS). Alternately it can be used for manufacturing different products, process called *utilisation of CO_2*. The latter process and its application are detailed throughout this book.

Different strategies are employed for energy optimisation and operational control purposes. Lean solvent leaving the reboiler is hot that needs to be cooled before entering the absorber so is passed through a cross exchanger and heat is transferred to the rich solvent before it enters the desorber. The lean solvent is further cooled in a lean cooler to adjust its temperature to be suitable for the absorption process usually 35–45 °C.

The absorber and striper are normally packed towers but can also be tray towers. Various different types of packings are available in the market, broadly categorised into two categories, random and structured. Both of these have their own advantage and disadvantages. Structured packings generally have lower pressure drop, thus saving costs on gas blowers but are relatively difficult to install and are expensive.

7.2.1 Degradation – impact on costs and process

The major disadvantage of the technology is the consumption of energy during stripping. The process is very complex and sensitive to operational conditions. Changes in stripper temperature and pressure, lean and rich CO_2 loadings, concentration of CO_2 in the flue gas, solvent concentration, regeneration efficiency, capture efficiency are amongst the factors that influence stripping energy consumption. Different strategies are employed to tackle this issue, including flow sheet optimisation, new solvents and parametric optimisation. Loading of CO_2 into the solvent is one of the operational parameters, which is optimised for a specific process. To minimise energy consumption during monoethanolamine (MEA) regeneration, lean and rich loadings need to be maintained in the range of 0.35–0.49 mol/mol [4, 5].

Figure 7.2 shows change in reboiler duty as a function of capture efficiency. The heat duty increases sharply as the capture efficiency is increased beyond 90%. Therefore, it is generally recommended to maintain the capture efficiency at 90%. Lower capture efficiency requires less energy during stripping but at the cost of release of a higher amount of CO_2 into the atmosphere. Therefore, the relationship

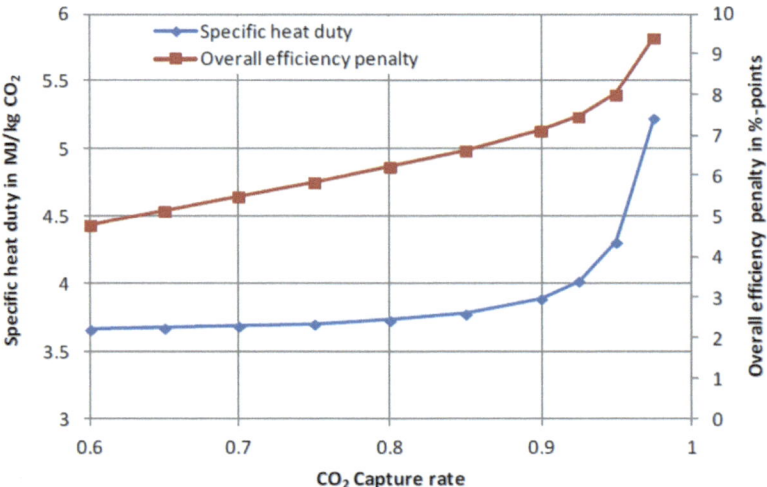

Figure 7.2: Energy penalty vs capture efficiency [8].

between the capture efficiency and the reboiler duty has to be optimised to get the maximum CO_2 capture at the least energy consumption.

The reboiler duty reported above does not consider solvent losses by degradation or evaporation. Solvent degrades due to thermal and oxidative degradation as well as due to reaction of solvent with contaminants in the flue gas, make up water, anti-foams and corrosion inhibitors [6]. Thermal degradation mainly occurs in the stripper depending upon stripper conditions, CO_2 loadings and solvent concentration [2]. Oxazolidin-2-one, MEA urea, 1-(2-hydroxyethyl) imidazolidin-2-one and N-(2-hydroxyethyl)-ethylenediamine are the main degradation products of MEA [7]. Oxidative degradation mainly occurs in the absorber due to the presence of oxygen in the flue gas and results in the production of, typically, ammonia [6] and heat-stable salts [7]. Unless removed, heat-stable salts accumulate in the process with time and reduce efficiency of the solvent to absorb CO_2 resulting in loss of solvent capacity and increased reboiler duty [2]. This phenomenon can lead to process problems such as fouling, corrosion and foaming [6], resulting in increased operational costs. If situation is not controlled, the cost of replacing degraded solvent can be enormous. For example, average annual cost for solvent make up for the Boundary Dam capture plant is estimated to be $17million. However, it went up due to higher than expected degradation of the solvent and they spent $32million on solvent maintenance in 2015–16.

7.3 Applications

Because of its flexible nature, post-combustion capture by absorption has a wide range of applications in the power and industrial sector, which will be briefly discussed in the following section.

7.3.1 Power

Power sector is expected to be the biggest beneficent of this technology in the near future. The technology has been demonstrated at commercial scale. There are only a few full-scale plants operational around the globe. Petra Nova and Boundary Dam are two prominent commercial operational plants capturing CO_2 from power plants. Petra Nova Carbon Capture plant (W. A. Parish power plant near Houston, Texas) is capable of capturing 1.4 $MtCO_2$ per year. It has been operational since January 2017 and is claimed to be the world's largest post-combustion CO_2 capture system currently in operation. The plant is capturing CO_2 from a slip stream of a 240 MW coal power plant at 90% capture efficiency using Mitsubishi Heavy Industries' KS-1 solvent [9]. Boundary Dam Carbon Capture (Sask Power), operational since 2014, is one such facility in Canada capturing over 2 $MtCO_2$ per year from a lignite fired power plant. The plant captures CO_2 at 90%

capture efficiency from flue gases using the Shell Cansolv process [10]. Operations at the Boundary Dam are supported by Shand carbon capture test laboratory. The captured CO_2 from both of these plants is being used for enhanced oil recovery.

Most of the effort has been focussed on the CO_2 capture from coal-fired plants followed by gas-CCS. Commercial plants operational on coal power plants are mentioned earlier. In addition to these, there are number of research facilities researching separation of CO_2 from coal using absorption-based CO_2 capture. PACT has 250 kW pulverised coal/biomass firing rig integrated to 1 t/day CO_2 capture plant.

Recently, after a surge in gas-fired plants, gas-CCS has been investigated widely. There are some research centres working in this field. TCM in Norway has 20 ktCO_2/year capture plant for separating CO_2 from the flue gases of a natural gas-fired combined heat and power (CHP) plant [11]. The PACT pilot plant is integrated with two CHP micro gas turbines of 330 kWth capacity each. The micro turbines are very lean combustion systems and produce flue gases having very low CO_2 concentration. In order to increase CO_2 concentration in the flue gases to represent commercial power plant gas turbine flue gases and conditions applicable to exhaust gas recycle (EGR) and Selective-EGR, CO_2 from cryogenic storage is injected into the slip stream of flue gases [12]. The Sulzer pilot plant treats 150 kg/h of flue gases from a commercial gas fired burner designed for heating homes [13], while the SINTEF pilot plant captures CO_2 at 50kg/h from a propane burner [14].

Capturing CO_2 from biomass power plants (BECCS) is considered desirable as it potentially can offer negative CO_2 emissions. According to a report published by Energy Technologies Institute [15], BECCS could deliver ca. 55 Mt/yr of negative emissions, roughly half of the UK emissions target, by 2050. However, these figures appear not to take into account the emissions during the life cycle of the biomass, that is, production, treatment, drying, transportation and so on.

There is almost no open literature information on practical CO_2 capture from biomass using absorption technology. Flue gases from biomass combustion can have many different types of contaminants that can impact the performance of the capture plant and can lead to increased emissions. Drax power station in the UK is claimed to be one of the biggest facilities for power produced from sustainable biomass. It is planning to install a number of capture projects based on post-combustion technology. The first pilot plant of this series is expected to capture 1t/day CO_2 from biomass flue gases. These projects when operational will open new venues for achieving, arguably, negative emissions and thus accelerating the path to achieve greenhouse gases emissions reduction targets.

7.3.2 Industry

Besides deploying CCS on power plants, it is also necessary to decarbonise the industrial sector if the planet is to be saved. Energy-intensive industries include steel, cement, aluminium and paper. Post-combustion capture seems to be the only technology to de-carbonise the cement, steel and aluminium sectors [16].

The Quest industrial CCS project captures more than 1 $MtCO_2$ per year from a steam methane reforming process. Captured CO_2 is transported through an underground pipeline to Basal Cambrian Sands for its permanent storage about 2 km beneath the ground. There are a number of planned CCS projects in the pipeline due to be operational post-2020. Up to date details of currently operational full-scale, demonstration and pilot-scale capture plants are available on Global CCS Institute website (https://www.globalccsinstitute.com/). However, combined global CCS projects represent less than 50 Mt/yr and a step-change in their roll-out is needed.

7.4 Flexibility of capture plants

Demand for electricity varies with season, climate on the day and time. Recently, UK has installed significant renewables capacity, including solar and wind. However, because of the intermittent nature of the renewables, thermal power plants should be designed to cover the short fall created by non-availability of the renewables. Moreover, coal power plants in the UK are aging and are expected to be replaced by gas-fired power plants. It is expected that installed capacity of renewable will also grow in the future. Therefore, gas-fired power plant capacity also needs to be increased to keep the lights on when there is no sun or wind.

Figure 7.3 shows a comparison of electricity demand variation on a typical working day and on a bank holiday in May 2018. The plot shows that on a working day the highest electricity demand was in the morning at around 7 am, while on a bank holiday the demand peaked in the evening at around 8:15 pm. The most important feature of the plot is to show the variation in load over a period of time during the day, which varied from 22 GW to 31 GW and from 24 GW to 34 GW on bank holiday and working day, respectively. Moreover, the demand is higher on a working day due to consumption in offices, class rooms and so on. The demand increased between 4:40 and 6:55 am very rapidly at a rate of 4 GW/hr (67 MW/min). The plot also shows CCGT generation on the working day. As can be witnessed from the plot, the variation in CCGT generation is in line with the demand. In order to compensate for the sharp increase in demand as witnessed on the working day morning, CCGTs have to undergo rapid load changes. Variations in the load on a CCGT power plant will also have a huge impact on the performance of the integrated CO_2 capture plant.

Capture plants have to be flexible enough to accommodate process variations in the source plant. In the power sector, particularly, flexibility of capture plant is very important to respond to power plant load variations. Gas-fired power plants operate on load following basis and undergo turndown and shutdown/start up scenarios quite frequently. At peak load demands, it is desirable to maximise the power output from the integrated power plant–capture plant setup. Different

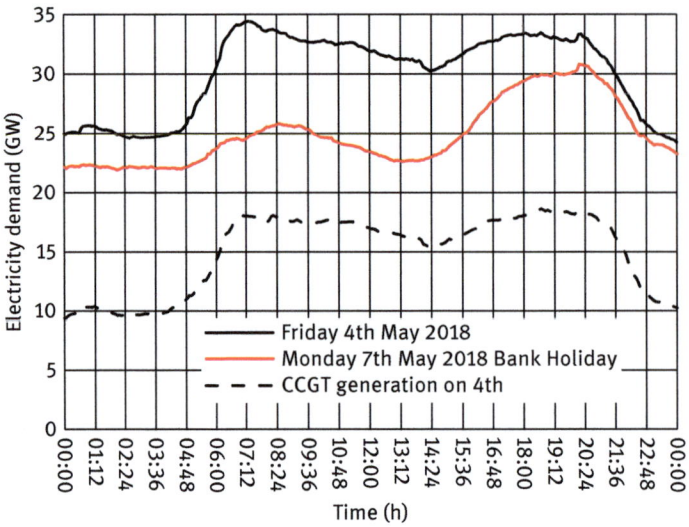

Figure 7.3: UK load variation during a day – comparison of a working day and a bank holiday [17].

technique are proposed to achieve this, namely, solvent storage and capture plant turn down or turn off. In the former case, no steam is used for stripping and rich solvent from the absorber is being stored in a storage tank. The stored solvent can be generated later when load demand is reduced. Although the technique can be used to enhance the power output during peak demands, it requires large tanks for storing the solvent that increases capital and operational costs. The latter technique involves reducing the capture plant load or turning down the capture plant based on the load demand. During this period, all or part of CO_2 will be emitted to the atmosphere. It may be required to capture CO_2 at a higher rate later on to compensate for the emitted CO_2 during peak demand period.

Moreover, shutting down the capture plant or stripper for short periods may have implications on CO_2 transport system due to intermittent supply of CO_2. In order to ensure steady supply of CO_2 to the pipeline, it may be required to store CO_2 onsite during periods of full load operation of CO_2 capture plant and release the stored CO_2 into the transport network when capture plant is turned down or shut off [18].

7.5 Recent developments

Despite its benefit of being well understood and the flexibility of integration with sources of CO_2 emissions, the process has a number of drawbacks. One of the major issues with this process is that it is very energy intensive. The major energy-consuming item of the process is the stripper, although the gas blower, pumps and coolers also contribute to the total energy cost. The regeneration of solvent in the stripper accounts for about 70% of the total operating cost of a gas-treating process [19], the majority of which is used for heating the water present in the solution. The other major problem is the degradation of the solvent being used for capturing of CO$_2$. The process works at elevated temperatures (above 120 °C) and pressures (above 1.2 bars), which lead to thermal degradation of the solvent. Moreover, the presence of oxygen in the flue gases can result in oxidative degradation of the solvent. Degradation and replacement of solvent can result in millions of pounds of revenue loss at commercial scale plants as mentioned in Section 7.2.

To address these issues, a number of the techniques are being proposed including process improvement through modifications and optimisation, alternative more stable solvents and process optimisation. These options are briefly discussed hereunder.

7.5.1 Process optimisation

One of the ways to enhance plant performance is optimisation by employing process modifications and optimisation. A number of modifications have been investigated and some of them have claimed to be better than the others.

Process modification can be differentiated in three main categories, namely, absorption enhancement, heat integration and heat pump effects [20]. Absorber inter-cooling and absorber inter-heating [21], rich solvent recycle and inter-heated stripper [22], split flow arrangement, multi-pressure stripper and rich solvent preheating [23], double loop absorber/stripper [24], rich solvent splitting and lean vapour compression [25], rich solvent flashing [26], rich vapour compression [27] and heat integrated stripper [28] amongst others have been reported as modifications for process enhancement.

The modifications can be applied individually or in combination. Combination of inter-cooling, rich solvent splitting, and semi-lean solvent extracting when applied to post-combustion capture plant integrated with a supercritical coal fired power plant have been shown to reduce the reboiler duty by around 10% [29].

Inter-cooled absorber, lean vapour compression or rich solvent split has been the focus of most of studies [20]. In some cases, the capture plant can be designed in such a way that it can be operated independent of the modification, but this is not always the case. Moreover, the modifications introduce additional complexity to the process; however, almost all of the modification result in the performance enhancement of the process [20].

Figure 7.4 shows a schematic of a double-loop absorber configuration as an example of process modification. In the process, the absorber is split into two section, each using a different solvents. The bottom section, where most of the absorption takes place, is fed by a partially stripped solvent, while the top section of the absorber is used for polishing the lean gas. Solvent from the lower section of the absorber is regenerated in the primary stripper, while that from the upper section is regenerated in the secondary stripper [24].

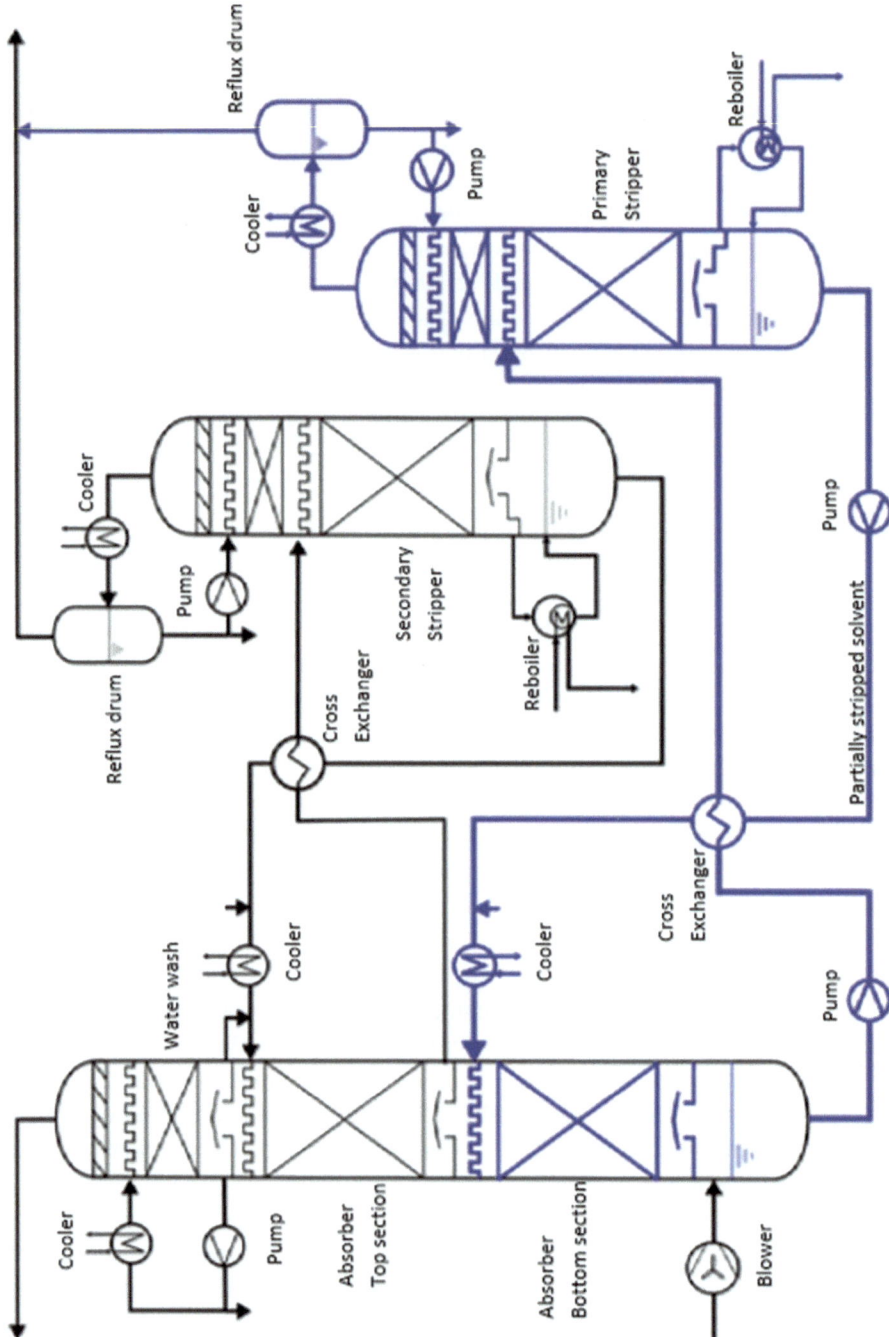

Figure 7.4: Double-loop absorber configuration [20].

Using lean vapour compression and absorber inter-cooling for a coal flue gas composition, Flour claimed to have reduced the reboiler duty to 2.9 MJ/kgCO_2 as opposed to 3.8–4 MJ/kgCO_2 in a conventional process using 30% MEA [30] Lean vapour recompression, even after considering compression energy, is found to be more effective than inter-cooling, showing 6% drop in reboiler duty at atmospheric flash pressure [16].

Flue gas from natural gas-fired power plants has a lower concentration of CO_2, which increases energy penalty per unit of CO_2 captured. To make the capture process energy efficient, by increasing the concentration of CO_2 in the flue gas, different technologies such as exhaust gas recycling (EGR), selective EGR (S-EGR) and humidified gas turbine cycles amongst others are proposed [31].

EGR involves the recycle of exhaust gas to join the air stream, while S-EGR involves the separation of CO_2 from the exhaust stream using a different technology such as membranes (see Figure 7.5) and recycle the CO_2 back to join the air stream. S-EGR can give higher CO_2 concentration in the exhaust gas as compared to EGR, which is limited due to the negative impact of reduction in oxygen concentration in the oxidant as the recycle ratio is increased. S-EGR, which can potentially increase the concentration of CO_2 in the flue gas by up to 18% [32], has been shown to decrease the energy penalty caused by the capture process by up to 40% [33].

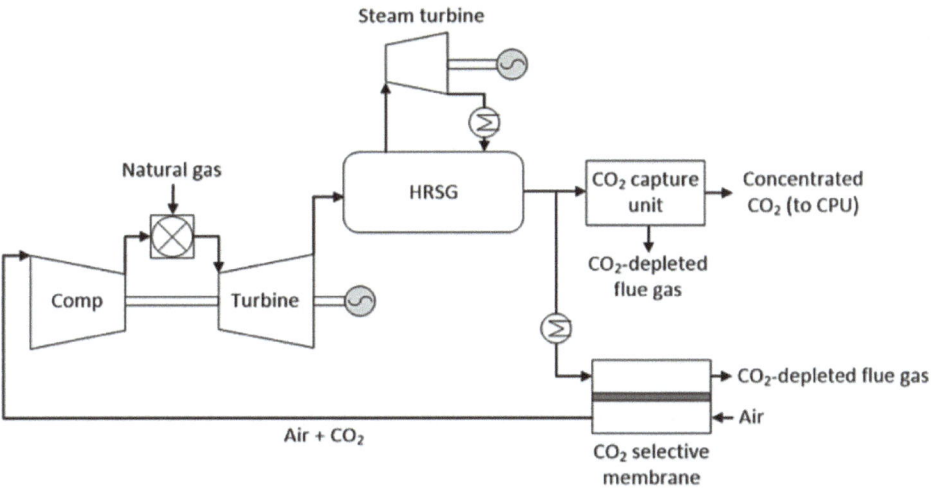

Figure 7.5: S-EGR with CO_2 capture – parallel configuration [31].

7.5.2 New solvents

Although MEA is the best known and a well-researched solvent for this technology, it has some drawbacks. The conventional solvent suffers from oxidative and thermal degradation, corrosion and low absorption capacity [34]. Therefore, considerable effort has been devoted to develop more stable and efficient alternative solvents. Physical properties of solvents such as density, viscosity, vapour pressure, heat capacity, heat of absorption, diffusivity, absorption capacity, kinetics and regeneration capability [35] play a key role in their performance for CO_2 capture applications. Many companies are active in this field and some of them have made encouraging claims.

Many different types of amines are being investigated. Single amine or mixture of different amines in different proportions and additives are proposed to enhance the performance of the system. Piperazine, due to high reactivity with CO_2 and resistance to oxidative and thermal degradation, has been suggested as solvent for CO_2 capture [36]. Despite its advantages, piperazine also has some disadvantages such as limited water solubility and higher volatility than MEA [37]. According to Yang et al. [38], some cyclic diamine and triamine solvents have shown to have better performance than MEA and piperazine, improving the cyclic capacity by up to 273%.

Application of blended amines is being favoured to take advantage of higher reaction rate of primary amine and higher loading capacity and lower regeneration costs of tertiary amines [19]. For example, Cansolv Technologies, Inc. introduced a tertiary amine solvent that has demonstrated fast mass transfer, good chemical stability and relatively higher net capacity (twice that of MEA, 0.5 mol of CO_2/mole of amine per cycle) [39].

A sterically hindered amine, for example, 2-amino-2-methyl-1-propanol have shown to have higher CO_2 loading than those of other amines and tertiary amine, for example, triethanolamine (TEA) showed lower CO_2 loading at all temperature ranges when compared to other amines [40]. At a reboiler temperature of 120 °C, regeneration energy of 20 wt% diethylenetriamine DETA + 10 wt% piperazine PZ aqueous solution was shown to be reduced by 25% in comparison to the regeneration energy of the 30 wt% MEA [41].

The rate of CO_2 absorption increases if the water in a MEA solution is replaced by a physical solvent, the so-called partially aqueous amine solvent, due to reduced operating CO_2 loading (higher free MEA concentration), greater CO_2 physical solubility and greater MEA activity [42]. Figure 7.6 shows a typical example of increase in CO_2 absorption rate of MEA when part of water was replaced with N-methyl-2-pyrrolidone (NMP).

Aqueous alkaline salts of amino acids are an alternative to alkanolamine solvents due to their lower volatilities and environmentally friendly nature [43] and good resistance to oxidative degradation [44]. However, as with other sorbents, amino acid salts have their own drawbacks, such as precipitation at high concentrations or high CO_2 loading leading to lower mass transfer rate and a possibility of damaging the process equipment [45].

Ionic liquids are another class of solvents used as an alternative to amines. Ionic liquids have exceptional physicochemical properties but also have some disadvantage such as hygroscopic nature, high viscosity and high cost. Combining amines with ionic liquids have been proposed as a better route to take advantage of superior qualities of the parent solvent [19].

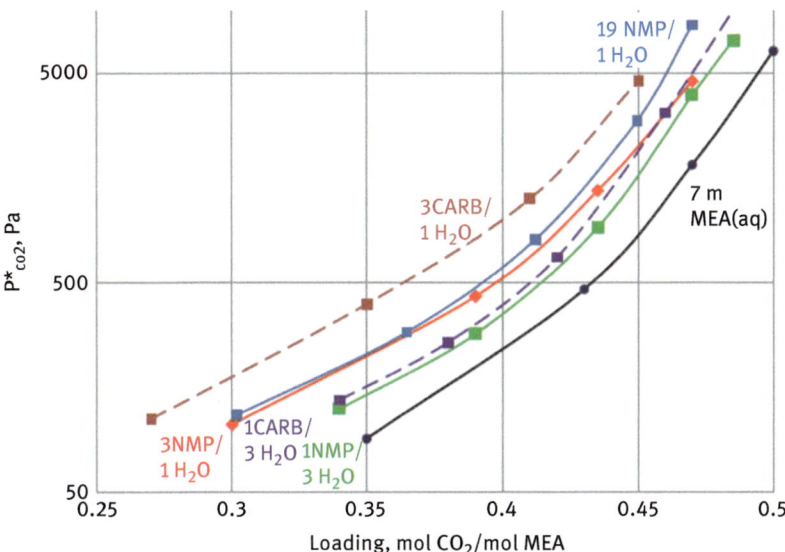

Figure 7.6: CO$_2$ solubility of 7 m MEA in NMP/water at 40 °C [42].

Binding organic liquids or switchable solvents are another class of solvents that have tunable physicochemical properties like ionic liquids [19]. Organic solvents cannot be used in amine scrubbers due to their higher volatility [42]. More studies are required to assess the performance of these potential solvents at optimised process conditions, but the limitation is that the chemicals are not available in bulk quantities to be tested at a higher scale [38].

7.5.3 Process intensification

Process intensification is a technique widely employed in process industries for process improvements and economic benefits. Replacing a conventional packed bed absorber with a rotating packed bed (RPB) absorber is one of the methods proposed to improve mass transfer coefficient of the solvent/CO$_2$ reaction. RPB uses centrifugal force to boost mass transfer coefficient by improving the slip velocity, flooding characteristics and interfacial shear stress [46]. A simplified layout of the capture plant designed with intensified absorber and stripper is shown in Figure 7.7.

A coal fired sub-critical power plant of 500 MWe if equipped with conventional CO$_2$ capture plant will need two absorbers having 17 m packing each and 9 m diameter each [47]. Process intensification can reduce capital costs significantly and improve process dynamics [48]. Volume of packing required for a traditional packed

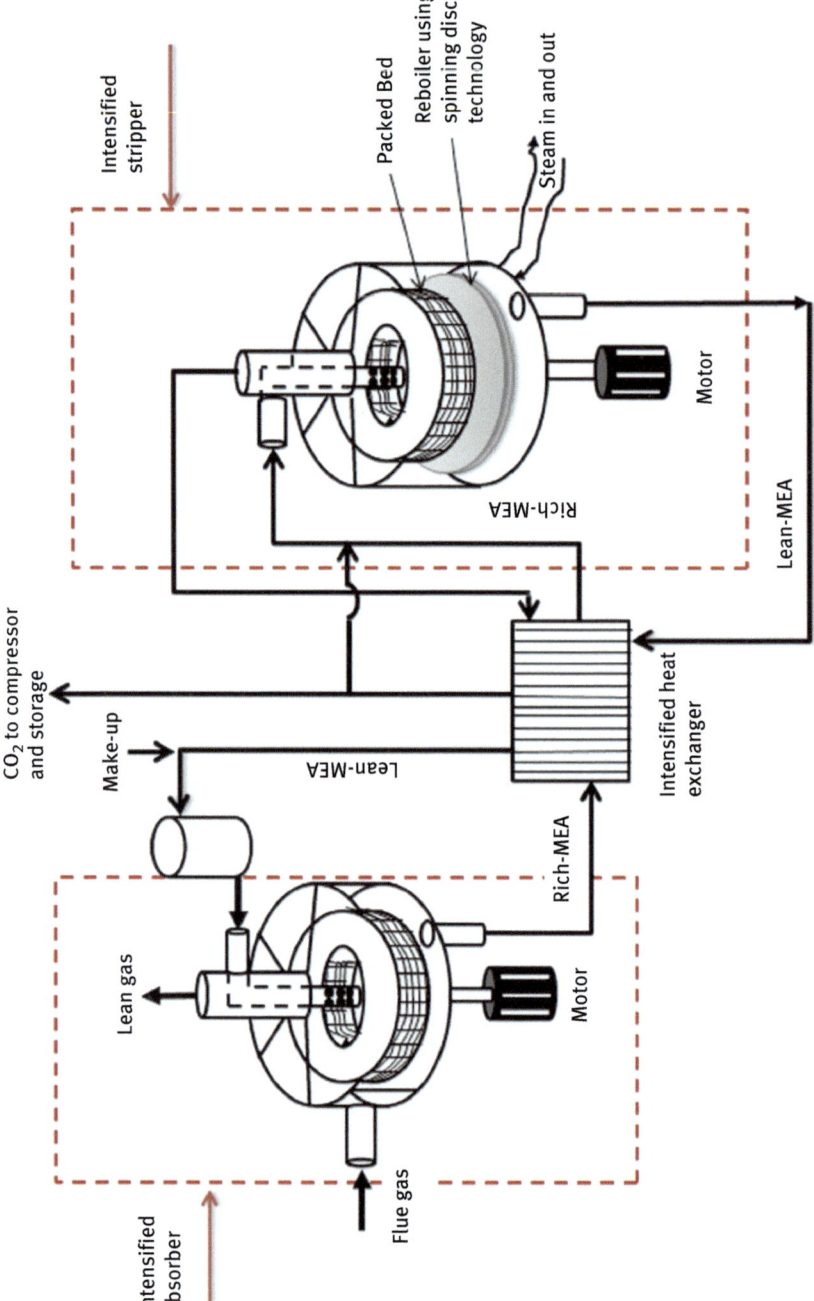

Figure 7.7: Simplified PFD of an RPB CO_2 capture process [37].

bed absorber as compared to an RPB absorber capturing 100 t/d of CO_2 is around 6 times higher for 30% MEA and 30 times higher for 3% aqueous ammonia [49]. Jassim et al. [46] have shown that height and diameter of stripper can be reduced by a factor of 8.4 and 11.3, respectively, by process intensification.

RPB technology reduces energy consumption per unit of CO_2 captured and it is possible to achieve same regeneration efficiency as a conventional packed bed by using 1/10th volume in a RPB due to higher heat and mass transfer coefficients and thus lower height of transfer unit in the case of RPB [41]. Of course, RPBs require mechanical energy to rotate [50], which is not the case for traditional packed beds, but the amount of energy is negligible when compared to the stripping energy costs [41].

A combination of two RPBs in series can decrease the regeneration energy by up to 9.5% and increase rich loading as compared to a single RPB of the same total volume [51]. The scale-up of RPBs could be a challenge that needs to be investigated. To predict the hydraulic and mass transfer behaviour of the RPB accurately, CFD and process modelling studies should be coupled together [37].

A cross exchanger, an integral part of absorption-based CO_2 capture process, also needs to be intensified [37]. Printed circuit heat exchanger and Marbond heat exchangers due to their compact design, high efficiency, low pressure drop and better economics [52–54] are better candidates to replace conventional heat exchangers being used currently in such plants [37].

7.6 Techno-economics

Capital and operational costs of a CO_2 capture plant can add significant expenditure to the power plant costs. Optimised integration of capture process with power plant can reduce OPEX. One of the parameters to determine the cost of CO_2 absorber, which contributes to around 30% of the CAPEX of the capture plant, is the rate of CO_2 absorption [55].

Increasing CO_2 concentration of NGCC flue gas from 4% to 6%, using EGR, decreases CAPEX by about 35% [56]. Figure 7.8 shows the impact of inlet flue gas CO_2 concentration on the total yearly cost (TOTEX) of post-combustion CO_2 capture plant assuming a plant life time of 20 years at an interest rate of 10%.

Levelised cost of electricity (LCOE) is usually used as an overall cost comparison parameter for different electricity generation technologies. The LCOE is defined as the total cost of generating electricity divided by the amount of electricity generated and is represented as £/kWh. LCOE increases by around 42% when a NGCC plant is equipped with a post-combustion CO_2 capture plant using MEA but drops to 37% by using EGR and to 31% if MEA is replaced with a proprietary solvent [57].

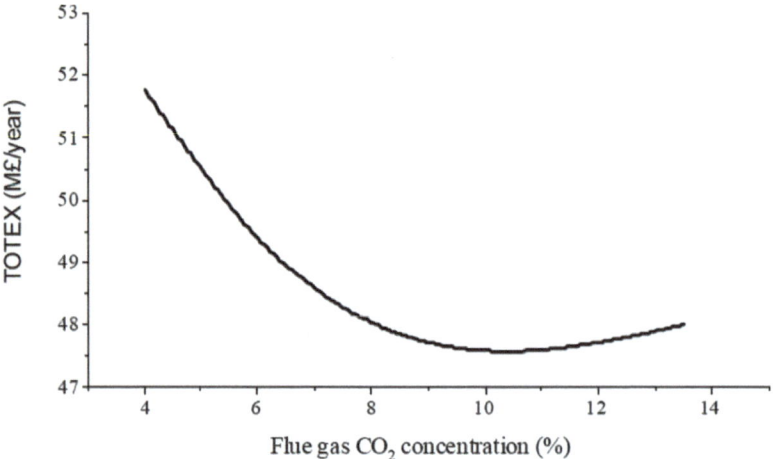

Figure 7.8: Variation of the plant total annual cost (TOTEX) in million pounds sterling (M£) with flue gas CO_2 concentration [55].

7.7 Future trends

In order to meet greenhouse gas emissions reduction targets, carbon capture utilisation and storage (CCUS) has to be deployed at a wide scale at least in the near future. Post-combustion capture, due to its flexible nature and well-known technology, is a leading candidate for the application in the near future. Therefore, it is expected that a number of these plants will be installed and operational by 2030 and beyond. During this period, emerging and advanced technologies with lower costs need to be available to replace the conventional technology.

7.8 Concluding remarks

CCUS, at the moment, is expected to be the first choice technology as without CCUS decarbonisation will have huge cost implications. Governments' needs to make policies to deploy CCUS at scale on gas- and biomass-fired power plants immediately and then in the industrial sector. There are some technical barriers, but there are much greater financial constraints to the deployment of the technology. Delay in deployment will make it harder to meet the emissions reduction targets and will cost more.

References

[1] Figueroa JD., Fout T., Plasynski S., McIlvried H., Srivastava RD. 2008. "Advances in CO_2 capture technology – The U.S. Department of Energy's Carbon Sequestration Program." Int. J. Greenh. Gas Control, 2: 9–20.

[2] Flø N.E., Faramarzi L., de Cazenove T., et al., 2017, "Results from MEA Degradation and Reclaiming Processes at the CO_2 Technology Centre Mongstad." Energy Procedia 114: 1307–1324.

[3] Tait P., Buschle B., Milkowski K., Akram M., Pourkashanian M., Lucquiaud M., 2018. "Flexible operation of post-combustion CO_2 capture at pilot scale with demonstration of capture-efficiency control using online solvent measurements." Int. J. Greenh. Gas Control, Volume 71: 253–277.

[4] Abu-Zahra, M.R., Niederer, J.P., Feron, P.H., Versteeg, G.F., 2007a. "CO_2 capture from power plants: Part II. A parametric study of the economical performance based on mono-ethanolamine." Int. J. Greenh. Gas Control 1: 135–142.

[5] Abu-Zahra, M.R., Schneiders, L.H., Niederer, J.P., Feron, P.H., Versteeg, G.F., 2007b. "CO_2 capture from power plants: Part I. A parametric study of the technical performance based on monoethanolamine." Int. J. Greenh. Gas Control 1: 37–46.

[6] Abdi MA, Golkar MM. 2011. "Improve contaminant control in amine systems." Hydrocarbon Processing. 63: 102C–102I.

[7] Dumée L, Scholes C, Stevens G, Kentish S. 2012. "Purification of aqueous amine solvents used in post combustion CO2 capture: A review." Int. J. Greenh. Gas Control 10: 443–455.

[8] Mletzko J., Ehlers S., Kather A., 2016. "Comparison of natural gas combined cycle power plants with post combustion and oxyfuel technology at different CO_2 capture rates". Energy Procedia 86: 2–11.

[9] NRG. NRG Energy, JX Nippon complete world's largest post-combustion carbon capture facility on-budget and on-schedule. [Online]. Available at: http://investors.nrg.com/phoenix.zhtml?c=121544&p=irol-newsArticle&ID=2236424 [28 April 2018].

[10] Stéphenne K.2014. Start-up of world's first commercial post-combustion coal fired CCS project: Contribution of Shell Cansolv to SaskPower Boundary Dam ICCS Project." Energy Proc 63: 6106–6110.

[11] de Cazenove T., Bouma RHB., Goetheer ELV., van Os PJ. and Hamborg ES. 2016. "Aerosol measurement technique: Demonstration at CO_2 Technology Centre Mongstad." Energy Proc 86:160–170.

[12] Akram M., Ali U., Best T., Blakey S., Finney KN. and Pourkashanian M., 2016. "Performance evaluation of PACT Pilot-plant for CO2 capture from gas turbines with Exhaust Gas Recycle." Int. J. Greenh. Gas Control 47:137–150.

[13] Notz R, Mangalapally HP. and Hasse H. 2012. "Post combustion CO_2 capture by reactive absorption: Pilot plant description and results of systematic studies with MEA." Int. J. Greenh. Gas Control 6: 84–112.

[14] Mejdell T. Vassbotn T. Juliussen O., et al. 2011. "Novel full height pilot plant for solvent development and model validation." Energy Procedia 4:1753–1760.

[15] ETI. 2016. The Evidence of Deploying Bioenergy with CCS (BECCS) in the UK, An insights report from the Energy Technologies Institute. Downloaded on 17-12-2018 (https://www.eti.co.uk/insights/the-evidence-for-deploying-bioenergy-with-ccs-beccs-in-the-uk)

[16] Kvamsdal HM., Haugen G., Svendsen HF., Tobiesen A., Mangalapally H., Hartono A., Mejdell T., 2011. Modelling and simulation of the Esbjerg pilot plant using the Cesar 1 solvent, Energy Procedia, Volume 4, 2011, Pages1644–1651.

[17] National Grid, 2018. Data downloaded from National Grid website http://www.gridwatch.templar.co.uk/ on 08/05/2018.

[18] IEA GHG 2012. Operating flexibility of power plants with CCS. IEA, Cheltenham, UK.

[19] Kumar S. Cho JH. Moon I. 2014. "Ionic liquid-amine blends and CO_2 BOLs: Prospective solvents for natural gas sweetening and CO2 capture technology – A review." Int. J. Greenh. Gas Control 20: 87–116.

[20] Le Moullec Y., Neveux T., Al Azki A., Chikukwa A. and Hoff KA., 2014. "Process modifications for solvent-based post-combustion CO_2 capture." Int. J. Greenh. Gas Control 31: 96–112

[21] Aroonwilas, A., Veawab, A., 2011. Heat recovery gas absorption process. University of Regina. US 7906087.

[22] Madan, T., Sachde, D., Lin, Y.-J., Frailie, P., Rochelle, G.T., 2013. "Improved process configurations for amine scrubbing." In: 7th Trondheim Conference on CO_2 Capture, Transport and Storage, Trondheim (NO), 4–6 June.

[23] Ahn, H., Luberti, M., Liu, Z., Brandani, S., 2013. "Process configuration studies of the amine capture process for coal-fired power plants." Int. J. Greenh. Gas Control 16: 29–40.

[24] Towler, G.P., Shethna, H.K., Cole, B., Hajdik, B., 1997. Improved absorber–stripper technology for gas sweetening to ultra-low H_2S concentrations. In: Proceedings of the 76th GPA Annual Convention, Tuka (OK). 93–100.

[25] Amrollahi, Z., Ertesvag, I.S., Bolland, A., 2011. "Optimized process configurations of post-combustion CO_2 capture for natural-gas-fired power plant-exergy analysis." Int. J. Greenh. Gas Control 5: 1393–1405.

[26] Akiyama, T., 2010. Method and device for removing CO_2 and H_2S. Hitachi.US2010/0101416.

[27] Koch, M., Naumovitz, J.P., 2012. Method and system for reducing energy requirement of a CO_2 capture system. Alstom Technology. WO2012/036878.

[28] Oyenekan, B.A., Rochelle, G.T., 2005. "Energy performance of stripper configurations for CO_2 capture by aqueous amines." Ind. Eng. Chem. Res. 45 (8): 2457–2464.

[29] Oh SY. Yun S. Kim JK. 2018. "Process integration and design for maximizing energy efficiency of a coal fired power plant integrated with amine-based CO_2 capture process." Appl. Energ. 216:311–322.

[30] Reddy, S., 2008. Econamine FG Plus[SM] Technology for Post-combustion CO_2 capture, presentation at the 11th meeting of the International Post-Combustion CO_2 Capture Network, May 20–21, 2008, Vienna, Austria.

[31] Diego ME., Akram M., Bellas JM., Finney KN. and Pourkashanian M. 2017. "Making gas-CCS a commercial reality: The challenges of scaling up." Greenhouse Gases: Science and Technology 7 (5): 778–807.

[32] Herraiz L., 2016. "Selective Exhaust Gas Recirculation in Combined Cycle Gas Turbine Power Plants with Post-combustion Carbon Capture." PhD Thesis. University of Edinburgh, Scotland, UK.

[33] Merkel TC. Wei X., He Z., White LS. Wijmans JG. Baker RW. 2013. "Selective exhaust gas recycle with membranes for CO_2 capture from natural gas combined cycle power plants." Ind. Eng. Chem. Res. 52: 1150–1159 (2013).

[34] Rochelle, G.T., Bishnoi, S., Chi, S., Dang, H., Santos, J., 2001. Research Needs for CO_2 Capture From Flue Gas by Aqueous Absorption/Stripping. US Department of Energy, Pittsburgh, PA.

[35] Bougie, F., Iliuta, M.C., 2012. "Sterically hindered amine-based absorbents for the removal of CO_2 from gas streams." J. Chem. Eng. Data 57: 635–669.

[36] Freeman SA, Dugas R, Van Wagener DH, Nguyen T, Rochelle GT. 2010. "Carbon dioxide capture with concentrated, aqueous piperazine." Int. J. Greenh. Gas Control 4: 119–24.

[37] Wang M., Joel AS., Ramshaw C., Eimer D., Musa NM. 2015. "Process intensification for post-combustion CO_2 capture with chemical absorption: A critical review." Appl. Energ. 158: 275–291.

[38] Yang Q., Puxty G., James S., Bown M., Feron P., Conway W.2016. "Toward intelligent CO$_2$ capture solvent design through experimental solvent development and amine synthesis." Energy Fuels 30: 7503–7510.

[39] Hakka, L., 2007. Cansolv Technologies Inc., private communication.

[40] Kim, Y.E., Lim, J.A., Jeong, S.K., Yoon, Y.I., Bae, S.T., Nam, S.C., 2013. "Comparison of carbon dioxide absorption in aqueous MEA, DEA, TEA and AMP solutions." Bulletin of Korean Chemical Society 34: 783–787.

[41] Cheng H., Lai C., Tan C. 2013. "Thermal regeneration of alkanolamine solutions in a rotating packed bed." Int. J. Greenh. Gas Control 16: 206–16.

[42] Yuan Y. Rochelle GT. 2018. "CO$_2$ absorption rate in semi-aqueous monoethanolamine." Chem. Eng. Sci. 182: 56–66.

[43] Aronu, U.E., Svendsen, H.F., Hoff, K.A., 2010. "Investigation of amine amino acid salts for carbon dioxide absorption." Int. J. Greenh. Gas Control 4, 771–775.

[44] Kumar, P.S., Hogendoorn, J.A., Timmer, S.J., Feron, P.H.M., Versteeg, G.F., 2003a. "Equilibrium solubility of CO$_2$ in aqueous potassium taurate solutions: Part 2. Experimental VLE data and model." Ind. Eng. Chem. Res. 42, 2841–2852.

[45] Kumar, P.S., Hogendoorn, J.A., Feron, P.H.M., Versteeg, G.F., 2003b. "Equilibrium solubility of CO$_2$ in aqueous potassium taurate solutions: Part 1. Crystallization in carbon dioxide loaded aqueous salt solutions of amino acids." Ind. Eng. Chem. Res. 42: 2832–2840.

[46] Jassim MS., Rochelle G., Eimer D., Ramshaw C. 2007. "Carbon dioxide absorption and desorption in aqueous monoethanolamine solutions in a rotating packed bed." Ind. Eng. Chem. Res. 46: 2823–33.

[47] Lawal A., Wang M., Stephenson P., Obi O. 2012. "Demonstrating full-scale post combustion CO2 capture for coal-fired power plants through dynamic modelling and simulation." Fuel 101: 115–28.

[48] Joel AS., Wang M., Ramshaw C. 2015. "Modelling and simulation of intensified absorber for post-combustion CO$_2$ capture using different mass transfer correlations." Appl. Therm. Eng. 74: 47–53.

[49] Kang JL., Wong DSH., Jang SS., Tan CS. 2016. "A comparison between packed beds and rotating packed beds for CO$_2$ capture using monoethanolamine and dilute aqueous ammonia solutions." Int. J. Greenh. Gas Control 46: 228–239.

[50] Keyvani, M., Gardner, N.C., 1992. Operating Characteristics of Rotating Beds. Technical Progress Report DOE # DF-FG22–87PC79924.

[51] Yu CH., Chen MT., Chen H., Tan CS. 2016. "Effects of process configurations for combination of rotating packed bed and packed bed on CO$_2$ capture." Appl. Energ. 175: 269–276.

[52] Reay, D. 1999. Learning from experiences with compact heat exchangers. Sittard, Netherlands: Centre for the Analysis and Dissemination of Demonstrated Energy Technologies, Caddet Analysis Support Unit.

[53] Hesselgreaves JE. 2001. Compact heat exchangers: Selection, design and operation. Gulf Professional Publishing.

[54] Pierres RL. 2013. Diffusion bonded compact heat exchangers – compact reactors presentation to PIN, 23rd May, 2013. Downloaded on 2nd of May 2018 from www.pinetwork. org/pubs/PIN21/lepierres.pdf

[55] Frailie P.T. 2014. Modeling of Carbon Dioxide Absorption/Stripping by Aqueous Methyldiethanolamine/Piperazine, PhD Thesis, University of Texas at Austin.

[56] Rezazadeh F. 2016. Optimal integration of post-combustion CO$_2$ capture process with natural gas fired combined cycle power plants, PhD Thesis, University of Leeds.

[57] Smith N. Miller G. Aandi I., Gadsden R., Davison J., 2013. "Performance and costs of CO$_2$ capture at gas fired power plants." Energy Procedia, 37, 2443–2452.

Melis S. Duyar, Shuoxun Wang, Martha A. Arellano-Treviño
and Robert J. Farrauto

8 CO$_2$ capture and catalytic conversion using solids

8.1 Introduction

The use of solid adsorbent materials for CO$_2$ capture enables the possibility to integrate catalytic materials into the process, and to convert captured CO$_2$ directly to value-added products, once concentrated in the solid material. Such dual-function materials (DFMs) offer a unique technological solution to the CO$_2$ capture and utilisation challenges for mitigating fossil fuel-based emissions. By combining an energy-consuming process (CO$_2$ capture/release cycle) with an energy-releasing process (CO$_2$ hydrogenation to a chemical), DFMs can potentially drive down the energy requirements of both CO$_2$ capture and utilisation by achieving a favourable energy balance within a single reactor. This chapter introduces the concept of DFMs for CO$_2$ capture and presents the first demonstrations producing synthetic natural gas directly from CO$_2$ emissions. Kinetic and cyclic tests for materials development are presented, and an outlook for the future of this concept is discussed.

Solids for CO$_2$ capture can be functionalised to capture CO$_2$ from dilute streams and in lieu of a typical sorbent regeneration cycle, they instead perform a desirable reaction on captured CO$_2$ to yield a concentrated carbon-containing product. Since there are a variety of exothermic reactions of CO$_2$ with a number of co-reactants, this mode of operation can take advantage of heat generation through the reactions of CO$_2$ to drive the endothermic desorption from sites of adsorption.

DFM addresses several of the weakness associated with the state-of-the-art CO$_2$ capture technology using liquid monoethanolamine as a solvent. This absorbent is highly corrosive and can be used only when diluted with water. Hence, in the operation of industrial CO$_2$ capture units, a large energy penalty arises from the regeneration of the absorbent (to release captured CO$_2$) due to its high water content and its energy-intensive volatilisation. This ultimately renders it too costly in post-combustion CO$_2$ capture and utilisation applications. Figure 8.1 shows the schematic for the operation of a DFM process for post-combustion CO$_2$ capture. The

Melis S. Duyar, SUNCAT Center for Interface Science and Catalysis, SLAC National Accelerator Laboratory, USA Chemical and Process Engineering University of Surrey Guildford, Surrey, GU2 7XH, UK; Department of Chemical Engineering, Stanford University, Stanford, USA
Shuoxun Wang, Martha A. Arellano-Treviño, Robert J. Farrauto, Department of Earth and Environmental Engineering, Columbia University, New York, USA

https://doi.org/10.1515/9783110563191-008

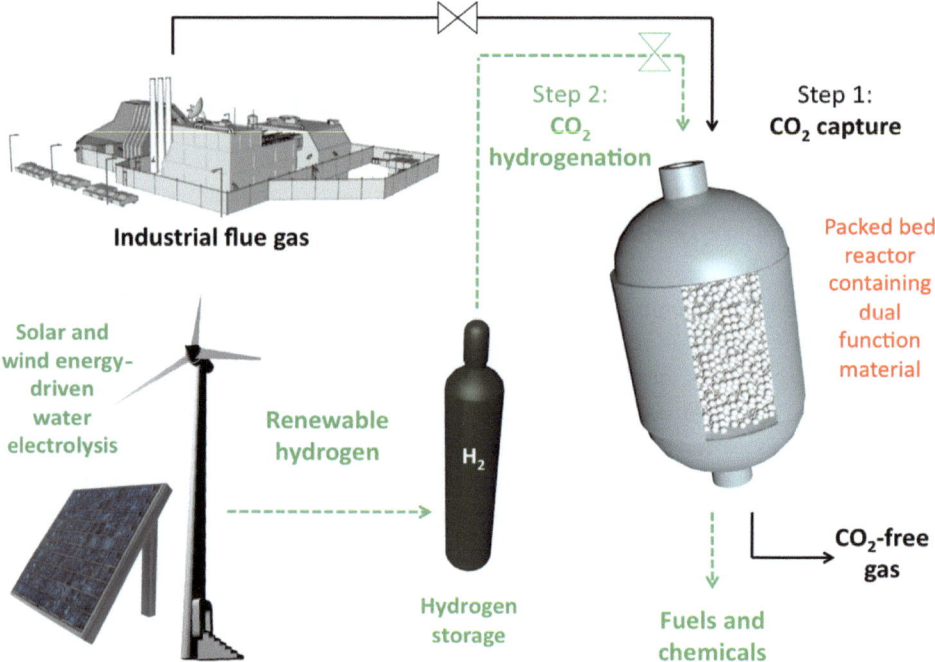

Figure 8.1: The dual-function material (DFM) concept offering CO_2 capture and utilisation in a single-process unit. Since the reactor will be operated in capture and hydrogenation cycles, parallel reactors are needed to run this system continuously. Both CO_2 capture and hydrogenation steps can operate at the same temperature, if the heat generation from reaction matches the requirement for CO_2 desorption and migration to catalytic sites.

solid DFM captures CO_2 and converts it to a synthetic fuel or a high-value chemical in one reactor at one temperature and at the site of CO_2 generation utilising renewable H_2. The sensible heat of flue gas exhaust itself provides the energy necessary for capture and conversion.

8.2 Proof of concept for DFMs: Synthetic natural gas production from flue gas

For a DFM to work under realistic operating conditions, it is essential to demonstrate selective capture of CO_2, and production of a value-added chemical or fuel from captured CO_2. In most applications, dilute CO_2 will be emitted as a mixture with nitrogen, oxygen, steam and additional species depending on the source of emissions. In light of these constraints, material optimisation, cyclic stability and kinetic considerations for the development of DFMs are discussed below.

The DFM concept was first demonstrated by combining a reversible intermediate temperature (300–650 °C) adsorbent (CaO/γ-Al$_2$O$_3$) [1, 2] and high mass-based activity methanation catalyst (Ru) [3–5]. Initial proof of concept studies employed a testing protocol where the DFM was first saturated with CO$_2$ from a dilute simulated post-combustion flue gas stream, then purged with inert gas to remove any remaining CO$_2$ in the gas lines and finally by hydrogenating captured CO$_2$ under isothermal conditions (Figure 8.2). In this first DFM, the following steps were hypothesised to occur, with CO$_2$ capture taking place on both adsorbent (CaO) and catalytic (Ru) sites and CO$_2$ hydrogenation taking place on only the Ru sites.

Step 1: CO$_2$ capture from simulated flue gas

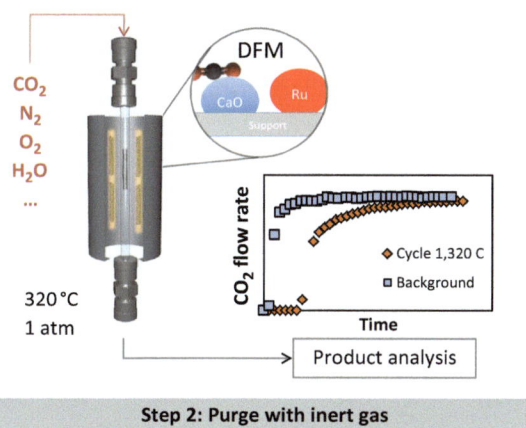

Step 2: Purge with inert gas

Step 3: Hydrogenation of captured CO$_2$

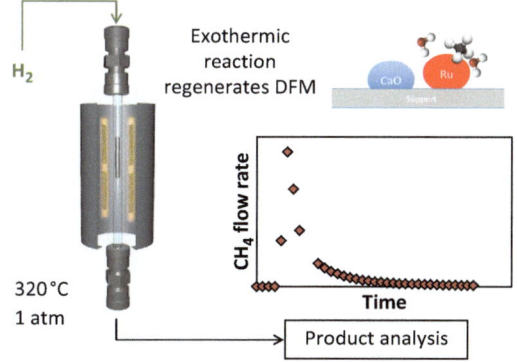

Figure 8.2: DFM proof of concept experimental protocol for Ru-CaO/Al$_2$O$_3$. Testing was performed using a quartz-packed bed reactor operating at atmospheric pressure and 320 °C. Product analysis shown is from data obtained using an online micro-GC. A three-step test protocol is used to demonstrate that captured CO$_2$ is the source of CH$_4$ formed. Time resolution is an important factor for accurately determining CO$_2$ capture capacity and conversion of captured CO$_2$ to desired products.

$$CO_2 + CaO^* \rightarrow CO_2 \cdots CaO \; \Delta H_{rxn} < 0 \tag{8.1}$$

$$CO_2 + Ru^* \rightarrow CO_2 \cdots Ru \; \Delta H_{rxn} < 0 \tag{8.2}$$

$$CO_2 \cdots CaO \rightarrow CO_2 + CaO^* \; \Delta H_{rxn} > 0 \tag{8.3}$$

$$CO_2 + 4H_2 \rightarrow CH_4 + 2H_2O \; \Delta H_{rxn} = -164 \, kJ/mol \tag{8.4}$$

The initial work showed that CO_2 adsorbs on both CaO and Ru (eqs. 8.1, 8.2), and is hydrogenated to methane on Ru sites (eq. 8.4) [5]. The exothermic methanation reaction (eq. 8.4) supplies all the heat required for the desorption of CO_2 from CaO sites (eq. 8.3) and their migration to Ru sites where they become hydrogenated; this allows the system to operate at the same temperature during both steps [5]. Moreover, the whole process works in a simulated flue gas environment, containing dilute CO_2, and with a stable performance in the presence of water and oxygen, making the system a realistic solution for post combustion CO_2 capture and utilisation [5].

The DFM concept has since been further developed and scaled-up using the alkali adsorbent/methanation catalyst system and adopted by other researchers [6–11]. There is tremendous room for materials research to develop high surface area, thermally stable, Earth crust abundant materials with high CO_2 adsorption capacity and catalytic activity towards valuable products such as alcohols, oxygenates, olefins and more, in the presence of a suitable co-reactant. Moreover, there is a need to understand the processes occurring at the atomic scale such as CO_2 adsorption, spillover to catalytic sites, reaction and desorption to improve the performance of DFMs.

8.2.1 Kinetic considerations and material optimisation for CO_2 capture and methanation

DFM development begins with choosing a high-capacity adsorbent and active and selective catalyst. Optimisation of the material composition relies on balancing the rate of reaction with the rate of CO_2 desorption during CO_2 conversion (Figure 8.2, Step 3). Since the DFM relies on heat generated from the exothermic reaction on the catalyst to drive the endothermic desorption of CO_2 from the adsorbent, mismatched rates can result in unreacted CO_2 desorption, or inadequate heat generation leading to low working capacity of the DFM and low yield of product. Transport limitations might also arise due to support morphology and choice of catalyst, adsorbent or operating conditions. To address these issues, it is necessary to optimise the loadings of adsorbent and catalyst based on kinetic insight on different processes taking place in the DFM.

Wang determined an empirical rate law for a Ru-"Na_2O"/Al_2O_3 DFM powder [12], by measuring the methanation rate dependence on hydrogen partial pressure (Figure 8.3A), CO_2 coverage (Figure 8.3B) and temperature (Figure 8.3C). These

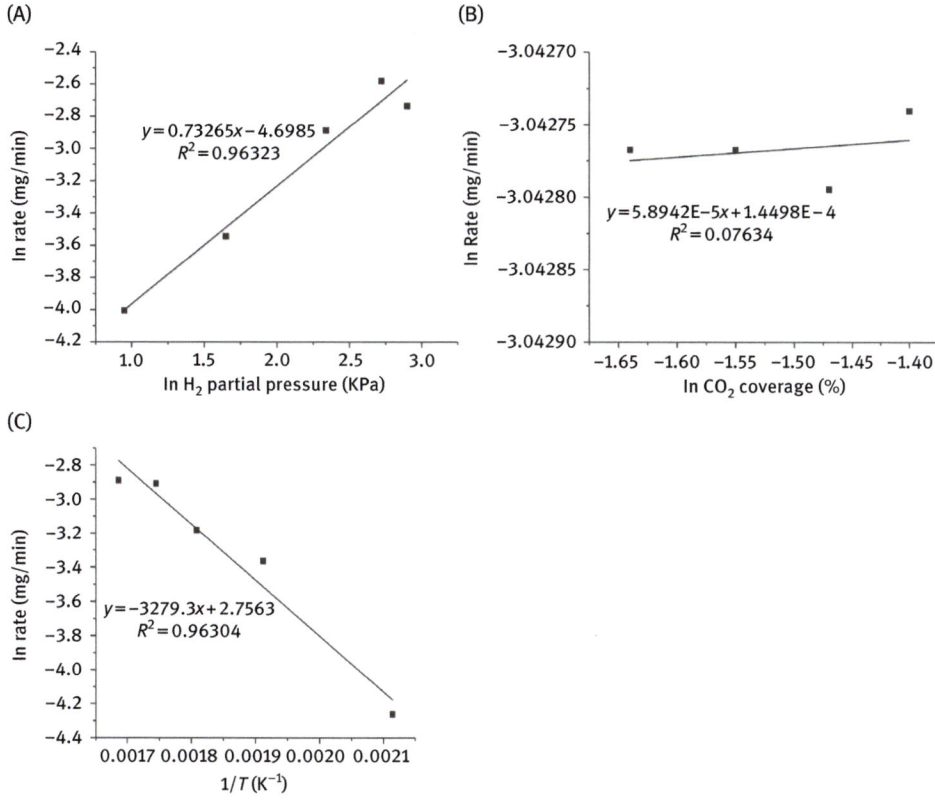

Figure 8.3: Methanation rate dependence on H$_2$ partial pressure (A), CO$_2$ coverage (B) and temperature (Arrhenius plot) (C) derived from TGA tests on Ru-"Na$_2$O" DFM powder.

kinetic studies indicate a significant dependence of methanation rate on H$_2$ partial pressure, approaching 1, and essentially zero dependence on CO$_2$ coverage, in accordance with previous work by Duyar et al. for methanation over Ru/Al$_2$O$_3$ [4].

Kinetic data can be collected using thermal gravimetric analysis (TGA), by monitoring the weight change on the DFM upon adsorption of CO$_2$ and subsequent hydrogenation rates under different conditions (Figure 8.4). Particularly interesting was the determination of apparent activation energy for this system (27.26 kJ/mol CO$_2$), which was half of that reported for CO$_2$ hydrogenation on Ru/Al$_2$O$_3$, in an analogous purely catalytic system [4]. From comparison of the activation energies of the catalyst and the DFM system, the methanation step for DFM was determined to be limited by slow spillover of adsorbed CO$_2$ to the Ru metallic sites. This insight can be used to engineer new materials with the use of a different support or increased interface between the catalyst and adsorbent, to overcome the observed transport limitation.

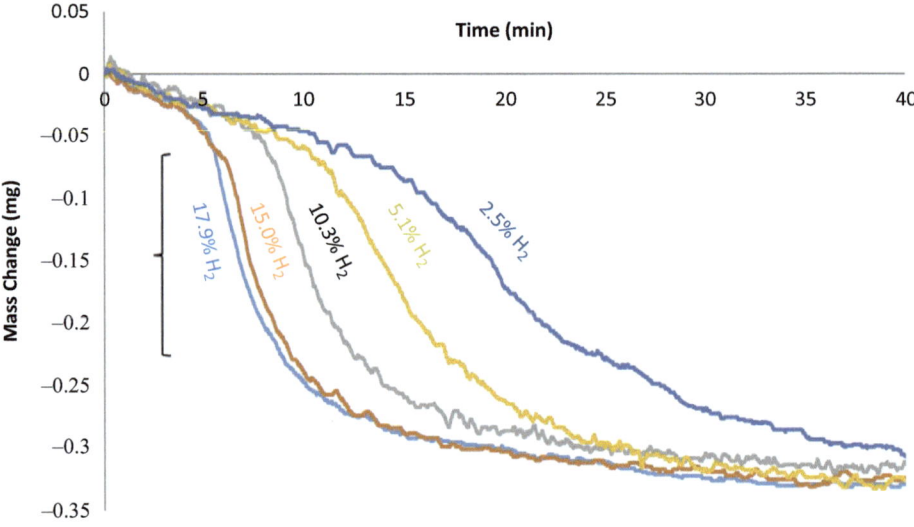

Figure 8.4: TGA profile for calculating reaction orders with respect to hydrogen. The region indicated in the figure was used for calculating hydrogenation rates. CO_2 adsorption conditions: 320 °C, 5% CO_2/N_2 for 1 h. Methanation conditions: 320 °C, 2.5–17.9% H_2/N_2.

8.2.2 Cyclic stability considerations

Switching from oxidising (flue gas) to reducing (reaction) conditions is expected to put stress on DFMs, and can lead to degradation for extended cyclic operation. Degradation of activity can occur by sintering and deactivation of the catalyst. For example, Ru catalysts are known to be active for methanation only in a reduced state and hence cyclic stability studies also reveal whether the catalyst will deactivate under CO_2 capture conditions. The presence of water in post-combustion emissions sources puts an added stress on the solid material as this can accelerate sintering.

Figure 8.5 displays results from cyclic stability tests performed on the Ru-CaO DFM [5]. Figure 8.5 shows that upon repeatedly switching from oxidising CO_2 capture conditions (10% CO_2/air) to reducing hydrogenation conditions over 20 cycles, the material undergoes only a slight decline in methanation activity. Under CO_2 capture conditions, where simulated flue gas also contains water (8%CO_2/21%H_2O/air), the same DFM displays a slight decrease in methanation capacity over 20 cycles compared to the case with no water present, and fluctuating behaviour. The fluctuations are possibly due to the complicated adsorption/desorption dynamics in the presence of water, which can compete with CO_2 for adsorption on CaO sites [1]. The repeated oxidation and reduction of the Ru can also lead to a loss in performance. However, the proper H_2 partial pressure during methanation eliminates this potential problem [12].

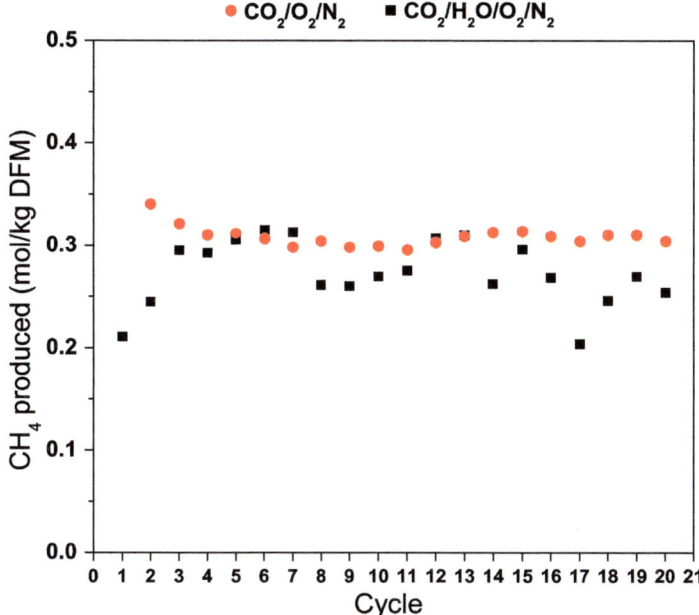

Figure 8.5: Stability of the optimised DFM with 5% Ru 10% CaO/Al$_2$O$_3$ for CO$_2$ capture and methanation over 20 cycles of operation. Conditions: 320 °C and 1 atm for all cycles with CO$_2$ capture from a 10% CO$_2$/air mixture (•) and from an 8% CO$_2$/21%H$_2$O/air mixture (▪) and hydrogenation with 5% H$_2$/N$_2$. Data extracted from [5] with permission.

Remarkably, the Ru-"Na$_2$O"/Al$_2$O$_3$ DFM developed by Wang [12] demonstrates increasing methanation activity over 50 cycles of operation (Figure 8.6). This unexpected improved result is attributed to re-dispersion of both Na and Ru due to the cycling conditions, based on CO$_2$ and H$_2$ chemisorption measurements as well as TEM on pre- and post-test DFM samples [12]. Results from both Figures 8.5 and 8.6 highlight the significant influence of cyclic oxidising and reducing conditions on DFM activity, and suggest that consequences depend on structural stability of the DFM.

8.2.3 Catalysts and carrier materials

Catalyst selection for DFM should be based on activity, selectivity and stability under cyclic oxidising/reducing conditions. For the case of synthetic natural gas (CH$_4$) production, a comparison of the methanation activity of various catalysts, including precious metals (Ru, Rh, Pt and Pd) and base metals (Ni and Co), is listed in Table 8.1. The data show that Ru is the most active catalyst, producing methane even at low temperatures (< 200 °C). Rhodium follows with a comparable activity at temperatures above 275 °C. While Nickel, the most commonly used methanation

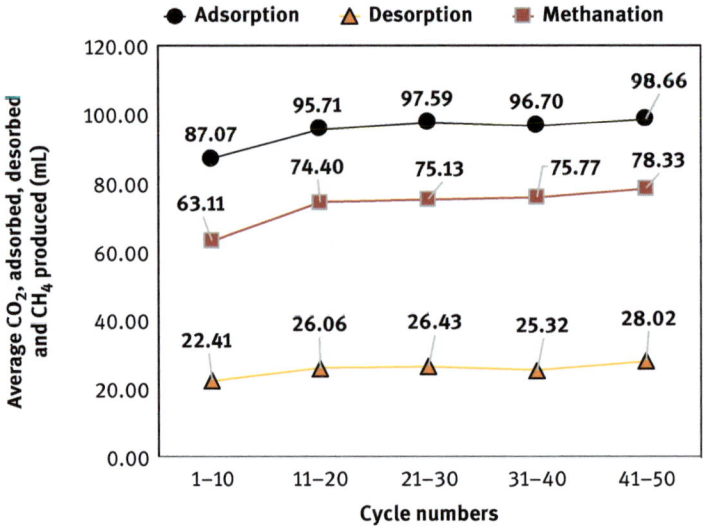

Figure 8.6: Stability of the Ru "Na$_2$O" DFM (10 g, 5 × 5 mm tablets) for CO$_2$ capture and methanation over 50 cycles of operation [12]. CO$_2$ adsorption, CO$_2$ desorption and CH$_4$ production are averaged for every 10 cycles. Conditions: CO$_2$ capture from simulated flue gas (7.5% CO$_2$/15% H$_2$O/4.5% O$_2$/N$_2$), 15 min for methanation in 15% H$_2$ / N$_2$, with 3 min of pure N$_2$ purge between each step, 300 °C, 1 atm.

Table 8.1: Methane concentration (shown as % of equilibrium) produced by different supported catalysts at different temperatures (150–320 °C). About 0.1000 g of the catalysts was tested in a fixed bed reactor at atmospheric pressure with a stoichiometric feed (1:4) of 4% CO$_2$, 16% H$_2$/ He balance. Only base metal catalysts with Ni and Co were pre-reduced at 450 °C for 3 h with 100% H$_2$ prior to the experiment. All other catalysts were reduced during startup by the reactant feed that contains 16% H$_2$.

	150 °C	200 °C	250 °C	275 °C	320 °C
10%Ru/Al$_2$O$_3$	2.1%	47.8%	83.3%	90.9%	100%
10%Rh/Al$_2$O$_3$	0%	8.69%	51.1%	79.5%	100%
10%Ni/Al$_2$O$_3$	0%	0%	8.9%	21.6%	53.6%
10%Pd/Al$_2$O$_3$	0%	0%	2.2%	5.7%	14.6%
10%Co/Al$_2$O$_3$	0%	0%	0%	0%	14.6%
10%Pt/Al$_2$O$_3$	0%	0%	0%	0%	0%

catalyst for industrial applications, is far less expensive than Ru or Rh, it has moderate methanation activity at 320 °C (~50% of equilibrium). Moreover, since a critical issue for DFM applications is maintaining the catalyst in its active state during exposure to cycling oxidising/reducing conditions, Ni may also present challenges due to rapid oxidation during the CO$_2$ capture process, which can result in deactivation [13, 14]. Pd, Pt and Co were found to show poor activity for methane production under atmospheric pressure hydrogenation conditions [8].

To date, γ-Al$_2$O$_3$ has been used as a carrier for the methanation applications of DFM due to its high surface area and commercial availability. Infrared studies have shown that alumina is capable of weakly adsorbing CO_2 to form bicarbonates [15] and can also form surface interactions with the alkaline adsorbents to promote CO_2 adsorption [15–24]. Other high surface carriers such as zeolites, CeO_2, CeO_2-ZrO_2, ZrO_2 and SiO_2 are also attractive candidates either for their intrinsic CO_2 adsorption capacities (due to their alkalinity) [25–29], or due to their oxygen storage capability [30] that may help the redox properties of the catalyst, which is necessary for restoring the catalytic activity of DFM when switching from O_2-containing flue gas to hydrogenation conditions. When choosing carrier materials for DFM development, stability, mass transfer, effects on catalytic activity and adsorbent/catalyst separation become important considerations.

References

[1] M.S. Duyar, R.J. Farrauto, M.J. Castaldi, T.M. Yegulalp, Industrial & Engineering Chemistry Research 53 (2014) 1064–1072.

[2] P. Gruene, A.G. Belova, T.M. Yegulalp, R.J. Farrauto, M.J. Castaldi, Industrial & Engineering Chemistry Research 50 (2011) 4042–4049.

[3] C. Janke, M.S. Duyar, M. Hoskins, R. Farrauto, Applied Catalysis B: Environmental 152–153 (2014) 184–191.

[4] M.S. Duyar, A. Ramachandran, C. Wang, R.J. Farrauto, Journal of CO2 Utilization 12 (2015) 27–33.

[5] M.S. Duyar, M.A.A. Treviño, R.J. Farrauto, Applied Catalysis B: Environmental 168–169 (2015) 370–376.

[6] L.F. Bobadilla, J.M. Riesco-García, G. Penelás-Pérez, A. Urakawa, Journal of CO2 Utilization 14 (2016) 106–111.

[7] Q. Zheng, R. Farrauto, A. Chau Nguyen, Industrial & Engineering Chemistry Research 55 (2016) 6768–6776.

[8] M.S. Duyar, S. Wang, M.A. Arellano-Trevino, R.J. Farrauto, Journal of CO2 Utilization 15 (2016) 65–71.

[9] S. Wang, E.T. Schrunk, H. Mahajan, Catalysts 7 (2017) 88.

[10] S.M. Kim, P.M. Abdala, M. Broda, D. Hosseini, C. Copéret, C. Müller, ACS Catalysis 8 (2018) 2815–2823.

[11] J.V. Veselovskaya, P.D. Parunin, O.V. Netskina, A.G. Okunev, Topics in Catalysis 1–9.

[12] S. Wang, A Study of Carbon Dioxide Capture and Catalytic Conversion to Methane using a Ruthenium," Sodium Oxide" Dual Functional Material: Development, Performance and Characterizations, Columbia University, 2018.

[13] B. Mutz, H.W.P. Carvalho, S. Mangold, W. Kleist, J.-D. Grunwaldt, Journal of Catalysis 327 (2015) 48–53.

[14] B. Mutz, A. Gänzler, M. Nachtegaal, O. Müller, R. Frahm, W. Kleist, J.-D. Grunwaldt, Catalysts 7 (2017) 279.

[15] Keturakis, C. J., Ni, F. , Spicer, M. , Beaver, M. G., Caram, H. S. and Wachs, I. E. (2014), Monitoring Solid Oxide CO2 Capture Sorbents in Action. ChemSusChem, 7: 3459-3466. doi:10.1002/cssc.201402474

[16] Choi, S. , Drese, J. and Jones, C. (2009), Adsorbent Materials for Carbon Dioxide Capture from Large Anthropogenic Point Sources. ChemSusChem, 2: 796-854. doi:10.1002/cssc.200900036

[17] K.B. Lee, M.G. Beaver, H.S. Caram, S. Sircar, Adsorption 13 (2007) 385–397.

[18] K.B. Lee, M.G. Beaver, H.S. Caram, S. Sircar, Industrial & Engineering Chemistry Research 47 (2008) 8048–8062.

[19] S.C. Lee, B.Y. Choi, T.J. Lee, C.K. Ryu, Y.S. Ahn, J.C. Kim, Catalysis Today 111 (2006) 385–390.

[20] L. Li, X. Wen, X. Fu, F. Wang, N. Zhao, F. Xiao, W. Wei, Y. Sun, Energy & Fuels 24 (2010) 5773–5780.

[21] N.T. Son, L. Leon, S.S.G.K. Babu, S. Kulathuiyer, ChemCatChem 7 (2015) 1833–1840.

[22] J.V. Veselovskaya, V.S. Derevschikov, T.Y. Kardash, O.A. Stonkus, T.A. Trubitsina, A.G. Okunev, International Journal of Greenhouse Gas Control 17 (2013) 332–340.

[23] J.V. Veselovskaya, V.S. Derevschikov, T.Y. Kardash, A.G. Okunev, Renewable Bioresources 3 (2015) 1.

[24] M. Zhao, A.I. Minett, A.T. Harris, Energy & Environmental Science 6 (2013) 25–40.

[25] J.A.H. Dreyer, P. Li, L. Zhang, G.K. Beh, R. Zhang, P.H.L. Sit, W.Y. Teoh, Applied Catalysis B: Environmental 219 (2017) 715–726.

[26] P. Frontera, A. Macario, M. Ferraro, P. Antonucci, Catalysts 7 (2017) 59.

[27] C. Fukuhara, K. Hayakawa, Y. Suzuki, W. Kawasaki, R. Watanabe, Applied Catalysis A: General 532 (2017) 12–18.

[28] J.M. Rynkowski, T. Paryjczak, A. Lewicki, M.I. Szynkowska, T.P. Maniecki, W.K. Jóźwiak, Reaction Kinetics and Catalysis Letters 71 (2000) 55–64.

[29] S. Sharma, Z. Hu, P. Zhang, E.W. McFarland, H. Metiu, Journal of Catalysis 278 (2011) 297–309.

[30] F. Ocampo, B. Louis, A.-C. Roger, Applied Catalysis A: General 369 (2009) 90–96.

Matthew O'Brien

9 Polymer membranes in CO_2 separation and continuous flow processing

9.1 Introduction

With a specific enthalpy of formation of -394 kJ mol^{-1}[1], the stability of CO_2 makes a significant contribution to the thermodynamic driving force for the combustion of flammable organic materials, one of the most important chemical processes in human history. The strong and very polar C=O bonds in CO_2 also provide an efficient chromophore for the absorption and emission of infrared radiation. In this way, atmospheric CO_2 (along with other compounds) contributes to the greenhouse effect by facilitating retention of solar energy (as heat), which would otherwise be radiated into space [2]. Without this phenomenon, terrestrial surface temperatures would be much lower (ca. -18 °C) [3], perhaps too low to sustain any form of life. However, very large post-industrial increases in anthropogenic CO_2 emissions have been strongly linked to rises in atmospheric temperature and associated climate change phenomena. Left unchecked, future temperature rises predicted as a result of CO_2 emissions are likely to have profound and potentially disastrous consequences for society. Although efforts to curb CO_2 emissions (e.g., by developing renewable energy sources as an alternative to fossil fuel combustion) are intensifying, it seems unlikely that CO_2 production will be reduced to adequate levels in sufficient time to prevent major climatic impact [4].

As a potential means of mitigation, several strategies including carbon capture and storage (CCS) [5] and carbon capture and utilisation (CCU) [6] have been proposed to prevent industrial CO_2 from being released into the atmosphere, or to remove the existing atmospheric CO_2.

At the same time, there is a growing recognition that CO_2 is potentially a useful and plentiful chemical feedstock, and a range of new reactor technologies have been developed that facilitate its efficient use in synthetic processes [7].

This chapter will focus on some recent developments in the use of polymer membrane materials in both these contexts.

Matthew O'Brien, Lennard-Jones Laboratories, Keele University, Keele, Staffordshire, UK.

https://doi.org/10.1515/9783110563191-009

9.2 Polymer membranes in CO_2 separation

To capture, store or utilise CO_2, it is advantageous to separate it from other components. Clearly, the suitability of separation methods will depend, in part, on the other components of the mixture.

The principle source of industrial CO_2 production is in fossil fuel combustion in power plants and also in other industries such as cement production [8]. The flue gas that is released during these processes consists largely of N_2, water vapour and CO_2. Natural gas mining is another major source of CO_2, and the ratio of CH_4 (the principle hydrocarbon component) to CO_2 that is released during mining can vary considerably depending on the location of the mine [9]. For industrial viability as a fuel, the CO_2 needs to be almost completely removed (at least to levels of 2% or below) [10]. Another major source of CO_2 is in the production of hydrogen in the syngas and water–gas–shift processes [11]. The levels of the other components (H_2, CO, H_2O) vary depending on the exact nature of the process and the hydrocarbon feedstocks.

Non-membrane methods of CO_2 separation include cryogenic distillation and chemical absorption [12]. Although these techniques can be extremely efficient in terms of selectivity and the amount of CO_2 that can be removed/separated, they tend to be relatively expensive and therefore increase the cost of energy generated by power plants. Membrane separation [10, 13], which avoids the need for the CO_2 to undergo a phase change, offers a potentially very simple and cost-effective means of CO_2 separation. A recent yardstick target given by the DOE suggests that the cost of CO_2 separation/capture should be less than \$30 per ton for any process to be viable [14].

Membrane phenomena have been known for a long time. Nollet was the first to record osmosis through a semipermeable membrane in 1748 [15]. Early pioneering investigations were also carried out by Dutrochet [16], Fick [17], Graham [18] and others. More recently, in the 1960s the development of asymmetric membranes, primarily by Loeb and Surirajan [19], led to the industrial application of reverse osmosis. Partly due to the use of these processes in water desalination and other industrially important applications, the intervening decades have witnessed a massive expansion in membrane research, both into the nature of the materials as well as efficient fabrication methods.

For porous membranes, gasses can physically pass through the pores (the pore flow model). More commonly, for non-porous gas separation polymer membranes, the gas molecules must first enter into the polymer material and then move through it. This is known as the *solution-diffusion model* [20]. If a gas cannot first dissolve into the polymer membrane material, it obviously cannot pass through it. If it cannot diffuse through the membrane to the other side, again it cannot pass through it. A simple relationship for the solution-diffusion model equates permeability as the product of diffusion and solubility:

$$P = DS$$

where P is the permeability expressed as the quantity of gas passing through the membrane per unit area per unit time per unit of pressure difference (across the membrane). D and S are the diffusion coefficient and the solubility coefficient, respectively. The Barrer is the most common unit used for permeability (1×10^{-10} cm^3 (at STP) cm^{-2} s^{-1} cmHg^{-1}).

An important characteristic of any gas separation membrane is its selectivity, expressed as follows:

$$\alpha_{AB} = \frac{P_A}{P_B}$$

where α_{AB} is the selectivity factor and P_A and P_B are the permeabilities of the two gasses, A and B, respectively.

Combining both equations, the selectivity can be seen to depend both on the ratio of the diffusion coefficients as well as the solubility coefficients:

$$\alpha_{AB} = \frac{P_A}{P_B} = \frac{D_A S_A}{D_B S_B} = \alpha_D \alpha_S$$

In general terms, there is a trade-off between permeability and selectivity. Polymers that are more permeable tend to be less discriminating and therefore less selective. Robeson has carried out detailed mathematical analyses and derived a log–log linear relationship between permeability and selectivity, which correlates quite well with published data for a number of polymers [21].

A wide variety of polymeric types are permeable to CO_2 and can be used in CO_2 separation. These include, amongst others, poly(ethylene oxide) (PEO), polydimethysiloxane (PDMS), polysulfones (PS), polyimides (PI), polyacetylenes, polyamides and polycarbonates. In rubbery polymers, such as PDMS, the CO_2 permeability is high due to the high degree of polymer chain mobility. However, this leads to fairly low selectivity [10]. Many polymers are glassy, with relatively rigid polymer chains. Whilst this can retard the diffusion of gas molecules through the membrane material, it usually also increases the selectivity. Because of the very polar nature of CO_2, polymers containing polar functional groups often display high levels of CO_2 permeability.

9.2.1 PEO membranes

A number of membranes based on PEO have been used in CO_2 separation [22]. The basic structure of PEO, also known as poly(ethylene glycol) (PEG), particularly at lower molecular weights, is shown below (Figure 9.1):

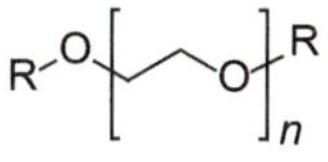

Figure 9.1: Basic structure of PEO.

The ether oxygen groups, having available lone pairs, are relatively electron rich. Because of the highly polar nature of the bonds in CO_2, the central carbon atom is relatively electron deficient. Spectroscopic studies suggest that CO_2 can behave as a Lewis acid in its interactions with polymer materials and that polyether materials interact more strongly than polyester materials [23]. A schematic representation of a Lewis-acid–Lewis-base pair between CO_2 and PEO is shown in Figure 9.2.

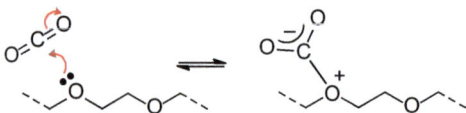

Figure 9.2: Schematic representation of a possible Lewis-acid–Lewis-base interaction between CO_2 and PEO.

The calculated electrostatic potential of CO_2 [24], plotted on its surface, is shown in Figure 9.3a. Recent calculations have shown that the relatively electron-rich oxygen atoms might also partake in binding to molecules similar to PEO. For example, weak C–H⋯O hydrogen bonds have been predicted in CO_2 complexes of ethers such as dimethoxyethane, which is structurally analogous to PEO (Figure 9.3b) [25].

(a)   (b)

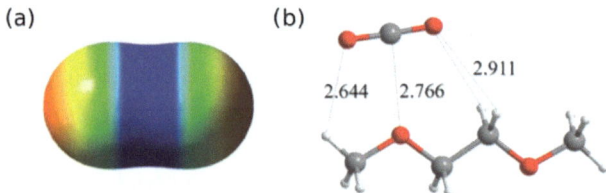

Figure 9.3: (a) Electrostatic surface potential of CO_2, calculated at the at the PBE1PBE/aug-cc-pVTZ level of theory (Reproduced from Ref. 24 with permission from The Royal Society of Chemistry). (b) Calculated geometry of a complex between CO_2 and DME, at the MP2 level of theory using the 6-31+G(d) and aug-cc-pVDZ basis sets (Reprinted with permission from Ref. 25: Kim, K. H.; Kim, Y. *J. Phys. Chem. A* 2008, *112*, 1596–160. Copyright (2008) American Chemical Society).

The mechanical strength and crystallinity of PEO increases with molecular mass. Very high molecular mass PEO (e.g., $> 10^5$) has very good mechanical properties but lower gas permeability. As low molecular mass PEO is either liquid or low melting solid, it cannot generally be used for gas separation membranes even though it is highly permeable. Several approaches have been investigated in order to improve the mechanical properties of PEO whilst retaining high levels of permeability.

These include the use of copolymers, polymer blends (e.g. with low molecular mass PEO) and cross-linking.

Many PEO copolymers have been developed with the goal of combining the high affinity of PEO with the enhanced mechanical rigidity of other materials [26]. The generic structure of the poly(ether-ester) block copolymer of PEO and poly(butylene terephthalate) (PEO-PBT) is shown in Figure 9.4.

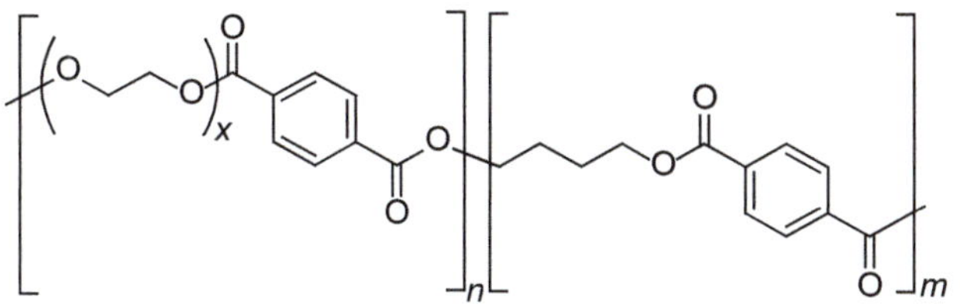

Figure 9.4: Generic structure of the PEO-PBT block copolymer.

Poly(butylene terephthalate) is a thermoplastic engineering polymer. The copolymer obtains significant mechanical rigidity from the PBT segments whilst retaining CO_2 permeability from the more flexible PEO segments. Additionally, the PBT segments also impart increased solubility in a range of organic solvents, thus facilitating a wider range of membrane manufacturing options. Commercial forms of PEO-PBT are available under the tradename of PolyActive and have found significant use as biodegradable polymers in medical applications [27]. A number of types of this polymer have been studied for use as CO_2 separation membranes, including several industrial scale investigations. Depending on the molecular mass range of the polymers, the permeability is temperature dependent. For PolyActive 1,500 (molecular mass of ca. 1,500 g mol^{-1}), the CO_2 permeability increases approximately linearly with temperature from ambient temperatures upward (from below 200 Barrer at 30 °C to almost 500 Barrer at 90 °C). However, PolyActive 3,000 and PolyActive 4,000 (molecular masses of ca. 3,000 and 4,000 g mol^{-1}) exhibit relatively low permeabilities below about 30 and 45 °C, respectively, but undergo phase transitions near these temperatures, above which they obtain significantly greater permeability than PolyActive 1,500 [28].

Another type of PEO copolymer that has received significant study for CO_2 separation is the poly(ether amide) block copolymers [29]. Generic structures are shown in Figure 9.5.

These polymers can vary according to the length of the alkyl section of the amide as well as the size of the amide and polyether blocks. A range of commercially available polymers of this type are known under the trade name of PEBAX® (Poly Ether Block Amide Extreme). A number of PEBAX polymers have poly(tetramethyene oxide) (PTMO) polyether segments rather than PEO.

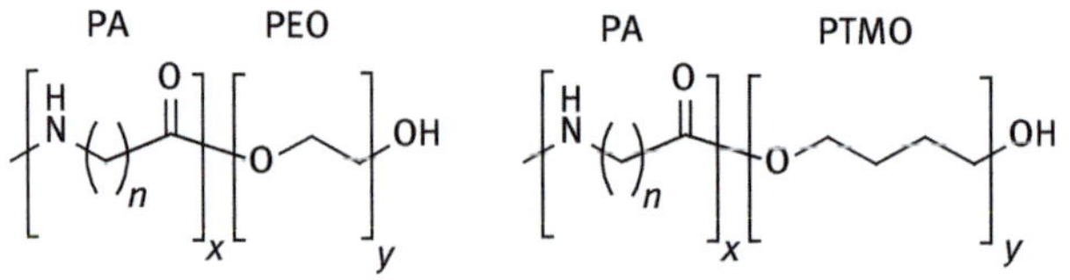

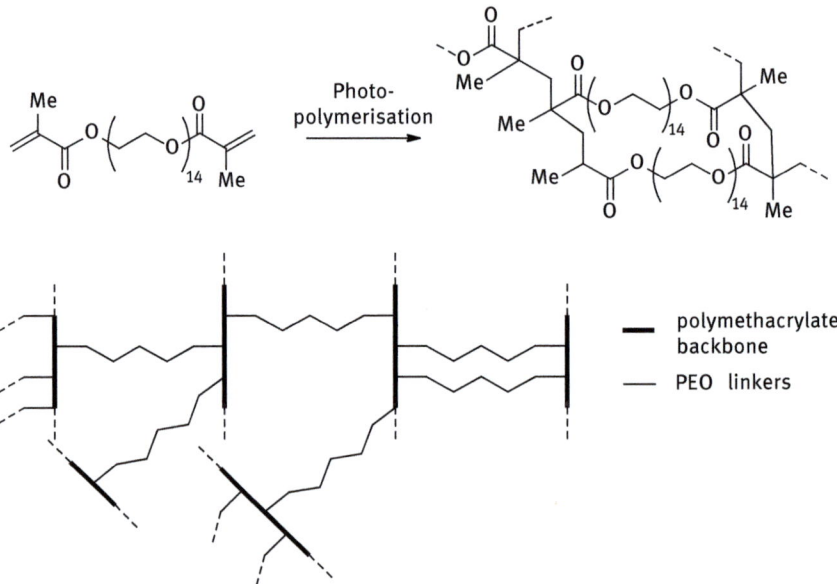

Figure 9.5: Generic structures of poly(ether amide) block copolymers (PEBA/PEBAX).

PEBAX 1074 contains PA12 (nylon 12) and PEO units. It has a CO_2 permeability of 120 Barrer, and a CO_2/N_2 selectivity of 51.4 at 35 °C. PEBAX 4011 also contains PEO as the polyether component but has PA6 (nylon 6) as the polyamide. PEBAX 5533 does not contain PEO but has PTMO as the polyether along with PA12 (nylon 12) as the polyamide. Those polymers with PEO as the polyether tend to have higher CO_2/N_2 and CO_2/CH_4 solubility selectivities than those with PTMO (with similar ratios of PE to PA). In general, a greater concentration of polyether segments increases the CO_2 permeability [22].

Cross-linking strategies have been used with PEO polymers to effectively reduce the crystallinity of the material and increase the amount of free volume [30]. Cross linking can also improve the mechanical properties of the material. A common method of producing cross-linked PEO materials is to polymerise oligomeric PEO precursors containing suitable functional groups. An example is shown in Figure 9.6.

Figure 9.6: Cross-linked polymer with poly(methacrylate) main chains linked via oligomeric PEO linkers [31].

In this case, oligomeric 14-mer PEO units capped at each end with methacrylate were subjected to photochemical activation (plasma UV) [31]. The resulting polymer had polymethacrylate chains cross-linked with PEO linker units. The PEO content by mass was about 80%. This material had a CO_2 permeability of 45 Barrer and CO_2/N_2 selectivity of 68 (at 25 °C) and a CO_2 permeability of 107 Barrer and CO_2/N_2 selectivity of 38 at 50 °C. By varying the lengths of the oligomeric PEO units, and by adding various quantities of non-cross-linking monomer (e.g., with acrylate/methacrylate only at one end), improved permeabilities could be obtained. For instance, with 90% of a 23-mer bis-methacrylate PEO unit and 10% of a non-cross-linking 9-mer mono methacrylate PEO unit, a CO_2 permeabilities of 145/290 Barrer (at 25 °C/50 °C) and CO_2/N_2 selectivities of 66/38 (at 25 °C/50 °C) were obtained.

9.2.2 Polyimide membrane materials

The structure of the imide functional group, along with a polyimide structure commonly encountered in CO_2 separation membranes [32], is shown in Figure 9.7.

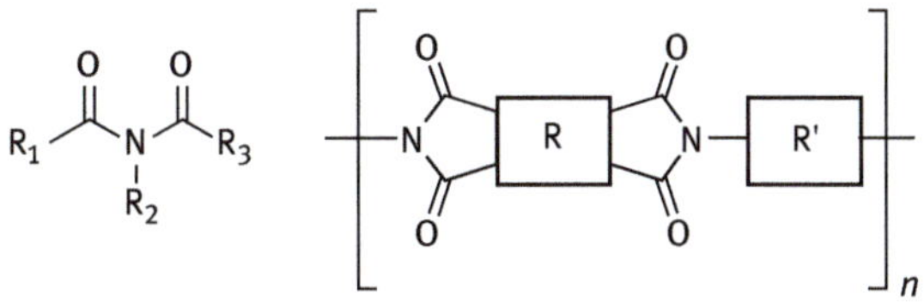

Figure 9.7: The structure of the imide functional group (left) and a generic structure of common polyimides used in CO_2 separation.

Many common polyimide materials result from the condensation of diamine monomers with bis-phthalic acid (or anhydride) monomers. They often have excellent thermal and mechanical qualities and are resistant to a number of chemical environments and reagents (including acids).

Polyimides containing the hexafluoroisopropylidene diphthalic anhydride moiety (6FDA) have seen intensive investigation as membranes for the separation of CO_2 from natural gas. The CF_3 groups in 6FDA are thought to confer enhanced chain stiffness and a reduction in chain-packing efficiency as a result of steric hindrance effects [33]. Four representative examples of such 6FDA polyimides are shown in Figure 9.8 [34].

These examples highlight the profound effects that relatively subtle differences on monomer structure can have on the gas permeation properties of the material. In each pair (upper, lower, Figure 9.8), the only change is in the substitution pattern on the aromatic rings in the diamine moiety. In each case, the meta substitution pattern

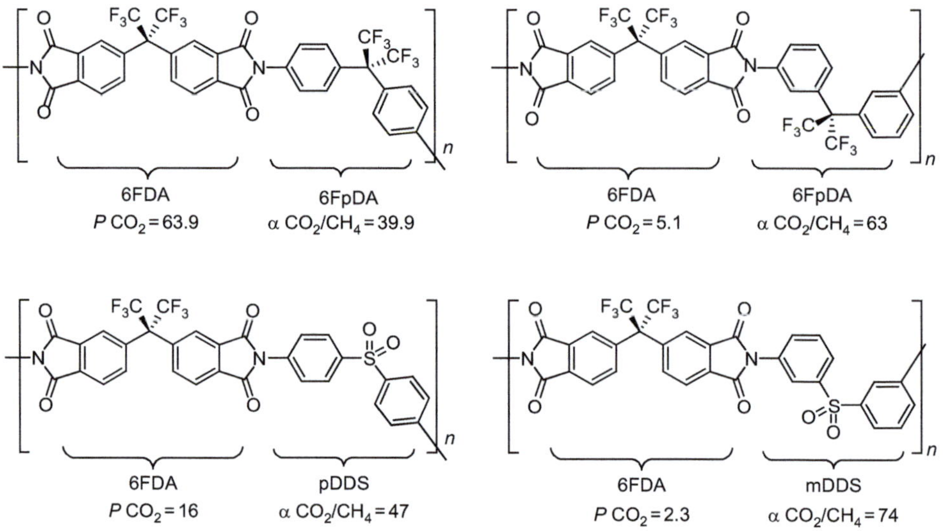

Figure 9.8: Polyimides containing the 6FDA moiety. 6FpDA, 2,2-bis-(4-aminophenyl) hexafluoropropane; 6FpDA, 2,2-bis-(3-aminophenyl) hexafluoropropane; pDDS, 4,4-diaminodiphenyl sulfone; mDDS, 3,3-diaminodiphenyl sulfone [34].

leads to lower CO_2 permeability, but this is accompanied by a significant increase in CO_2/CH_4 selectivity. This has been ascribed to a greater chain-packing efficiency and restricted rotation, which results in a reduction of fractional free volume.

Another type of polyimide structure that has been used for CO_2 separation from natural gas is Matrimid® [35], whose general structure is shown in Figure 9.9.

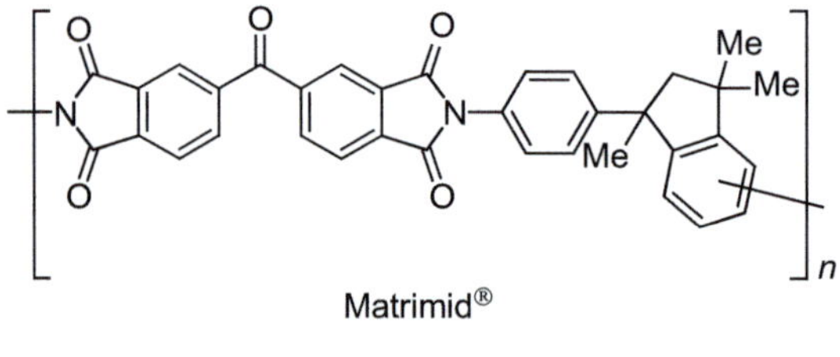

Figure 9.9: Generic structure of the Matrimid® polymers.

At high CO_2 concentrations and pressures, plasticisation and swelling can significantly affect the performance of polyimide membranes. Cross-linking can be used to impart resistance to plasticisation, and also to affect changes to the mechanical properties of polyimides. Many cross-linking strategies involve ring opening of the cyclic imide moieties with bis-nucleophilic compounds (e.g., diamines). However, these reactions appear to be reversible and reformation of the cyclic imides can occur following heat treatment, although some of the characteristics imparted by cross-linking remain (Figure 9.10) [36].

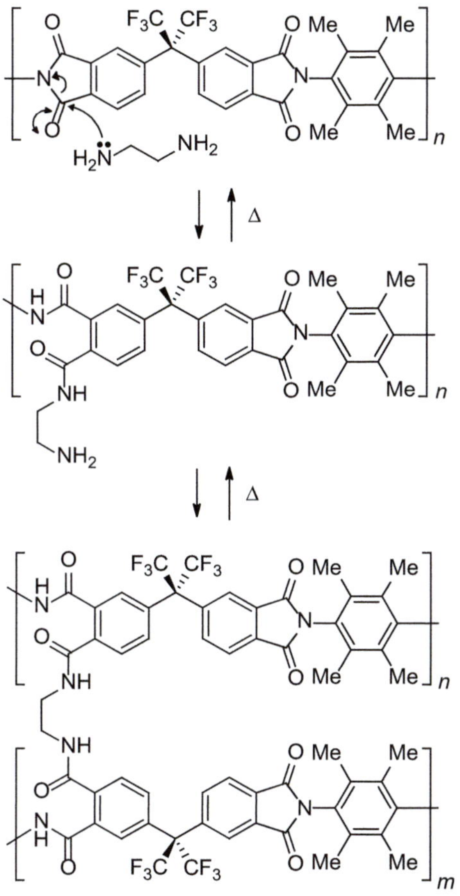

Figure 9.10: Reversible diamine cross-linking in a 6-FDA-durene polyimide [36].

A cross-linking strategy that left the cyclic imide units intact was reported by Hillock and Koros (Figure 9.11). The polyimide chains were based on the 6FDA moiety with some having mono-propylene glygol mono-ester side groups. Under heating, these appendages formed bis-ester linkages with other polymer chains [37].

For material that was cross-linked at 200 °C, the CO_2 permeability was significantly increased (57.5 Barrer compared to 17.1 for uncross-linked material). The CO_2/CH_4 selectivity saw a slight increase (from 34 to 37). Importantly, the cross-linking imparted resistance to CO_2 plasticisation at pressures up to at least 450 psia.

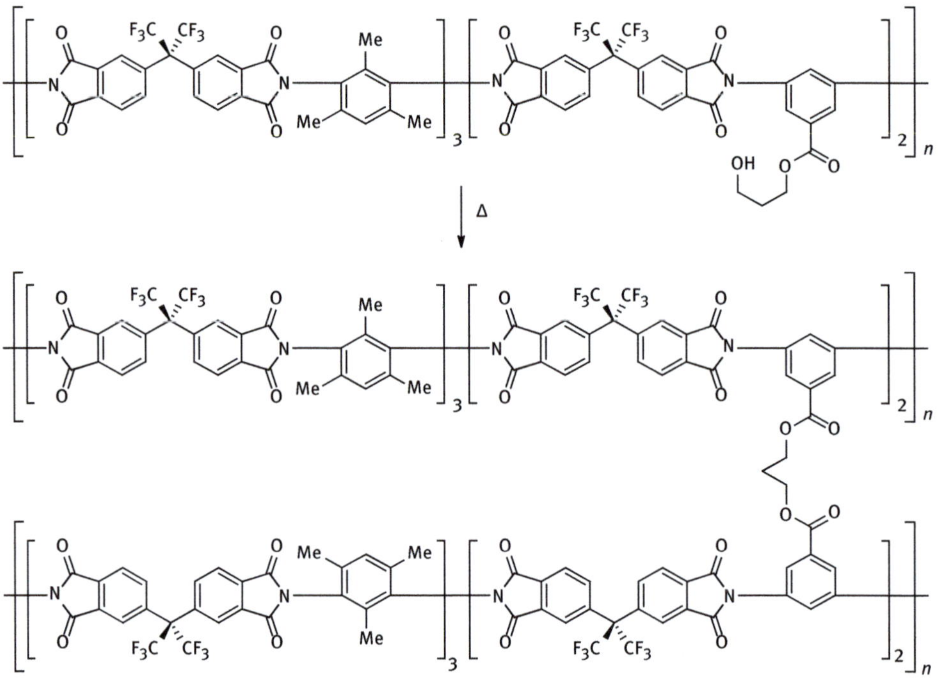

Figure 9.11: 6FDB-based polyimide cross-linked with propylene oxide bis-ester linkages [37].

If polyimide materials with suitable nucleophilic ortho substituents are heat treated, the ortho substituent can effect a dehydrative cyclisation onto the cyclic imide, creating another ring fusion. With suitable treatment, the imide ring can then be cleaved, facilitating decarboxylation. The resulting materials are called *thermally rearranged polyimides*, and tend to have significantly higher permeability than the native material [38]. For instance, when the 6FDA-based polyimide shown in Figure 9.12 (with an amino group in the ortho position) was heated to between 300 and 450 °C, a fused tetracyclic structure was generated through the formation of an imine bond between the ortho-amino group and one of the imide carbonyls [39]. Subsequent treatment with alkali initiated ring opening of the 5-membered cyclic amide. Heating the resulting carboxylic acid at 450 °C resulted in loss of CO_2 to form the benzimidazole product shown. Compared to the starting material (P CO_2 = 24 Barrer), the thermally rearranged material displayed significantly higher permeability (P CO_2 = 1624 Barrer). This was accompanied by a noticeable but only relatively small loss in CO_2/CH_4 selectivity (from 93 to 46). The very high permeabilities seen in such thermally rearranged polymimides has been ascribed to an increase in the fractional free volume. Many of the materials have permeability/selectivities above the Robeson "upper bound". Thermally rearranged polyimides have also been shown to have very high resistance to plasticisation, even at high pressures.

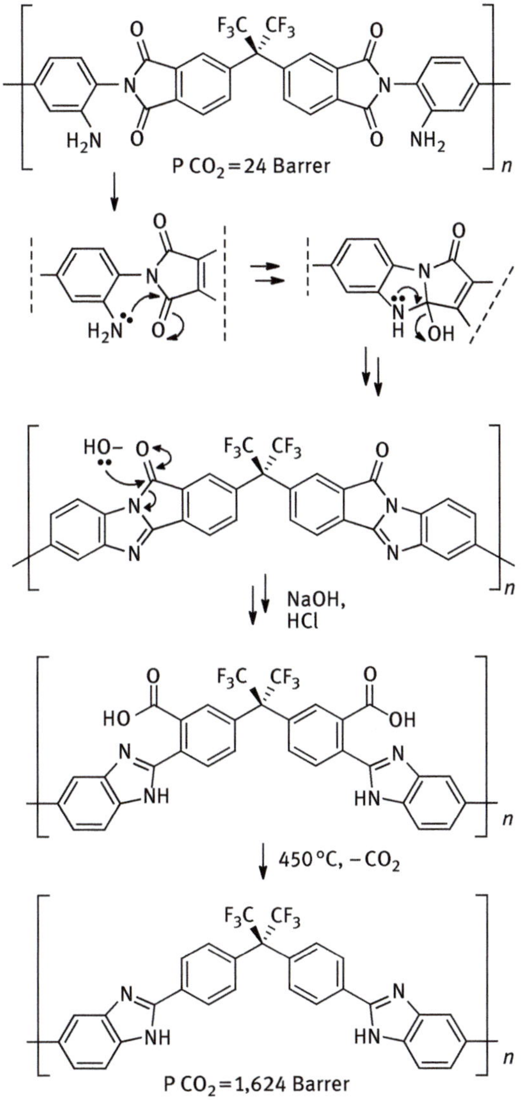

Figure 9.12: Heat/alkali treatment of an ortho-amino polyimide to create a thermally rearranged membrane material. Only one of several possible structural isomers (e.g., *meta/para*) is shown [39].

9.2.3 Facilitated transport membranes

As mentioned for the PEO-containing membranes, CO_2 selectivity can be enhanced if the membrane material has functionality that can interact with CO_2. In facilitated transport membranes, the polymer material contains functional groups (particularly amines) that can strongly and reversibly interact with CO_2 [40].

An early example used pyridine moieties attached to the polymer chains and, under the dry conditions used, acid–base interactions between the pyridine nitrogen and CO_2 was thought to facilitate transport [41].

Poly(vinylamine), poly(ethyleneimine) and poly(allylamine) membranes have been extensively studied. The presence of water within amino-functionalised membranes has been shown to have a profound positive effect on the rate of CO_2 transport and is thought to influence the type of transport mechanism. In the absence of water, the main interaction available to the amino groups and CO_2 molecules is an acid–base mechanism through the formation of carbamate species (e.g., direct nucleophilic reaction of the amine with CO_2) [42]. However, with significant levels of hydration, it is thought that an indirect bicarbonate mechanism is operative, whereby the amine groups activate water molecules towards nucleophilic attack at CO_2 (Figure 9.13) [43]. In certain membranes with very high levels of water content (e.g., hydrogels), immobilised water is thought to form essentially aqueous passageways that facilitate CO_2 permeation [44].

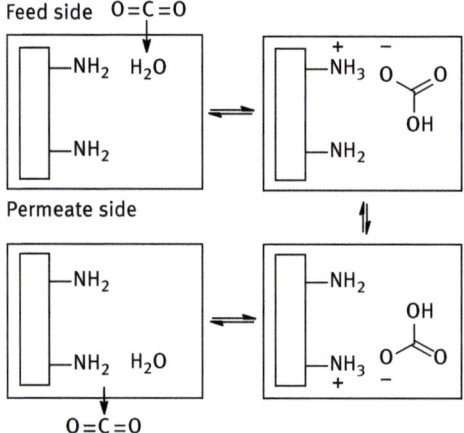

Figure 9.13: Simplified schematic for a bicarbonate-mediated facilitated transport mechanism in a hydrated amino-functionalised membrane.

In a recent study, a series of alkyl substituted poly(allylamine) materials were synthesised that had varying degrees of steric hindrance at the amine site (Figure 9.14) [45]. In blended membranes consisting of 70% alkylated poly(allylamine) and 30% crosslinked poly(vinyl alcohol) (by weight), the more hindered amine membranes (particularly those incorporating the PAA-C_3H_7 structure) displayed both significantly enhanced CO_2 permeability and also superior selectivities (for both CO_2/H_2 and CO_2/N_2).

The greater permeabilities seen with the hindered amines was ascribed to a possible switch in mechanism. Although the unhindered amines may be able to

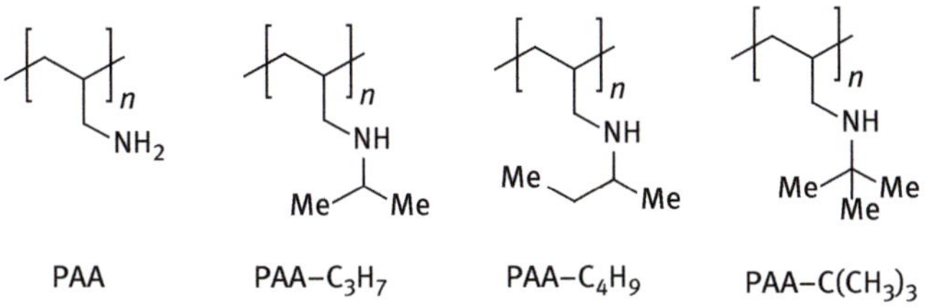

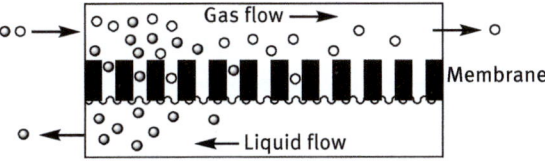

Figure 9.14: Poly(allylamine) and alkylated forms with varying degrees of steric hindrance [45].

bind to CO_2 covalently, forming carbamate species in addition to bicarbonate anions, the hindered amines were thought to operate exclusively through the bicarbonate mechanism.

9.2.4 Polymer membranes as gas–liquid contactors in CO_2 separation

An alternative to using membranes as a selective transport layer is to use the membrane as a means to create a very high contact area between flowing gas and liquid phases [46]. In this case, the liquid phase can be chosen to have selective absorption properties, rather than relying on the membrane to provide selectivity. A particularly practical configuration, which provides very high interfacial surface areas, is the hollow fibre membrane. These contactors often have significantly higher surface-to-volume ratios than more conventional contactors, such as packed bed absorption columns.

A general schematic for selective gas–liquid membrane contacting is shown in Figure 9.15.

Figure 9.15: Simplified schematic of a membrane gas–liquid contactor separating two gasses.

A significant issue that arises with the highly porous materials used in hollow fibre contactors is wetting [47]. This occurs when the liquid (which is usually aqueous) enters the pores, and can cause significantly increased resistance to mass transfer across the membrane. A common strategy to mitigate this is to increase the surface hydrophobicity of the membrane. PTFE and other fluoropolymers have intrinsic hydrophobicity, but they are relatively expensive. Surface modification of cheaper polymers may provide a more economical alternative. In one study, CO_2 absorption by a mixed piperazine/2-amino-2-methyl-propanol aqueous absorbent was investigated using both native polypropylene (PP) and PP that was plasma treated in the presence of CF_4 (forming a hydrophobic fluorous surface) [48]. The water contact angle was found to increase with plasma treatment time up to a certain point after which it reached a plateau (ascribed to establishment of equilibrium between making and breaking surface C–F bonds). The wetting ratio was significantly reduced from 0.0674% to 0.027% following plasma treatment. Increases in CO_2 transfer rates and improved membrane durability were also observed for the plasma-treated membrane, which was found to have a similar mass transfer coefficient to PTFE membranes.

A related approach to avoid unwanted wetting is to use composite membranes in which the standard membrane material is coated with a thin layer of a denser, hydrophobic polymer material. In a recent example, a standard PP hollow fibre membrane was coated with a thin layer of Teflon AF-2400, a glassy amorphous fluoropolymer that has very high permeability to a range of gasses along with excellent chemical resistance [49]. Surprisingly, the Teflon AF-2400 coating, whilst protecting against wetting, caused almost no additional resistance to CO_2 transfer through the membrane (into monoethanolamine solution). For the other two coatings assayed (PDMS and poly (trimethylsilylacetylene)), the composite membrane had significantly poorer CO_2 transport properties when compared with the native PP hollow fibre membrane material.

Although this has been only a rather brief survey, and has not included many polymer materials that show great promise in CO_2 separation technologies (including Polymers of Intrinsic Porosity, PIMs) [50], it is hopefully clear that polymer membranes have the potential to make a huge contribution to mitigation of the effects of anthropogenic CO_2 emissions in future. The next section highlights another aspect of polymer membrane technology, but focuses on CO_2 utilisation rather than CO_2 separation.

9.3 Continuous flow synthetic utilisation of CO_2 using Teflon AF-2400 membrane reactors

There is a growing realisation that CO_2 is a cheap and readily available C1 feedstock for chemical synthesis and recent years have seen a significant growth of research in this area [7]. In this section we will look at some recent developments in the use of Teflon AF-2400 membranes to facilitate the utilisation of CO_2 in continuous flow chemical synthesis applications.

In terms of process considerations, gases (such as CO_2) are often advantageous to use in chemical synthesis as, due to their kinetic mobility, they are readily separated from other components that are in the condensed phase. However, the use of gasses in chemical synthesis is accompanied by several issues that need to be addressed.

To obtain adequate concentrations of gas in solution, high pressures are often required. This brings with it significant safety concerns and frequently necessitates the use of expensive specialist equipment. As the energy stored through pressurisation is proportional to the volume of the vessel, this issue is exacerbated on scale up.

Another scale-related issue is the nature of the contact between the gas and liquid phases (assuming that the reaction will take place in a liquid phase, as is common). As surface scales with the square of dimension but volume scales with its cube (for a fixed reactor geometry), the surface area-to-volume ratio for gas–liquid contact will naturally grow smaller as a reaction is scaled up. This obviously leads to unwanted scale variant reaction kinetics and, at very large scales, the transport of gas molecules into the liquid phase can become rate limiting.

In recent years, continuous flow chemical synthesis has emerged as an alternative paradigm to more traditional batch protocols that, under some circumstances, can offer some attractive benefits [51]. As only a small amount of material is being processed, continuously, at any one time, this alleviates many safety concerns particularly when hazardous conditions are being used (e.g., high pressures of gas). Additionally, as the dimensions of the reaction/contact zone are relatively small, this provides enhanced (and well-defined) interfacial surface contact. As reactions are scaled over time rather than through increasing reactor dimensions, this leads to highly predictable and scale-invariant processes.

For continuous flow processes using gaseous reagents such as CO_2, a key factor is the nature of the gas–liquid contact. A straightforward option is to simply mix two flowing streams of gas and liquid at some kind of fluidic junction. This leads to biphasic flow (Figure 9.16).

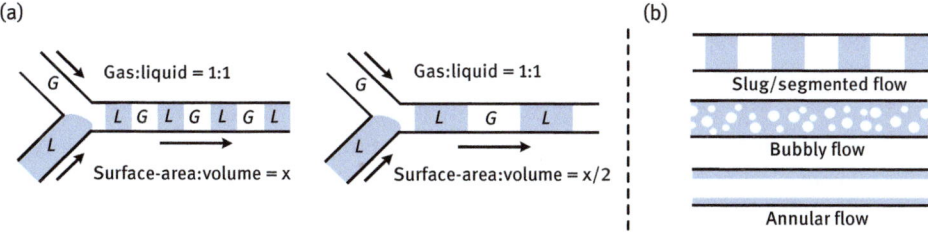

Figure 9.16: Schematic diagrams of biphasic gas–liquid flow patterns.

A number of different flow regimes can be obtained (including slug/segmented flow, bubbly flow and annular flow) depending on a number of factors, including the flow rates of the gas and liquid phases as well as the dimensions of the tubing. This is not always easy to control and a change in parameters may trigger a transition from one flow regime to another (e.g., at high gas flow rates annular flow can be obtained). Even if we are able to keep within a segmented flow regime (typically obtained with the tubing dimensions used in laboratory-scale continuous flow processes), control of the segment sizes is difficult. As shown in Figure 9.16a, whilst the relative amounts of gas and liquid phase (here 1:1) can be controlled by varying their relative flow rates, controlling the sizes of the segments is not at all straightforward. In the example shown, an increase in the segment size by a factor of 2 causes a corresponding decrease in the surface area-to-volume ratio by the same factor. This is difficult to control and/or predict and can actually be unstable over time even with fixed flow rates.

A number of engineering strategies have been developed aimed at maximising and controlling the interfacial surface area by mechanically breaking up the phase segments [52]. An alternative approach, which avoids many of the issues associated with biphasic flow, is to use a semi-permeable membrane to facilitate gas–liquid contact, thereby generating a homogenous solution of gas.

In 2010, O'Brien and Ley developed a new type of gas–liquid continuous flow reactor based on the amorphous glassy fluoropolymer Teflon AF-2400 [53]. The polymeric structure of Teflon AF-2400 is shown in Figure 9.17.

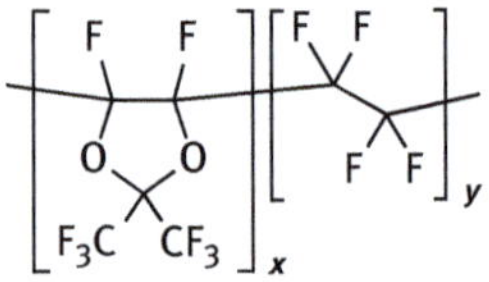

Teflon AF-2400
x:y = 83 : 17

Figure 9.17: The chemical structure of Teflon AF-2400.

It is a copolymer of tetrafluoroethylene and a perfluorodioxolane moiety [54] – developed by DuPont (now Chemours) – which is commercially available in tube form from Biogeneral [55]. The sterically bulky dioxolane component significant reduces crystallinity compared with PTFE and this leads to a very high degree of free volume and porosity. The material is extremely permeable to a number of gasses, and is practically impermeable to the liquid phase. For CO_2, it has a permeability of 2,800 Barrer. Although its modest permselectivity (5.7 for CO_2/N_2) means that it has not been widely used as a directly selective layer for CO_2 separation (although see above for an example of its use in a composite membrane) [49], the fact that it is practically impermeable to the liquid phase makes it an ideal contact membrane for gas–liquid flow processes using CO_2 as a single gas. The first synthetic application of the membrane by O'Brien and Ley was in an ozonolysis application and it has since been used by a number of research groups with a range of reactive gasses including ozone [53, 56], hydrogen [57], carbon monoxide [58], syngas [59], ethylene [57c, 59a, 60], ammonia [61], dimethylamine [58a], oxygen [57c, 62], diazomethane [63], formaldehyde [64] and fluoroform [65], as well as with trifluoroacetic acid [64]. In the first use with CO_2, the Ley group used a tube-in-tube configuration, where a tube of Teflon AF-2400 is contained within a slightly wider tube that contains pressurised gas. In this way, the volume of pressurised gas is kept to a minimum (Figure 9.18). The liquid can also flow in the outer tube (with the gas contained in the inner tube). For elevated gas pressures, it is important to use a back-pressure regulator downstream of the reaction zone in order to maintain homogeneity of the flow stream. In the absence of a back-pressure regulator, premature out-gassing occurs, whereby bubbles of gas form as a separate phase in the flow stream.

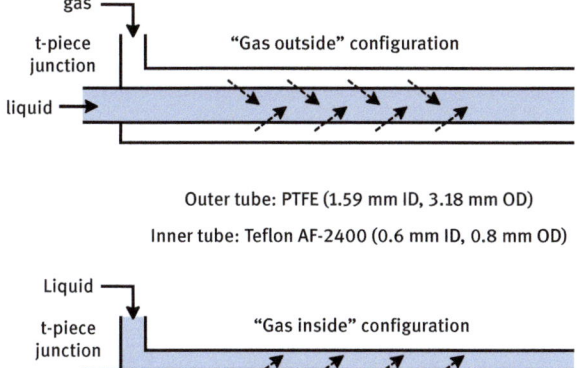

Figure 9.18: Tube-in-tube configurations for Teflon AF-2400-based gas–liquid membrane contactors.

The reaction the Ley group studied was the carboxylation of Grignard reagents [66]. A schematic of the flow system is shown in Figure 9.19.

Initially, the simple system shown in Figure 9.19a was used to gauge the effect of variables such as flow rate and CO_2 pressure. The Grignard reagent 3,5-dimethoxyphenylmagnesium bromide was pumped (via an injection loop) as a solution in THF into a T-piece where it was joined by a stream of THF (this provided the necessary dilution to avoid the formation of precipitates). The combined flow stream then entered the tube-in-tube device where it became enriched with CO_2. After passing through the terminal back-pressure regulator, the magnesium carboxylate product was subjected to aqueous workup and product analysis. Conversion of the Grignard reagent was very high at low flow rates but was found to decrease in a linear manner with increased flow rate. This was attributed to the flow stream taking on insufficient CO_2 when it passed through the tube-in-tube device too quickly. Increasing the CO_2 pressure to 3 bar led to significant increases in conversion. With optimised conditions, a series of Grignard reagents were processed using a system that incorporated inline quench/workup and "catch and release" stages (Figure 9.19b). In this configuration, the initial magnesium carboxylate product passed through a cartridge

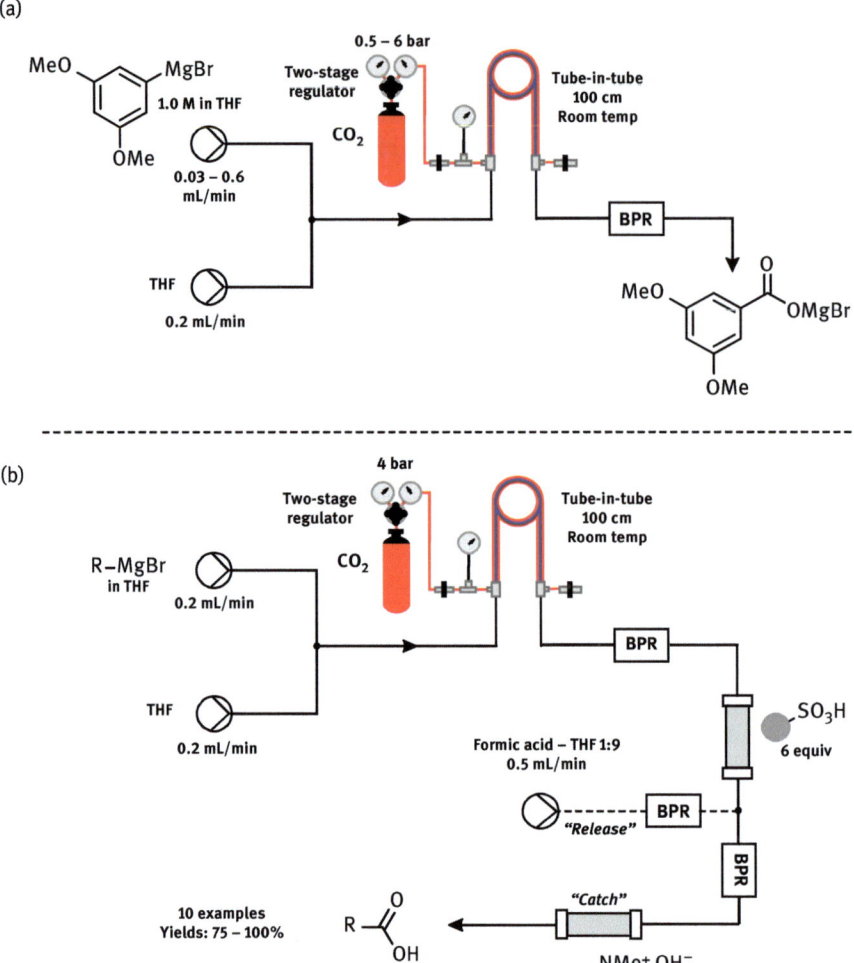

Figure 9.19: Schematic diagrams of the flow systems for carboxylation of Grignard reagents using a Teflon AF-2400 tube-in-tube reactor: a) simple system and b) system incorporating "catch and release" purification strategy.

of sulfonic acid resin, which protonated the carboxylate (forming the acid) and scavenged the magnesium salts. The carboxylic acid was then "caught" by passing through an alkali-loaded ion-exchange resin (where it formed the carboxylate and became ionically attached to the resin). Subsequent washing with a formic acid solution protonated the carboxylate, releasing it from the resin. This "catch and release" protocol led to the isolation of analytically pure carboxylic acids by removing impurities that were present in the Grignard reagent solution (e.g., alkyl bromide and other contaminants).

The Ley group also carried out a series of deprotonation/carboxylation reactions on a set of aryl fluorides (Figure 9.20) [67]. Streams of aryl fluoride in THF and *n*-butyllithium in hexanes were pumped through precooling coils, via a junction, into a 2.5 mL reaction loop where the *n*-butyllithium removed the proton *ortho* to the fluoride. The resulting aryllithium species then met a stream of CO_2-enriched THF (also precooled) emerging from a tube-in-tube device. These streams then went on to react in another 2.5 mL reaction loop before exiting the system through a back-pressure regulator. All the cooling and reaction coils were kept at the correct temperature (−50 °C) using a refrigerated mandrel type system. It was particularly important to keep the intermediate *ortho*-fluoro organolithium compound cool as these species are highly susceptible to elimination of LiF to form reactive benzyne species.

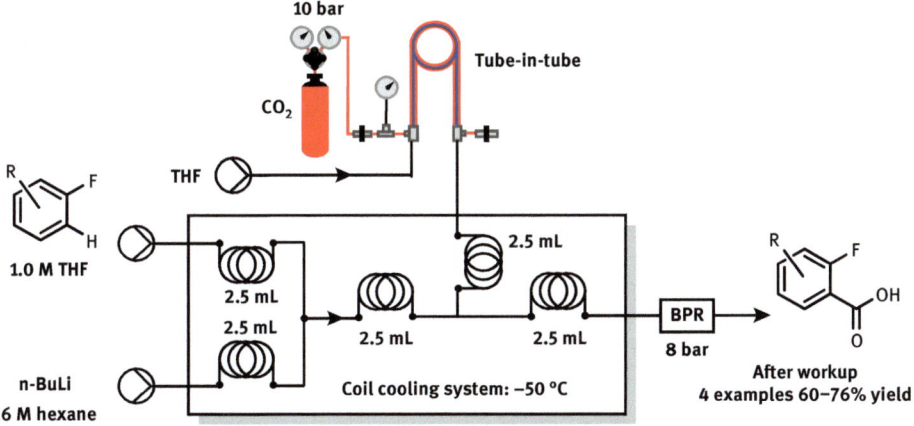

Figure 9.20: Sequential deprotonation/carboxylation reaction of aryl fluoride compounds using a tube-in-tube device.

Recently, Rehman et al. carried out extensive optimisation and kinetic studies on the conversion of styrene oxide to styrene carbonate using a tube-in-tube device (Figure 9.21) [68]. By delivering homogeneous solutions of CO_2 into solution, in a controllable and efficient manner, the device facilitated a rapid investigation of key reaction parameters (CO_2 pressure, catalyst loading, etc.), and complete conversion was able to be achieved in a significantly reduced reaction time.

9.3.1 Gas permeation properties of Teflon AF-2400 in continuous flow

With a number of gasses, the solution concentration of gas in the liquid phase seems to have a very simple relationship to the residence time in the section of AF-2400

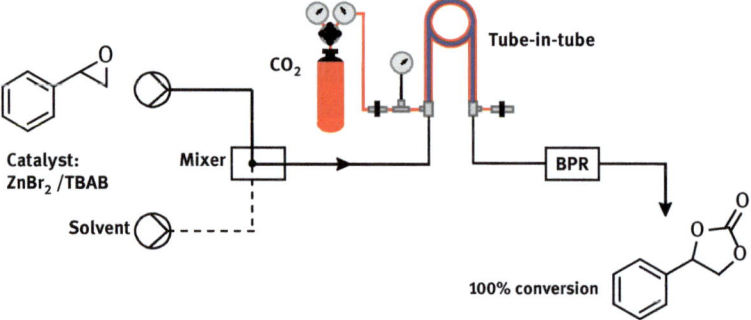

Figure 9.21: Catalytic conversion of styrene oxide to styrene carbonate using a Teflon AF-2400 tube-in-tube device.

tubing. To investigate the permeation/concentration of CO_2 in acetonitrile, O'Brien developed an automated colorimetric inline titration system [69]. The flow system used is shown in Figure 9.22.

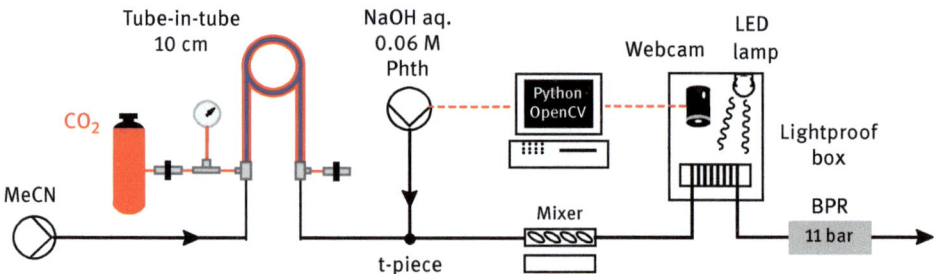

Figure 9.22: Flow system used in an automated colorimetric inline titration to determine concentrations of CO_2 in acetonitrile emerging from a Teflon AF-2400 tube-in-tube device. Reproduced from Ref 69 with permission from Elsevier [69].

At a given set of tube-in-tube conditions (e.g., flow rate of acetonitrile, CO_2 pressure), the concentration of CO_2 was obtained by adjusting the flow rate of a solution of aqueous sodium hydroxide (of known concentration) that also contained phenolphthalein indicator, so that the end point was just reached. This was done in an automated manner. A Python script, incorporating the OpenCV computer vision library [70], was used to control the titrant pumping rate and also to detect the colorimetric endpoint via processing of digital images. To increase accuracy of endpoint determination, an equation was used to model the observed colour intensity. Two approaches to data gathering were used. In one, titrant flow rates were varied uniformly across a range. In the other, a bisection search algorithm [71] was used to "home in" to the correct endpoint flow rate (this would also provide a greater number of data points around the true endpoint). Plots of calculated CO_2 concentrations against residence time (in the tube-in-tube device) are shown in Figure 9.23 for two different pressures of CO_2.

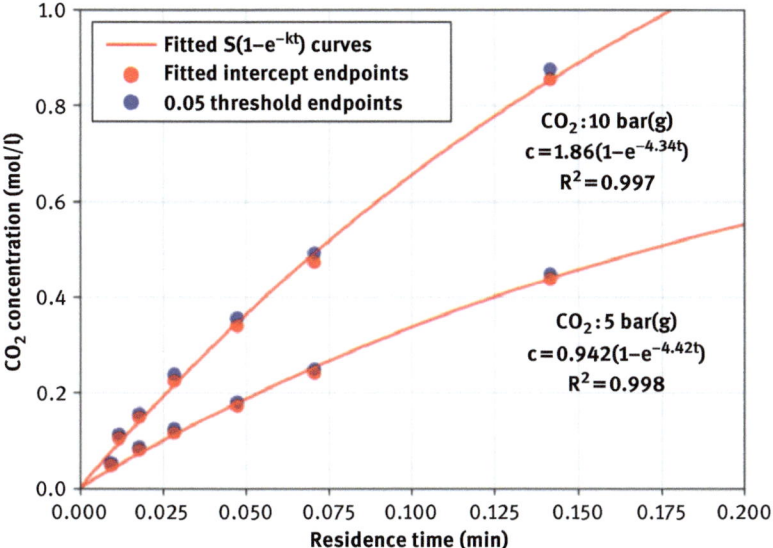

Figure 9.23: Plots of derived CO_2 concentrations against residence time for acetonitrile solutions in a tube-in-tube device at 5 and 10 bar of CO_2. Reproduced from Ref 69 with permission from Elsevier [69].

The data are consistent with a simple model where the rate of increase in concentration is proportional to the difference in between the current concentration and the saturation concentration:

$$\frac{dC}{dt} = k(S - c)$$

$$c = S\left(1 - e^{-kt}\right)$$

where c is the concentration, S is the saturation concentration and k is a rate factor. Although, at the residence times investigated, saturation has not been reached, the calculated saturation concentrations for the two pressures are consistent with an approximate Henry's law [72] relationship.

The Jensen group has also carried out automated experimental work to determine the rate of CO_2 permeation into organic solvents [73]. Additionally, they have performed detailed numerical calculations, including finite element analysis, to model the permeation and reaction of gasses in Teflon AF-2400 tube-in-tube reactors [74].

Although gas–liquid contacting devices based on semipermeable membranes such as Teflon AF-2400 have only been used in continuous flow chemical synthesis for a relatively short period, results reported so far are extremely encouraging. In particular, a number of demonstrations of their use with CO_2 suggest a promising future for their further exploitation with this gas.

References

[1]   Chase, M. W., Jr. *J. Phys. Chem. Ref. Data, Monograph 9* 1998, 1–1051.
[2]   a) Zhong, W.; Haigh, J. D. *Weather* 2013, *68*, 100–105; b) Augustsson, T.; Ramanathan, V. *J. Atmos. Sci.* 1977, *34*, 448–451; c) Held, I. M.; Soden, B. J. *Ann. Rev. Energ. Env.* 2000, *25*, 441–475; d) Anderson, T. R.; Hawkins, E.; Jones, P. D. *Endeavour* 2016, *40*, 178–187.
[3]   Lindzen, R. S. *Bull. Amer. Meteor. Soc.* 1990, *71*, 288–299.
[4]   *IPCC, 2014: Climate Change 2014: Synthesis Report. Contribution of Working Groups I, II and III to the Fifth Assessment Report of the Intergovernmental Panel on Climate Change [Core Writing Team*, R.K. Pachauri and L.A. Meyer *(eds.)]. IPCC, Geneva, Switzerland, 151 pp.* https://www.ipcc.ch/report/ar5/syr/.
[5]   *IPCC, 2005: IPCC Special Report on Carbon Dioxide Capture and Storage. Prepared by Working Group III of the Intergovernmental Panel on Climate Change [Metz, B., O. Davidson, H. C. de Coninck, M. Loos, and L. A. Meyer (eds.)]. Cambridge University Press, Cambridge, United Kingdom and New York, NY, USA, 442 pp.*
[6]   a) Cuellar-Franca, R. M.; Azapagic, A. *J. CO2 Util.* 2015, *9*, 82–102; b) Markewitz, P.; Kuckshinrichs, W.; Leitner, W.; Linssen, J.; Zapp, P.; Bongartz, R.; Schreiber, A.; Muller, T. E. *Energy Env. Sci.* 2012, *5*, 7281–7305; c) Hunt, A. J.; Sin, E. H. K.; Marriott, R.; Clark, J. H. *ChemSusChem* 2010, *3*, 306–322.
[7]   a) Styring, P.; Armstrong, K. *Chim. Oggi-Chem. Today* 2011, *29*, 34–37; b) Alissandratos, A.; Easton, C. J. *Beilstein J. Org. Chem.* 2015, *11*, 2370–2387; c) Liu, A.-H.; Yu, B.; He, L.-N. *Greenhouse Gases-Sci. Technol.* 2015, *5*, 17–33; d) Taherimehr, M.; Pescarmona, P. P. *J. Appl. Polym. Sci.* 2014, *131*; e) Kondratenko, E. V.; Mul, G.; Baltrusaitis, J.; Larrazabal, G. O.; Perez-Ramirez, J. *Energy Env. Sci.* 2013, *6*, 3112–3135; f) Manjolinho, F.; Arndt, M.; Gooßen, K.; Gooßen, L. J. *ACS Catalysis* 2012, *2*, 2014–2021; g) North, M.; Pasquale, R.; Young, C. *Green Chem.* 2010, *12*, 1514–1539; h) Aresta, M.; Dibenedetto, A. *Dalton Trans.* 2007, 2975–2992.
[8]   Tzimas, E.; Georgakaki, A. *Energy Policy* 2010, *38*, 4252–4264.
[9]   Wind, J. D.; Paul, D. R.; Koros, W. J. *J. Memb. Sci.* 2004, *228*, 227–236.
[10] Wang, S.; Li, X.; Wu, H.; Tian, Z.; Xin, Q.; He, G.; Peng, D.; Chen, S.; Yin, Y.; Jiang, Z.; Guiver, M. D. *Energy Env. Sci.* 2016, *9*, 1863–1890.
[11] a) Voldsund, M.; Jordal, K.; Anantharaman, R. *Int. J. Hydrogen Energy* 2016, *41*, 4969–4992; b) Terrien, P.; Lockwood, F.; Granados, L.; Morel, T. *Energy Procedia* 2014, *63*, 7861–7866; c) LeValley, T. L.; Richard, A. R.; Fan, M. *Int. J. Hydrogen Energy* 2014, *39*, 16983–17000; d) Gradisher, L.; Dutcher, B.; Fan, M. *Appl. Energy* 2015, *139*, 335–349.
[12] a) Wang, Y.; Zhao, L.; Otto, A.; Robinius, M.; Stolten, D. *Energy Procedia* 2017, *114*, 650–665; b) Rochelle, G. T. *Science* 2009, *325*, 1652–1654; c) Hart, A.; Gnanendran, N. *Energy Procedia* 2009, *1*, 697–706; d) Rao, A. B.; Rubin, E. S. *Env. Sci. Technol.* 2002, *36*, 4467–4475.

[13] a) Du, N.; Park, H. B.; Dal-Cin, M. M.; Guiver, M. D. *Energy Env. Sci.* 2012, *5*, 7306–7322; b) Yong-Woo, J.; Dong-Hoon, L. *Env. Eng. Sci.* 2015, *32*, 71–85; c) Kim, S.; Lee, Y. M. *Curr. Opin. Chem. Eng.* 2013, *2*, 238–244; d) Norahim, N.; Yaisanga, P.; Faungnawakij, K.; Charinpanitkul, T.; Klaysom, C. *Chem. Eng. Technol.* 2018, *41*, 211–223; e) Zhang, Y.; Sunarso, J.; Liu, S.; Wang, R. *Int. J. Greenhouse Gas Control* 2013, *12*, 84–107; f) Han, Y.; Ho, W. S. W. *Chin. J. Chem. Eng.* 2018.

[14] U.S. Department of Energy, Report 2016, "Carbon Capture, Utilization, and Storage: Climate Change, Economic Competitiveness, and Energy Security".

[15] Nollet, J. A. *Recherches sur les causes du bouillonnement des liquides. Hist Acad Roy Sci 1748, 1:57–104*. 1748.

[16] Dutrochet, H. *Mémoires pour servir à l'histoire anatomique et physiologique des végétaux et des animaux, 2 Vols. + Atlas. Paris & London: J.-B. Baillière*. 1837.

[17] Fick, A. *On Liquid Diffusion, The London, Edinburgh, and Dublin Philosophical Magazine and Journal of Science*, Vol. X 1855, 30–39.

[18] Graham, T. *On the absorption and dialytic separation of gases by colloid septa, Phil. Mag. 32, 401* 1866.

[19] a) Loeb, S. In *Synthetic Membranes:* AMERICAN CHEMICAL SOCIETY, 1981; Vol. 153, 1–9; b) Glater, J. *Desalination* 1998, *117*, 297–309.

[20] a) Wijmans, J. G.; Baker, R. W. *J. Memb. Sci.* 1995, *107*, 1–21; b) Koros, W. J.; Fleming, G. K.; Jordan, S. M.; Kim, T. H.; Hoehn, H. H. *Prog. Polym. Sci.* 1988, *13*, 339–401.

[21] a) Robeson, L. M. *J. Memb. Sci.* 1991, *62*, 165–185; b) Robeson, L. M. *J. Memb. Sci.* 2008, *320*, 390–400.

[22] Liu, S. L.; Shao, L.; Chua, M. L.; Lau, C. H.; Wang, H.; Quan, S. *Prog. Polym. Sci.* 2013, *38*, 1089–1120.

[23] a) Lee, B.-S. *J. CO2 Util.* 2018, *28*, 228–234; b) Newby, J. J.; Peebles, R. A.; Peebles, S. A. *J. Phys. Chem. A* 2004, *108*, 11234–11240; c) Kazarian, S. G.; Vincent, M. F.; Bright, F. V.; Liotta, C. L.; Eckert, C. A. *J. Amer. Chem. Soc.* 1996, *118*, 1729–1736; d) Madsen, L. A. *Macromolecules* 2006, *39*, 1483–1487; e) Nalawade, S. P.; Picchioni, F.; Marsman, J. H.; Janssen, L. P. B. M. *J. Supercrit Fluids* 2006, *36*, 236–244.

[24] Murphy, L. J.; Robertson, K. N.; Kemp, R. A.; Tuononen, H. M.; Clyburne, J. A. C. *Chem. Commun.* 2015, *51*, 3942–3956.

[25] Kim, K. H.; Kim, Y. *J. Phys. Chem. A* 2008, *112*, 1596–1603.

[26] Car, A.; Stropnik, C.; Yave, W.; Peinemann, K.-V. 2008, *18*, 2815–2823.

[27] a) Li, P.; Bakker, D.; van Blitterswijk, C. A. 1997, *34*, 79–86; b) Ramakrishna, S.; Mayer, J.; Wintermantel, E.; Leong, K. W. *Composite Sci. Technol.* 2001, *61*, 1189–1224.

[28] Brinkmann, T.; Lillepärg, J.; Notzke, H.; Pohlmann, J.; Shishatskiy, S.; Wind, J.; Wolff, T. *Engineering* 2017, *3*, 485–493.

[29] a) Bondar, V. I.; Freeman, B. D.; Pinnau, I. 2000, *38*, 2051–2062; b) Bondar, V. I.; Freeman, B. D.; Pinnau, I. 1999, *37*, 2463–2475.

[30] a) Barrer, R. M.; Barrie, J. A.; Wong, P. S. L. *Polymer* 1968, *9*, 609–627; b) Lin, H.; Freeman, B. D. *Macromolecules* 2005, *38*, 8394–8407; c) Lin, H.; Wagner, E. V.; Swinnea, J. S.; Freeman, B. D.; Pas, S. J.; Hill, A. J.; Kalakkunnath, S.; Kalika, D. S. *J. Memb. Sci.* 2006, *276*, 145–161.

[31] Hirayama, Y.; Kase, Y.; Tanihara, N.; Sumiyama, Y.; Kusuki, Y.; Haraya, K. *J. Memb. Sci.* 1999, *160*, 87–99.

[32] Favvas, E. P.; Katsaros, F. K.; Papageorgiou, S. K.; Sapalidis, A. A.; Mitropoulos, A. C. *React. Func. Polym.* 2017, *120*, 104–130.

[33] Xiao, Y.; Low, B. T.; Hosseini, S. S.; Chung, T. S.; Paul, D. R. *Prog. Polym. Sci.* 2009, *34*, 561–580.

[34] a) Coleman, M. R.; Koros, W. J. J. *Memb. Sci.* 1990, *50*, 285–297; b) Kawakami, H.; Anzai, J.; Nagaoka, S. 1995, *57*, 789–795.

[35] a) Sridhar, S.; Veerapur, R. S.; Patil, M. B.; Gudasi, K. B.; Aminabhavi, T. M. 2007, *106*, 1585–1594; b) Loloei, M.; Moghadassi, A.; Omidkhah, M.; Amooghin, A. E. 2015, *5*, 530–544; c) Falbo, F.; Brunetti, A.; Barbieri, G.; Drioli, E.; Tasselli, F. J. A. P. R. 2016, *6*, 439–450.

[36] Shao, L.; Chung, T.-S.; Goh, S. H.; Pramoda, K. P. *J. Memb. Sci.* 2005, *256*, 46–56.

[37] Hillock, A. M. W.; Koros, W. J. *Macromolecules* 2007, *40*, 583–587.

[38] a) Park, H. B.; Jung, C. H.; Lee, Y. M.; Hill, A. J.; Pas, S. J.; Mudie, S. T.; Van Wagner, E.; Freeman, B. D.; Cookson, D. J. 2007, *318*, 254–258; b) Sanders, D. F.; Guo, R.; Smith, Z. P.; Stevens, K. A.; Liu, Q.; McGrath, J. E.; Paul, D. R.; Freeman, B. D. *J. Memb. Sci.* 2014, *463*, 73–81; c) Guo, R.; Sanders, D. F.; Smith, Z. P.; Freeman, B. D.; Paul, D. R.; McGrath, J. E. *J. Mater. Chem. A* 2013, *1*, 6063–6072; d) Scholes, C. A.; Dong, G.; Kim, J. S.; Jo, H. J.; Lee, J.; Lee, Y. M. *Sep. Purif. Technol.* 2017, *179*, 449–454; e) Calle, M.; Chan, Y.; Jo, H. J.; Lee, Y. M. *Polymer* 2012, *53*, 2783–2791; f) Liu, Q.; Borjigin, H.; Paul, D. R.; Riffle, J. S.; McGrath, J. E.; Freeman, B. D. *J. Memb. Sci.* 2016, *518*, 88–99; g) Yeong, Y. F.; Wang, H.; Pallathadka Pramoda, K.; Chung, T.-S. *J. Memb. Sci.* 2012, *397–398*, 51–65.

[39] Han, S. H.; Lee, J. E.; Lee, K.-J.; Park, H. B.; Lee, Y. M. *J. Memb. Sci.* 2010, *357*, 143–151.

[40] a) Rafiq, S.; Deng, L.; Hägg, M.-B. 2016, *3*, 68–85; b) Tong, Z.; Ho, W. S. W. *Sep. Sci. Technol.* 2017, *52*, 156–167.

[41] Yoshikawa, M.; Ezaki, T.; Sanui, K.; Ogata, N. 1988, *35*, 145–154.

[42] a) Yamaguchi, T.; Boetje, L. M.; Koval, C. A.; Noble, R. D.; Bowman, C. N. *Ind. Eng. Chem. Res.* 1995, *34*, 4071–4077; b) Wang, Z.; Li, M.; Cai, Y.; Wang, J.; Wang, S. *J. Memb. Sci.* 2007, *290*, 250–258; c) Masakazu, Y.; Kiyoshi, F.; Hirokazu, K.; Toshio, K.; Naoya, O. 1994, *23*, 243–246.

[43] a) Matsuyama, H.; Teramoto, M.; Sakakura, H. *J. Memb. Sci.* 1996, *114*, 193–200; b) Kim, T.-J.; Li, B.; Hägg, M.-B. 2004, *42*, 4326–4336.

[44] Liu, L.; Chakma, A.; Feng, X. *J. Memb. Sci.* 2008, *310*, 66–75.

[45] Zhao, Y.; Winston Ho, W.S. *J. Memb. Sci.* 2012, *415–416*, 132–138.

[46] a) Cui, Z.; deMontigny, D. *Carbon Management* 2013, *4*, 69–89; b) Li, J.-L.; Chen, B.-H. *Sep. Purif. Technol.* 2005, *41*, 109–122; c) Mansourizadeh, A.; Ismail, A. F. *J. Haz. Mater.* 2009, *171*, 38–53; d) Marzouk, S. A. M.; Al-Marzouqi, M. H.; El-Naas, M. H.; Abdullatif, N.; Ismail, Z. M. *J. Memb. Sci.* 2010, *351*, 21–27.

[47] a) Mosadegh-Sedghi, S.; Rodrigue, D.; Brisson, J.; Iliuta, M. C. *J. Memb. Sci.* 2014, *452*, 332–353; b) Lv, Y.; Yu, X.; Tu, S.-T.; Yan, J.; Dahlquist, E. *J. Memb. Sci.* 2010, *362*, 444–452; c) Wang, Z.; Fang, M.; Ma, Q.; Yu, H.; Wei, C.-C.; Luo, Z. *J. Memb. Sci.* 2014, *455*, 219–228.

[48] Lin, S.-H.; Tung, K.-L.; Chen, W.-J.; Chang, H.-W. *J. Memb. Sci.* 2009, *333*, 30–37.

[49] Nguyen, P. T.; Lasseuguette, E.; Medina-Gonzalez, Y.; Remigy, J. C.; Roizard, D.; Favre, E. *J. Memb. Sci.* 2011, *377*, 261–272.

[50] a) Budd, P. M.; McKeown, N. B. *Polym. Chem.* 2010, *1*, 63–68; b) Budd, P. M.; Msayib, K. J.; Tattershall, C. E.; Ghanem, B. S.; Reynolds, K. J.; McKeown, N. B.; Fritsch, D. *J. Memb. Sci.* 2005, *251*, 263–269; c) McDermott, A. G.; Budd, P. M.; McKeown, N. B.; Colina, C. M.; Runt, J. *J. Mater. Chem. A* 2014, *2*, 11742–11752.

[51] a) Ley, S. V. *Chem. Record* 2012, *12*, 378–390; b) Lummiss, J. A. M.; Morse, P. D.; Beingessner, R. L.; Jamison, T. F. *Chem. Record* 2017, *17*, 667–680; c) Movsisyan, M.; Delbeke, E. I. P.; Berton, J.; Battilocchio, C.; Ley, S. V.; Stevens, C. V. *Chem. Soc. Rev.* 2016, *45*, 4892–4928; d) Pastre, J. C.; Browne, D. L.; Ley, S. V. *Chem. Soc. Rev.* 2013, *42*, 8849–8869; e) Gutmann, B.; Cantillo, D.; Kappe, C. O. *Angew. Chem. Int. Ed.* 2015, *54*, 6688–6728; f) Gemoets, H. P. L.; Su, Y. H.; Shang, M. J.; Hessel, V.; Luque, R.; Noel, T. *Chem. Soc. Rev.* 2016, *45*, 83–117; g) Hessel, V.; Kralisch, D.; Kockmann, N.; Noel, T.; Wang, Q. *ChemSusChem* 2013, *6*, 746–789; h) McQuade, D. T.; Seeberger, P. H. *J. Org. Chem.* 2013, *78*, 6384–6389;

i) Webb, D.; Jamison, T. F. *Chem. Sci.* 2010, *1*, 675–680; j) Wiles, C.; Watts, P. *Chem. Commun.* 2011, *47*, 6512–6535; k) Plutschack, M. B.; Pieber, B.; Gilmore, K.; Seeberger, P. H. *Chem. Rev.* 2017, *117*, 11796–11893; l) Wegner, J.; Ceylan, S.; Kirschning, A. *Adv. Synth. Catal.* 2012, *354*, 17–57; m) Fitzpatrick, D. E.; Battilocchio, C.; Ley, S. V. *ACS Central Science* 2016, *2*, 131–138.

[52] a) Steinfeldt, N.; Abdallah, R.; Dingerdissen, U.; JÄ¤hnisch, K. *Org. Proc. Res. Dev.* 2007, *11*, 1025–1031; b) Wada, Y.; Schmidt, M. A.; Jensen, K. F. *Ind. Eng. Chem. Res.* 2006, *45*, 8036–8042; c) Hübner, S.; Bentrup, U.; Budde, U.; Lovis, K.; Dietrich, T.; Freitag, A.; Küpper, L.; Jähnisch, K. *Org. Proc. Res. Dev.* 2009, *13*, 952–960; d) Chambers, R. D.; Fox, M. A.; Sandford, G.; Trmcic, J.; Goeta, A. *J. Fluorine Chem.* 2007, *128*, 29–33; e) Chambers, R. D.; Sandford, G.; Trmcic, J.; Okazoe, T. *Org. Proc. Res. Dev.* 2008, *12*, 339–344; f) Jähnisch, K.; Baerns, M.; Hessel, V.; Ehrfeld, W.; Haverkamp, V.; Löwe, H.; Wille, C.; Guber, A. *J. Fluorine Chem.* 2000, *105*, 117–128; g) Zanfir, M.; Gavriilidis, A.; Wille, C.; Hessel, V. *Ind. Eng. Chem. Res.* 2005, *44*, 1742–1751; h) Ziegenbalg, D.; Löb, P.; Al-Rawashdeh, M. m.; Kralisch, D.; Hessel, V.; Schönfeld, F. *Chem. Eng. Sci.* 2010, *65*, 3557–3566; i) Dietrich, T. R.; Freitag, A.; Scholz, R. *Chem. Eng. Technol.* 2005, *28*, 477–483.

[53] O'Brien, M.; Baxendale, I. R.; Ley, S. V. *Org. Lett.* 2010, *12*, 1596–1598.

[54] a) Pinnau, I.; Toy, L. G. *J. Memb. Sci.* 1996, *109*, 125–133; b) Alentiev, A. Y.; Yampolskii, Y. P.; Shantarovich, V. P.; Nemser, S. M.; Plate, N. A. *J. Memb. Sci.* 1997, *126*, 123–132; c) Polyakov, A.; Yampolskii, Y. *Desalination* 2006, *200*, 20–20; d) Hammoud, A. N.; Baumann, E. D.; Overton, E.; Myers, I. T.; Suthar, J. L.; Khachen, W.; Laghari, J. R. *NASA Technical Memorandum 1992 105753.*

[55] *Biogeneral Inc., 9925 Mesa Rim Road, San Diego, California, USA.* www.biogeneral.com

[56] Sun, C. X.; Zhao, Y. Y.; Curtis, J. M. *Anal. Chim. Acta* 2013, *762*, 68–75.

[57] a) O'Brien, M.; Taylor, N.; Polyzos, A.; Baxendale, I. R.; Ley, S. V. *Chem. Sci.* 2011, *2*, 1250–1257; b) Newton, S.; Ley, S. V.; Arce, E. C.; Grainger, D. M. *Adv. Synth. Catal.* 2012, *354*, 1805–1812; c) Lau, S.-H.; Bourne, S. L.; Martin, B.; Schenkel, B.; Penn, G.; Ley, S. V. *Org. Lett.* 2015, *17*, 5436–5439.

[58] a) Gross, U.; Koos, P.; O'Brien, M.; Polyzos, A.; Ley, S. V. *Eur. J. Org. Chem.* 2014, 6418–6430; b) Koos, P.; Gross, U.; Polyzos, A.; O'Brien, M.; Baxendale, I.; Ley, S. V. *Org. Biomol. Chem.* 2011, *9*, 6903–6908; c) Mallia, C. J.; Walter, G. C.; Baxendale, I. R. *Beilstein J. Org. Chem.* 2016, *12*, 1503–1511.

[59] a) Bourne, S. L.; O'Brien, M.; Kasinathan, S.; Koos, P.; Tolstoy, P.; Hu, D. X.; Bates, R. W.; Martin, B.; Schenkel, B.; Ley, S. V. *ChemCatChem* 2013, *5*, 159–172; b) Kasinathan, S.; Bourne, S. L.; Tolstoy, P.; Koos, P.; O'Brien, M.; Bates, R. W.; Baxendale, I. R.; Ley, S. V. *Synlett* 2011, 2648–2651.

[60] a) Battilocchio, C.; Iannucci, G.; Wang, S.; Godineau, E.; Kolleth, A.; De Mesmaeker, A.; Ley, S. V. *React. Chem. Eng.* 2017, *2*, 295–298; b) Schotten, C.; Plaza, D.; Manzini, S.; Nolan, S. P.; Ley, S. V.; Browne, D. L.; Lapkin, A. *ACS Sust. Chem. Eng.* 2015, *3*, 1453–1459; c) Bourne, S. L.; Koos, P.; O'Brien, M.; Martin, B.; Schenkel, B.; Baxendale, I. R.; Ley, S. V. *Synlett* 2011, 2643–2647.

[61] a) Pastre, J. C.; Browne, D. L.; O'Brien, M.; Ley, S. V. *Org. Proc. Res. Dev.* 2013, *17*, 1183–1191; b) Browne, D. L.; O'Brien, M.; Koos, P.; Cranwell, P. B.; Polyzos, A.; Ley, S. V. *Synlett* 2012, 1402–1406; c) Cranwell, P. B.; O'Brien, M.; Browne, D. L.; Koos, P.; Polyzos, A.; Pena-Lopez, M.; Ley, S. V. *Org. Biomol. Chem.* 2012, *10*, 5774–5779.

[62] a) Chaudhuri, S. R.; Hartwig, J.; Kupracz, L.; Kodanek, T.; Wegner, J.; Kirschning, A. *Adv. Synth. Catal.* 2014, *356*, 3530–3538; b) Petersen, T. P.; Polyzos, A.; O'Brien, M.; Ulven, T.; Baxendale, I. R.; Ley, S. V. *ChemSusChem* 2012, *5*, 274–277; c) Tomaszewski, B.; Lloyd, R. C.; Warr, A. J.; Buehler, K.; Schmid, A. *ChemCatChem* 2014, *6*, 2567–2576; d) Tomaszewski, B.; Schmid, A.; Buehler, K. *Org. Proc. Res. Dev.* 2014, *18*, 1516–1526; e) Brzozowski, M.; Forni,

J. A.; Savage, G. P.; Polyzos, A. *Chem. Commun.* 2015, *51*, 334–337; f) Sharma, S.; Maurya, R. A.; Min, K. I.; Jeong, G. Y.; Kim, D. P. *Angew. Chem. Int. Ed.* 2013, *52*, 7564–7568; g) Par k, J. H.; Park, C. Y.; Kim, M. J.; Kim, M. U.; Kim, Y. J.; Kim, G. H.; Park, C. P. *Org. Proc. Res. Dev.* 2015, *19*, 812–818.

[63] a) Dallinger, D.; Kappe, C. O. *Nature Prot.* 2017, *12*, 2138–2147; b) Dallinger, D.; Pinho, V. D.; Gutmann, B.; Kappe, C. O. *J. Org. Chem.* 2016, *81*, 5814–5823; c) Garbarino, S.; Guerra, J.; Poechlauer, P.; Gutmann, B.; Kappe, C. O. *J. Flow Chem.* 2016, *6*, 211–217; d) Mastronardi, F.; Gutmann, B.; Kappe, C. O. *Org. Lett.* 2013, *15*, 5590–5593; e) Pinho, V. D.; Gutmann, B.; Miranda, L. S. M.; de Souza, R.; Kappe, C. O. *J. Org. Chem.* 2014, *79*, 1555–1562; f) Koolman, H. F.; Kantor, S.; Bogdan, A. R.; Wang, Y.; Pan, J. Y.; Djuric, S. W. *Org. Biomol. Chem.* 2016, *14*, 6591–6595.

[64] Buba, A. E.; Koch, S.; Kunz, H.; Lowe, H. *Eur. J. Org. Chem.* 2013, *2013*, 4509–4513.

[65] Musio, B.; Gala, E.; Ley, S. V. *ACS Sust. Chem. Eng.* 2018, *6*, 1489–1495.

[66] Polyzos, A.; O'Brien, M.; Petersen, T. P.; Baxendale, I. R.; Ley, S. V. *Angew. Chem. Int. Ed.* 2011, *50*, 1190–1193.

[67] Newby, J. A.; Blaylock, D. W.; Witt, P. M.; Pastre, J. C.; Zacharova, M. K.; Ley, S. V.; Browne, D. L. *Org. Proc. Res. Dev.* 2014, *18*, 1211–1220.

[68] Rehman, A.; Fernandez, A. M. L.; Resul, M.; Harvey, A. *J. CO2 Util.* 2018, *24*, 341–349.

[69] O'Brien, M. *J. CO2 Util.* 2017, *21*, 580–588.

[70] a) *opencv.org/*; b) Bradski, G. Dr. Dobbs J. 2000, *25*, 120–125; c) O'Brien, M.; Cooper, D. A.; Dolan, J. *Tetrahedron Lett.* 2017, *58*, 829–834; d) O'Brien, M.; Cooper, D. A.; Mhembere, P. *Tetrahedron Lett.* 2016, *57*, 5188–5191; e) Ley, S. V.; Ingham, R. J.; O'Brien, M.; Browne, D. L. *Beilstein J. Org. Chem.* 2013, *9*, 1051–1072.

[71] Burden, R., L.; Faires, J. D. *Numerical Analysis, 3rd Edition (2.1 The Bisection Algorithm)*; Prindle, Weber and Schmidt: Boston, MA, US, 1985.

[72] Henry, W. *Phil. Trans. Royal Soc. London* 1803, *93*, 29–274.

[73] Zhang, J.; Teixeira, A. R.; Zhang, H.; Jensen, K. F. *Anal. Chem.* 2017.

[74] Yang, L.; Jensen, K. F. *Org. Proc. Res. Dev.* 2013, *17*, 927–933.

Part III: **General aspects of CO$_2$ chemistry**

CD Hills, N Tripathi, C Lake, PJ Carey, D Heap and AT Hills

10 Mineralisation of CO_2 in solid waste

10.1 Introduction

Carbonation is a natural process that takes place when rocks or wastes are exposed to CO_2 in the air. Natural carbonation proceeds over timescales that extend from decades to millennia.

Industry uses CO_2 as an additive for the production of carbonated drinks to the accelerated curing of concrete. Concrete carbonation has been cited by the industry as good for the environment [1].

The carbonation of mineral substrates can be managed and accelerated by low-energy-intensive processes that can harvest the reaction exotherm not generated in other carbon capture and utilisation (CCU)/transformation technologies.

Zevenhoven and Fagerlund [2] define the mineralisation of CO_2 as its reaction with materials containing alkaline-earth oxides, such as calcium or magnesium oxide (CaO and MgO, respectively). They suggest that for large-scale CO_2 capture and sequestration to take place, the vast resources represented by magnesium silicate-based rocks are ideal, because these rocks occur widely and their carbonated products are environmentally benign. Table 10.1 gives common geologically derived rocks and their mineral requirement to react with CO_2 gas [3–4].

The carbonation of rocks can be considered as natural chemical weathering. On contact with rain or moist air, CO_2 dissolves to produce carbonic acid. The carbonic acid dissociates to H^+ and HCO_3^- (bicarbonate), which degrades susceptible minerals liberating Ca and Mg. These cations then bind with bicarbonate to form carbonate salts. This natural weathering process is estimated to store around 0.3 Gt of atmospheric carbon every year in solid carbonates [6].

Table 10.1: Geologically derived materials with potential to react with CO_2.

Minerals	Chemical formula	Mineral requirement (kg/kg CO_2) *	Reaction completion in optimum conditions
Mg olivine	Mg_2SiO_4	1.6	49.5[+]
Mg serpentinite	$Mg_3Si2O_5(OH)_4$	2.1	73.5[+]
Wollastonite	$CaSiO_3$	2.6	81.8[+]
Basalt	Varies	4.9	15 [#]
Magnetite	Fe_3O_4	5.3	8 [#]

*Penner L., [3]; [+]Zevenhoven, [4]; [#]Sanna et al., [5]

CD Hills, N Tripathi, AT Hills, Engineering Science, University of Greenwich, UK.
PJ Carey, Carbon8 Systems Ltd. Medway Enterprise Hub, UK.
C Lake, Civil and Resource Engineering, Dalhousie University, Canada.

https://doi.org/10.1515/9783110563191-010

The interaction of carbonic acid with geological materials, such as basic igneous rock exposed at or near the Earth's surface, is illustrated below for (1) olivine and (2) serpentine [7–8].

Olivine (Mg_2SiO_4):

$$Mg_2SiO_4 + 2CO_2 + 2H_2O \rightarrow 2MgCO_3 + SiO_2 + 2H_2O = Mg_2SiO_4 + 2CO_2 \rightarrow 2MgCO_3 + SiO_2 \tag{10.1}$$

Serpentine ($Mg_3Si_2O_5(OH)_4$):

$$Mg_3Si_2O_5(OH)_4 + 3CO_2 \rightarrow 3MgCO_3 + 2SiO_2 + 2H_2O, \tag{10.2}$$

Over geological timescales, the reaction between geo-materials and CO_2 will remove large volumes (x 10 Gt) of this gas from the atmosphere, whilst producing voluminous mineral carbonate deposits. Indeed, it has been suggested that tectonic forcing influenced the late Cenozoic climate, when chemical weathering of the Tibetan plateau caused a drop in atmospheric CO_2 concentrations, resulting in a cooler World climate, and the growth of continental ice sheets e.g. [9].

The main issue facing the managed carbonation of geo-materials is that the reaction kinetics tend to be slow. To overcome this considerable engineering effort involving the use of intensive physical and chemical processing, catalysts and raised temperatures and pressures are required. These energy-intensive reaction conditions can significantly increase process yield [10–12]. However, processing rock in this way remains economically unfavourable, unless the value of the carbonate-based products exceeds treatment and transport costs [13].

An alternative approach is the accelerated weathering of geo-materials by spreading rock such as basalt that has been extracted, crushed and ground onto soil to enhance bicarbonate storage in the oceans [14]. A number of workers are currently active in this area e.g. [15], [16], Kohler *et al.* [17].

As discussed by Lackner et al. [18], the prospect of processing earth materials to sequestrate anthropogenic CO_2 in quantities that mitigate climate change is tantalising. Verduyn et al. [19] reviewed the options for CO_2 mineralisation with geologically derived materials; however, before mineralisation can proceed, suitable quantities of rock must be extracted and/or digested via complex chemical treatment processes. Majumdar and Deutch [20] present a high-level review of the research and development opportunities for the Gt-scale utilisation of CO_2, including its disposal into the geosphere. Sanna et al. [5] discuss mineralisation-based CCU with conventional carbon capture and storage (CCS). Nevertheless, despite nearly 25 years of activity in this area, progress is limited.

Table 10.2 gives global mineralisation potential in wastes and geo-materials [21], [22].

Table 10.2: Global CO_2 mineralisation potential.

Source	Potential Gt CO_2 equivalent
Bide et al. [21]	5.25
Global Carbon Initiative (GCI, [22])	3.60

As the status of mineral carbonation is not technically advanced and processing costs remain prohibitive [23], another approach is needed. Thus, for CO_2 mineralisation to be achieved in the shorter term, alternative high volume "mineral" feedstock is required, and contacted with CO_2 in an accelerated way. Industrial process residues have potential in this respect.

Many solid industrial residues have a mineralogy similar to geologically derived materials. The potential of solid residues to be diverted from landfill into feedstock for mineral carbonation has been discussed by several authors e.g. [24], including modelling the carbon emissions for particular waste streams [23].

Of particular interest are the solid residues from thermal processes. These wastes tend to have a high reactive surface area as they have been pulverised and contain calcium silicates or oxides suitable for reaction with CO_2. Fernandez-Bertos et al. [25] reviewed candidate residues and discussed the potential of accelerated carbonation not only as a means to manage the associated risks, but also as a means for the permanent sequestration of CO_2 in products with value. Table 10.3 gives a number of EU waste streams and their potential CO_2 uptake based on stoichiometry (see Steinour [26]) and by laboratory experimentation.

Solid industrial residues are often found as finely particulate materials that are relatively consistent in nature and available in relatively high volumes (Mt). Estimates for the global quantities of CO_2-reactive process residues vary considerably [27], but likely to be around 2 GT worldwide. The potential market for carbonated mineralised

Table 10.3: EU waste streams suitable for carbonation.

Waste type	Mean Steinour-calculated CO_2 uptake (% by weight)	Mean lab CO_2 uptake (% by weight)	Potential CO_2 storage in EU (Kt)
Air pollution control residue	34.3	11.4	172
Incinerator bottom ash	19.7	6.2	1074
Steel slag	27.4	16.5	3267
Coal pulverised fuel ash	16.7	9.7	2896
Cement kiln dust	39.2	22.3	437

products is often located near waste producing plants, which have an established transport infrastructure in place for the process feedstock and the process waste.

The risks associated with potentially carbonate-able residues are often related to their small particle size and high surface area, high pH and the mobility of priority contaminants [25]. When waste is treated by accelerated carbonation, the products are rapidly carbonate-solidified and pH-adjusted/neutralised, and this mitigates risk and promotes re-use.

10.2 History of development

The use of raised CO_2 levels in curing chambers used for the enhanced hydraulic hardening of concrete articles (e.g., concrete blocks) has been practiced for a hundred years or more. The amount of CO_2 employed rarely exceeded a few percent (v/v) due to safety concerns, as CO_2 is a recognised asphyxiant.

One of the first reported uses of CO_2 for treating cementitious material was in the mid-1800s for the conditioning of Portland cement to remove free lime and, therefore, improve the soundness of this hydraulic product [28]. In recognising that cementitious materials tend to react with CO_2, Steinour [26] examined the 'carbonate-ability' of these materials.

Young et al. investigated the carbonation of Portland cement-based compacted samples and showed that increased partial and elevated pressures of CO_2 could be used to induce rapid hardening [29–30]. Their work explored the decalcification of hydrated or poorly hydrated cement, the reduction in product porosity and formation of a finer capillary pore network within carbonated products. They noted that a water/solid (w/s) ratio of 0.125 was optimal for, for example, the carbonation of B-C_2S. The kinetics of carbonation of concrete have been discussed by Shi et al. [31].

The effect of a dominantly CO_2 curing environment for use in commercial processes, such as in the manufacture of concrete roofing tiles, illustrated the enhanced production rates that could be achieved [32]. The precipitation of carbonate salts in pores reduces drying shrinkage and water adsorption. In turn, strength and dimensional stability increase to levels comparable with steam-cured concrete [31].

When hydrated Portland cement is carbonated on exposure to CO_2 and moisture (equation 10.3), calcium hyoxide (10.4) and calcium silicate (10.5) form carbonates [33–35]. However, under saturated conditions, carbonation is suppressed by slow rate of diffusion of CO_2 in water compared to air [36]. The reactions with cementitious materials can be described as follows:

$$CO_2 + H_2O \rightarrow H_2CO_3 \tag{10.3}$$

$$Ca(OH)_2 + H_2CO_3 \rightarrow CaCO_3 + 2H_2O \tag{10.4}$$

$$3CaO.2SiO_2.3H_2O + H_2CO_3 \rightarrow CaCO_3 + 2SiO_2 + 6H_2O \qquad (10.5)$$

In the 1990s, the accelerated carbonation of hazardous waste was described by Lange et al. [37–38]. The hazardous wastes in question were difficult to solidify and stabilise as they interfered with the hydraulic hardening of the Portland cement binder [39]. A carbonation step was sufficient to overcome the inhibiting synergistic effects of heavy metals contained in the residues [40]. As a result, carbonation could be used to solidify the waste forms, enabling their disposal to landfill. Further work by the authors led to the first successful field application of carbonation to treat soil at a contaminated site in SE England [41].

The prevalence of CO_2-reactive minerals as found in many thermal residues has been discussed by Fernández-Bertos et al. [23], Li *et al.* [42] and Gunning et al. [35]. Candidate residues included municipal solid incinerator waste, paper incineration ashes, wood ash, pulverised fuel ash and steel slag [43].

The quantities of potential industrial residues for carbonation include legacy wastes. In countries where the regulatory framework for waste disposal is not very stringent, legacy residues are available in abundance. For example, a land-raise containing 1 Mt of metallurgical process waste in Rajasthan, India, has been shown to be able to be combined with 30 % CO_2 (v/v) with the product having potential for use in construction (unpublished work by the authors). It is suggested there may be > 7 Mt of these wastes in India.

Furthermore, with the potential of some wastes to be re-processed for metal(s) recovery, a treatment train involving carbonation may be of future importance. With respect to Argon Oxygen Decarburisation (AOD) and basic oxygen furnace (BOF) steel slags, a mechanism for the extraction/recovery of metal by ion exchange [44] or by ion-exchange resin [45] has been investigated. The recovery of significant trace metals including Cu, V, Zn, Ni, Mo and Cr prior to carbonation has been reported [46].

Many residues that are CO_2 reactive are alkaline in nature and upon carbonation, the precipitation of carbonate salts increases product density and strength, and reduces pH (typically by several units). As a result, the carbonate-hardened products can have re-use potential in engineering applications [47–48]. In combining carbonation with the manufacture of agglomerates, residues can be reconstituted into a lightweight aggregate e.g. [49].

Manufactured carbonated building materials made from waste have been commercially available in the UK for a number of years now, clearly illustrating that CCU-based technologies can successfully compete in the open market.

10.3 Research developments

There is increasing interest in legacy residues such as mine-tailing and other deposits for their potential to be passively carbonated or mined as a feedstock. Asbestos, copper, nickel, platinum deposits and bauxite residues (red mud from alumina processing) produce tailings with potential for mineral carbonation and CO_2 sequestration [50]. Ebrahini et al. [11] investigated calcium-rich alkaline waste/tailings including steel slag. Power et al. [51] reviewed the carbonation of alkali earth silicate and hydroxide minerals. Teir et al. [52] extracted magnesium from serpentinite (obtained from stockpile nickel tailings) and carbonated the Mg extracts with a carbonate conversion of 94%. Although the process produced individual precipitates of silica, iron oxide and hydromagnesite (93–99% purity), the cost was high at US$600-1600 t$CO_2$.

The mineralisation of flue gas CO_2 using electric arc furnace slag was investigated [53] to develop green construction materials to blend with cement mortar. Hamiliton et al. [54] investigated ultramafic tailings arising from mining crysotile and showed that the carbonation of gangue minerals helps sequester transition metallic contaminants rather than enhancing their release into surface or sub-surface mine waters.

There are the existing processes in China directly using flue gas to produce fertilisers [55]. The authors reviewed the mineralisation of Portlandite generated from industrial alkaline residues (including carbide, steel slag, paper mill waste, cement kiln dust and coal fly ashes) to produce $NaHCO_3$ and energy.

The effect of partial pressure of CO_2 and moisture content of wastes such as these has been examined by Sanna and Maroto-valer [56]. A separate study [5] noted that the carbonation reaction slowed when using diluted CO_2.

The enhancement of the carbonation reaction can be achieved by a pretreatment step including increasing reactant surface area, and the use of higher "energy" reaction conditions. Gerdemann et al. [57] reported the direct carbonation of ground magnesium (or calcium) silicates at 150–200 °C and 100–150 bar in the presence of 0.64 M $NaHCO_3$ and 1 M $NaCl$. With this approach, the extent of carbonation of olivine was >80% at 6h and >70% for Wollastonite at 1 h.

10.4 Development of commercial CO_2 mineralisation processes for solid waste treatment

The accelerated carbonation of industrial residues coupled with granulation can be used to manufacture engineered products. Gunning et al. [35] described lightweight aggregates (bulk density <1,000 kg/m^3 and compressive strength >0.10 MPa) made from alkaline residues including cement kiln dust, wood and paper ash. The process was commercialised under license in 2012 to produce aggregates from MSW air pollution control residue (APCr) for use in construction [58]. A number of production plants are in operation or planned, with current production being in the region of 200kt /pa [59].

Morone et al. [60] developed carbonate-bonded aggregates from BOF steel slags at pilot-scale. The products captured up to 10% CO_2 (v/v) after 28 days curing. Similar products from AOD slag were developed by Salman et al. [61]. A compressive strength of 34 MPa was developed after 3 weeks of curing in an atmosphere of 5% CO_2 (v/v) and 60 MPa when held at 8 bar and 80°C for 15 h. Quaghebeur et al. [62] described stainless steel waste-based monolithic compacts with compressive strengths of 55 MPa, when cured in pure CO_2 at 140 °C and 20 bar. In 2017, pre-cast building blocks made from moist carbonated steel slag were commercially available [63]. The products contain a blend of natural sand and finely milled steel slag.

Resulting from an increasing interest in waste mineralisation as a viable CCU technology, there are a number of commercial processes combining CO_2 with solid waste streams. Some of these are summarised in Table 10.4, and discussed below:

Alcoa Inc. (Australia) uses CO_2 directly from an ammonia plant to carbonate bauxite residue. The process involves conversion of flue-derived CO_2 into soluble bicarbonate and carbonate via an in-duct scrubber system with an enzyme catalyst. Alcoa's carbonation plant has been operational since 2007, treating bauxite residues/alkaline clay (a by-product of aluminium refining) to make construction fill and soil amendments. According to Evans et al. [64], multiple 10 kt quantities of red mud will be carbonated. The yield is 33 kg/t red mud according to the [22].

Calera (USA) produces a calcium carbonate cement using solid waste and flue gas emitted from power plants. The process involves aqueous precipitation of mineral carbonates and bicarbonates. Calera's process sources metal cations from hard water and brines. Sodium hydroxide is used to absorb CO_2 from coal-fired power plants (US Dept of Energy, [65] with acid gases such as SO_2 and heavy metals [66]. Each tonne of carbonated mineral/cement formed contains 500 kg CO_2. The International Energy Agency [68] evaluated the potential of this technology, although over the past 10 years it has not met commercial success.

Carbicrete (Montreal, Canada) manufactures a concrete made from waste and CO_2 (8% from flue gas) using a carbonation activation step. The process uses steel slag as the primary binder source. The manufactured construction materials meet specification for use as a concrete [69–70] and fixes 200 kg CO_2 per day [71]. It is reported that Carbicrete has approximately 50% greater compressive strength than conventional concrete [72].

Carbon8 Systems Ltd. (UK) developed accelerated carbonation technology (ACT) for the treatment of industrial residues and contaminated soils. The technology allows permanent capture of CO_2 under ambient temperature and pressure conditions in residues from the treatment of flue gas from Energy from Waste plants, to produce construction products that meet the requirements of "End of Waste" as agreed with the Environment Agency.

Table 10.4: Status of selected CO_2 mineralisation processes.

Process	Locations	Technology	TRL	Reaction environment	Product	Production status (production)	CO_2 used	Reference
Alcoa	Australia	Red mud treatment with flue gas CO_2 with enzymes	6	Raised temperature and pressure	Construction fill material, soil amendment/ fertiliser	Commercial	70,000 t	1
Calera	USA	Carbonate precipitates from CO_2 in water/brine	8–9	Aqueous-based, raised temperature and pressure	Cement	>700 t/yr	460 kg/t	2, 3
Carbicrete	Canada	Carbonation of steel slag	6–7	Raised temperature and pressure	Cement-free concrete	Pilot	200 kg/day	4
Carboncure	Canada, USA	Direct injection of CO_2 gas into green concrete/mortar	8–9	Ambient temperature and pressure	Accelerated cured concrete	Used in >40 pre-cast plants in USA/ Canada	3% CO_2 avoided	5
Carbon8 Systems	UK	Accelerated carbonation technology	9	Ambient temperature and pressure	Aggregates and fill, e.g., for blocks, and screed	Technology development/ licensor	100-200kg/t	6

Carbon8 Aggregates	UK	Accelerated carbonation technology	9	Ambient temperature and pressure	Aggregates	Ca. 200,000 t/yr (3 UK plants)	9000 t	7
Carbstone Innovation	Belgium	Carbonation of steel slag using CO_2 from flue gas	9	Autoclave-based (100% CO_2, 0.5-10 MPa and 20-140°C)	Construction materials including blocks and tiles	100,000 t	200 kg/t	8
Solidia	USA, Canada	CO_2 curing for cement concrete manufacturing	8	Chamber-based curing with 100% CO_2/24hr	Carbonate cement/concrete	Pilot (quantities unknown)	240 kg/t	9
CCm Research	UK	CO_2 combined ammonia-coated waste fibres	7	Waste fibres and CO_2 from exhaust gas	Fertiliser	Pilot (quantities unknown)	85% GHG reduction	10, 11
Blue Planet	California, USA	CO_2 sequestered coating over a substrate	6–7	Flue gas CO_2 and alkaline rock/waste	Lightweight aggregate	Pilot/demonstration (quantities unknown)	Not known	12
Carboclave	Ontario, Canada	Precipitation of nano-$CaCO_3$ crystals	7	CO_2 curing of concrete	Concrete blocks	Quantities unknown	300g CO_2 per 20cm block	13, 14

(continued)

Table 10.4 (continued)

Process	Locations	Technology	TRL	Reaction environment	Product	Production status (production)	CO_2 used	Reference
Green Minerals	Netherlands	Carbonation of olivine	3	Carbonation under pressure with CO_2 additive(s)	Building materials	Pilot	200 kg/t	15

1) https://www.globalccsinstitute.com/sites/www.globalccsinstitute.com/files/content/page/122975/files/Alcoa%20Kwinana%20Carbonation%20Plant.pdf | 2) http://www.calera.com/ 3) https://www.arb.ca.gov/cc/etaac/meetings/102909pubmeet/mtgmaterials102909/basicsofcaleraprocess.pdf | 4) http://www.carbicrete.com/ | 5) https://www.carboncure.com/ | 6) www.c8s.co.uk | 7) http://c8a.co.uk/ | 8) https://www.loesche.com/sites/default/files/list-content/brochure/2017-08/259_LOESCHE_EDS%20Steel%20Slag_EN.pdf | 9) http://solidiatech.com/ | 10) http://www.ccmresearch.co.uk/intro.html | 11) https://www.ktn-uk.co.uk/perspectives/ccm-research-develops-carbon-dioxide-utilisation-process-for-value-added-materials | 12) http://www.blueplanet-ltd.com/ | 13) http://www.carboclave.com/ | 14) https://www.on-sitemag.com/features/1003957345/| 15) www.ccmtechnologies.co.uk/fertiliser.html

CarbonCure (Nova Scotia, Canada) uses the direct injection of small amounts of CO_2 into green concrete mixtures to accelerate curing and enhance strength development. Increased product strength enables lower cement contents to be used, reducing the carbon footprint of the product [73].

Carbstone Innovation (Belgium) treats electric arc furnace and BOF slags with CO_2 for use as construction materials [46]. Following metals recovery from the slags, the process involves three steps: (1) pre-treatment of slags, (2) shaping of the products by compaction at pressures between 75 and 609 kgf/cm^2 and (3) CO_2 curing in an autoclave at a gas-feed pressure of 0.5-10 MPa and temperatures of 20 to 140 °C. The process sequesters 180-200 g CO_2/kg of slag.

Solidia Technologies (USA) produces a C3S-based cement that reacts with moist CO_2 to harden by carbonate cementation [74]. The concrete produced is white in colour. Solidia cement-based products such as bricks and blocks are cured for up to 24hr in a CO_2 curing environment. The concrete blocks produced from this process capture about 240 kg of CO_2 for every 1,000 kg of cement used in the mixture [75].

10.5 Future developments

Although much attention has been focussed on the sources of geologically derived and process-waste-based feedstock for CO_2 mineralisation, there has been less attention given to potential sources of CO_2. Ideally, the direct use of flue gas will reduce the cost of treatment of mineral feedstocks.

In the UK, as is the norm elsewhere, the infrastructure for the capture and transport of CO_2 is not in place, so the choice facing mineralisation processes is to locate near a source of CO_2, the mineral feedstock or the market for the mineralised products. This choice is an economic one. In the UK, road-tankered CO_2 derived, e.g., from fertiliser production provides flexibility in the location of a mineralisation plant.

However, high-purity CO_2 is expensive as it requires capturing, scrubbing, pressurisation, liquefaction and transportation in specialised road tankers. As mineralisation processes continue to develop, the requirement to capture CO_2 directly without purification and liquefaction and transportation will become essential.

In recent years, the use of catalysts (e.g., carbonic anhydrase (CA) enzymes) has gained attention as a means to catalyse mineral carbonation of reactive feedstocks. Catalysts accelerate the carbonation process, where the supply of CO_2 is limited, or the solvation/hydration of CO_2 is a rate limiting factor. Power et al. [76] reported an acceleration in carbonation by 240% in alkaline brucite slurry after amending the CO_2 supply-limited carbonation process with bovine CA. In another study [77], it was observed that the anhydrase enzyme promotes precipitation of calcium carbonate, under appropriate temperature and pH conditions of the reaction mixture. However, Lorenzo et al. [78] recently studied the carbonation of wollastonite using natural (e.g., CA) and biomimetic metal-organic framework catalysts. They reported that although CA accelerates the final carbonate precipitation step, it hinders the overall carbonation of wollastonite. They suggested the use of metal-organic framework alone or in combination with natural catalysts for enhancing silicate dissolution and accelerated carbonation. An enhanced rate of carbonation via enhanced CO_2 hydration has been reviewed by Maries and Hills [79]. The use of chemical enhancers (or homogeneous catalysts), including inorganic oxyanions (e.g., hypochlorite or sulphite); organic solutes (e.g., sugars and polyhydric alcohols) and amines and alkaloamines producing carbamates can enhance solvation and hydration of CO_2 in water.

In addition to catalyst-based accelerated carbonation, biologically enhanced mineralisation has also been explored. This involves the biologically engineered organisms to enhance the CO_2 hydration process. A yeast-based catalysis involving *Saccharomyces cerevisae* for enhanced production of $CaCO_3$ in coal fly ash indicated ca. 10% cost effectiveness for per ton capture of CO_2 [80]. Oliver et al. [81] reported a catalytic impact on the rate-limiting carbonate reactions by using an algal species *Scenedesmus* for biologically enhanced degassing and precipitation of carbonates.

Other recent advances in the transformation of CO_2 have been demonstrated by HeidelbergCement [82]. In one approach, flue gas has been used for the growth of micro-algae in bioreactors in three HeidelbergCement sites (Sweden, Turkey and France). The micro-algae produced meet all the necessary criteria for feed for cattle or fish.

The capacity of soil to hold carbon in organic matter is well recognised [83] and the managed enhancement of soil organic matter is one mitigation strategy being investigated [84]. NRC [85] also reviewed options for enhancing organic matter pool into agricultural soils. The inclusion of mineralised CO_2 in the remediation of contaminated soils has received less attention.

An indication of the advantages from the sequestration of CO_2 in soil via carbonate-based solidification treatment was obtained in a 'World first' experiment in 1999. Contaminated soil located in SE England was stabilised by accelerated carbonation. Instrumented test-cells containing the granular carbonated and un-carbonated soils were monitored for a period of nine years. The carbonate-cemented soil proved to be resilient to weathering and the leaching performance was well within prescribed limits. This work was reviewed by Antimer et al. [86]. Figure 10.1 shows the carbonated soil being sampled by coring in 2005.

An international collaborative project called PASSIFY involving the UK, USA and France investigated a number of cement-stabilised and solidified waste forms located in the partner countries. It was noted that the stabilised/solidified soils were carbonated to a lesser or greater amount and that this appeared to be occurring sub-aerially [87]. Figure 10.2 gives a backscattered electron micrograph showing carbonated stabilised-solidified waste.

Figure 10.1: Accelerated carbonated contaminated soil being sampled in 2005.

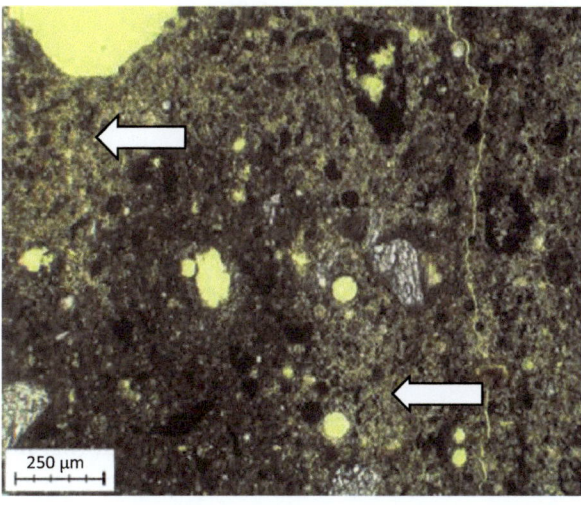

The lighter coloured matrix (arrowed), adjacent to a large void (upper left) and extending to lower right has been carbonated by subaerial reactions with CO_2

250 µm

Figure 10.2: Carbonated stabilised solidified waste (transmitted light photomicrograph).

Work by Lake et al. [88] and Choi et al. [89] have examined the potential of aggregated carbonated CKD wastes for geotechnical applications (i.e., roadway fills, etc.). In this work, CKD was aggregated and carbonated in the laboratory without the presence of cement or lime-based binders. Particle breakage during mechanical loading (i.e., compression, shear) and freeze-thaw cycling improved as the particle size of the aggregate decreased (2.5 mm vs 1.25 mm). It was shown that these aggregates had similar mechanical durability as natural calcareous sands that are used to support existing infrastructure.

Table 10.5 lists a summary of the particle breakage in terms of 'relative breakage' (B_r). A B_r value of zero would indicate no change in grain size distribution (i.e., no breakage) during testing, while a value of 1 would indicate that all particles (i.e., maximum breakage) were reduced to sizes <0.074 mm during testing.

Table 10.5: Comparison of relative breakage, B_r, between calcareous.

Reference	Mean stress (kPa)	Relative breakage, B_r
Calcareous Sand (Shipton and Coop, [91]	500 (triaxial compression)	0.1 (estimated from plot)
	3,000 (triaxial compression)	0.225 (estimated from plot)
	4,000 (triaxial compression)	0.1 (estimate from plot)
Shahnazari and Rezvani	866 (triaxial shear)	0.1 (estimated from plot)
Choi et al. [89]	1,024 (triaxial shear)	0.04
(1.25 mm–2.5 mm aggregate size)	1,289 (triaxial shear)	0.04
	1,585 (triaxial shear)	0.05
	1,988 (triaxial shear)	0.06
	2,254 (triaxial shear)	0.07
Choi et al. [89]	871 (triaxial shear)	0.07
(2.5 mm–5 mm aggregate size)	1,225 (triaxial shear)	0.08
	1,552 (triaxial shear)	0.10
	1,879 (triaxial shear)	0.10
	2,159 (triaxial shear)	0.12

The CO_2 capture potential of industrial process residues is difficult to quantify. However, in Europe and other parts of the world, there are significant and well documented, voluminous wastes with potential for CO_2 mineralisation. Table 10.6 gives example of European process waste streams, the amount landfilled (i.e. not already valorised) and their potential CO_2 uptake as determined from laboratory experimentation [90]. With the improvement of materials properties as described by Choi et al. [89], it is anticipated that more wastes will be valorised by mineralisation in the future.

Example EU waste streams suitable for carbonation.

	Production (Mt/yr)	Amount landfilled (Mt)	Potential CO$_2$ uptake (Mt)
MSW APCr	2.45	2	0.2
C&D	321	41	4.1
Red mud +	5 to 6	>5	>1.5
CKD/CBD ++	19.5	5.2	4.29 (22%)
Combustion and mineral wastes	192	83	–
Steel slag**	21.4	1.28	3.53
Paper ash	5.5	0.5	0.1

Based on total cement production in EU (260 Mt) and disposal calculated based on UK data (2010) of CKD disposal [90]. + [93] ++[94] ** Grubeša et al. [95] [96].

10.6 Summary

Pathways to mitigate industrial CO$_2$ emissions by CCU are being extensively explored in various sectors. A common impediment is that CCU processes are generally energy intensive and the infrastructure for handling large amounts of CO$_2$ remains to be established.

The mineralisation of CO$_2$ in waste is perhaps the most attractive option as process residues can be suitable feedstocks for mineralisation without significant further processing. Furthermore, the wastes often arise from processes that also emit CO$_2$ i.e., these two "feedstocks" are often co-located.

Construction products made from waste are already available in the market and they are demonstrably cost competitive. There is an established track record of successful commercial use of manufactured carbonated aggregates.

As high-volume wastes are often produced from facilities located near urban areas, the proximity of the 'market' is an advantage from a logistical point of view.

Recent research demonstrates that manufactured carbonated aggregates are comparable with virgin stone and are suitable for difficult applications, such as where a high level of freeze-thaw resistance is required.

Demand for construction aggregates is currently in the region of 50 Gt worldwide each year, including for the production of 10 Gt of concrete. The size of this market is advantageous as it has potential to utilise Gt of mineralised construction products without being significantly disrupted.

The EU-28 market including for aggregates in 2015 was 2.3 Gt, and with current technologies there is potential to sequestrate 5–20 Mt CO$_2$ [21] Going forward, the quantity of CO$_2$ sequestered worldwide in mineralised waste-based construction products is likely to contribute significantly to the circular economy and mitigation of climate change.

References

[1]   Andersson, R., Fridh, K., Kan S. H., Hä, M. (2013). Calculating CO_2 Uptake for Existing Concrete Structures during and after Service Life.

[2]   Zevenhoven, R. and Fagerlund, J. (2010). Mineralisation of CO_2. In: M. MarotoValer (Ed.). Developments and innovation in CCS technology. Woodhead Publishing Ltd., Cambridge (UK), p. 433-462 (Chapter 16). ISBN 10.1533/9781845699581.4.433.

[3]   Penner L., O. W. (2004). Mineral Carbonation: Energy Costs of Pretreatment Options and Insights Gained from Flow Loop Reaction Studies. 3th Annual Conference on Carbon Capture and Sequestration. Va.

[4]   Zevenhoven, R. (2010). Mineral carbonation for long-term CO_2 storage: an exergy analysis. International Journal of Applied Thermodynamics, 23.

[5]   Sanna, A., Uinbu, M. Caramanna, G., Kuusik, R. and Maroto-Valer, M.M. (2014). A review of mineral carbonation technologies to sequester CO_2. Chemical Society Reviews, 43, 8049–8080.

[6]   Beaulieu, E., Godđeris, Y., Donnadieu, Y., Labat, D., and Roelandt, C. (2012). High sensitivity of the continental-weathering carbon dioxide sink to future climate change, Nature Climate Change, 2, 346–349, 2012

[7]   Alexander, G. & Maroto-Valer, M. M. (undated) Serpentine and single stage mineral carbonation for the storage of carbon dioxide. Accessed July 2018 at: https://pdfs.semanti cscholar.org/1b97/91c844cdc5b7031734442244515fcac3bb69.pdf

[8]   O'Connor, W. K., Dahlin, D. C., Nilsen, D. N., Walters R. P. & Turner, P. C. (2000). Carbon Dioxide Sequestration by Direct Mineral Carbonation with Carbonic Acid. Proceedings of the 25th International Technical Conf. On Coal Utilization & Fuel Systems, Coal Technology Assoc., Clear Water, FL. Albany Research Center & U.S. Department of Energy. DOE/ARC-2000-008. [2] https://www.osti.gov/servlets/purl/896218

[9]   Raymo, M.E. and Ruddiman, W.F. (1992). Tectonic Forcing of Late Cenozoic Climate. Nature Sep 10, 1992; 359, 6391; Research Library Core p. 117

[10] Eikeland, E., Blichfield, A.B., Tyrsted, C., Jensen, A.P., Iversen, B.B. (2015). Optimized Carbonation of Magnesium Silicate Mineral for CO_2 Storage. ACS Applied Materials & Interfaces 7(9). http://dx.doi.org/10.1021/am508432w

[11] Ebrahini, A., Saffari, M., Hong, Y., Miliani, D., Montoya, A., Valix, M., Minett, A and Abass, A. (2018). Mineral sequestration of CO_2 using saprolite mine tailings in the presence of alkaline industrial wastes. Journal of Cleaner Production 188. http://dx.doi.org/10.1016/j.jcle pro.2018.04.046

[12] Balucan, R.D., Dluggogorski, B.Z., Kennedy, E.M., Belova, I.V. and Murch, G.E. Energy cost of heat activating serpentinites for CO_2 storage by mineralisation. January 2013. International Journal of Greenhouse Gas Control 17:225-239, http://dx.doi.org/10.1016/j.ijggc.2013.05.004

[13] Wang, J., Wu, H., Duan, H., Zillante, G., Zuo, J. and Yuan, H. (2018a). Combining life cycle assessment and building information modelling to account for carbon emission of building demolition waste: a case study. J. Clean. Prod., 172, 3154–3166.

[14] Sackler Forum (2017). Dealing with Carbon Dioxide at Scale. May 2018. www.royalsociety. org/sackler-forum

[15] Renforth, P., Pogge von Strandmann, PAE and Henderson, G.M. (2015). The dissolution of Olivine added to soil: Implications for enhanced weathering. Applied Geochemistry, 61, pp 109–118.

[16] Wilson, S.A., Harrison, A.L., Dipple G.M., Power, I.M., Barker, S.L.L, mayer, K.U., Fallon, S.J, Raudsepp, M. and Southam, G. (2014). Offsetting of CO_2 emissions by air capture in mine

tailings at the Mount keith Nickel Mine, Western Australia: Rates, controls and prospects for carbon neutral mining. International Journal of Greenhouse Gas Control 25, pp. 121-140.

[17] Kohler, P., Hartmann, J and Wolf-Gladrow, D.A. (2010). Geoengineering Potential of Artificially Enhanced Silicate Weathering of Olivine. PNAS, November 23, 2010 107 (47) 20228-20233. https://doi.org/10.1073/pnas.1000545107

[18] Lackner, K.S., Wendt, C.H., Butt, D.P., Joyce, E.L., Sharp, D.H. (1995) Carbon dioxide disposal in carbonate minerals. Energy 20 (11): 1153–1170. https://doi.org/10.1016/0360-5442(95) 00071-N

[19] Verduyn, M., Geerlings, H., van Mossel, G. and Jijayakumari, S. (2011). Review of the various CO₂ mineralization product forms. Energy Procedia 4, 2885–2892.

[20] Majumder, A. and Deutch, J. (2018). Research Opportunities for CO₂ Utilisation and Negative Emissions at the Gigatonne Scale. Joule 2 (5) 801–809. https://doi.org/10.1016/j.joule.2018. 04.018

[21] Bide, T.P., Styles, M. T. and Naden, J. (2014). An assessment of global resources of rocks as suitable raw materials for carbon capture and storage by mineralisation. *Applied Earth Science*, 179–195.

[22] GCI (2016). The Global Competitiveness Report 2016-2017. World Economic Forum Geneva. ISBN-13: 978-1-944835-04-0

[23] Wang, F., Dreisinger, D.B., Jarvis, M, and Hitchins, T. (2018b). The technology of CO₂ sequestration by mineral carbonation: current status and future prospects. Canadian Metallurgical Quarterly, Vol 57 (1), 46-58.

[24] Sanna, A., Dri, M., Hall, M. R., & Maroto-Valer, M. (2012). Waste materials for carbon capture and storage by mineralisation (CCSM) – A UK perspective. DOI: 10.1016/j. apenergy.2012.06.049

[25] Fernandez-Bertos, M. Simons, SJ, Hills, CD and Carey, PJ (2004). A Review of Accelerated Carbonation technology in the Treatment of Cement-Based materials and Sequestration of CO₂. J Hazar Mater, 112(3): 193-205. DOI: 10.1016/j.jhazmat.2004.04.019

[26] Steinour, H.H. (1959). Some Effects of Carbon Dioxide on Mortars and Concrete- discussion J. Am Concrete Inst., 30 (2) 905–907

[27] Gomes, H.I., Mayes, W., Rogerson, M., Stewart, D.I. and Burke, I.T. (2016). Alkaline residues and the environment: A review of impacts, management practices and opportunities. Journal of Cleaner Production 112(4). DOI: 10.1016/j.jclepro.2015.09.111

[28] Rowland, J. L. (1870). "Improvement in the manufacture of artificial stone", U.S. Patent No. 109669.

[29] Young J.F., Berger, P.L., Breese, J. (1974). Accelerated Curing of Compacted Calcium Silicate Mortars on Exposure to CO₂. J. Am. Ceram. Soc. Sept 1974. https://doi.org/10.1111/j.1151-2916.1974.tb11420.x

[30] Klemm, W.A., Berger, R.L. (1972). Accelerated Curing of Cementitious Systems by carbon Dioxide: Part 1. Portland cement. Cem Concr. Res., 2 (5), 567-576.http://dx.doi.org/10.1016/ 00008-8846(72)90111-1

[31] Shi, C.J., Liu, M., He, PP, Ou, ZH. (2012). Factors affecting the Kinetics of CO₂ Curing of Concrete. J Susuain Cement-based Mater, (1-2), 24-25. http://dx.doi.org/10.1080/21650373. 2012.727321

[32] Maries, A. and Hills, C.D. (1983). Improvement in Concrete Articles. UK Patent 2,192,392. Matter, J. M.; Keleman, P. B. (2009). Permanent storage of carbon dioxide in geological reservoirs by mineral carbonation. Nat. Geosci., 2, 837–841.

[33] Sulapha, P., Wong, S.F., Wee, T.H., Swaddiwudhipong, S. (2003). Carbonation of concrete containing mineral admixtures. Materials in Civil Engineering, 15, 134–143.

[34] Jiang, L., Lin, B., Cai, Y. (2000). A model for predicting carbonation of high-volume fly ash concrete. Cement and Concrete Research 30, 699–702.

[35] Gunning, P.J., Hills, C.D. and Carey, P.J. (2009). Production of lightweight aggregate from industrial waste and carbon dioxide. Waste Management, 29 (10), 2722–2728.

[36] Balen, V. (2005). Carbonation reaction of lime, kinetics at ambient temperature. Cement and Concrete Research 35 (4), 647–657.

[37] Lange, L.C., Hills, C.D. and Poole, A.B. (1997a). The Effect of Carbonation on the Properties of Blended and Non-Blended Cement Solidified Waste Forms. Journal of Hazardous Materials, Special Publication, 52, pp. 193–212.

[38] Lange, L.C., Hills, C.D. and Poole, A.B. (1997b). The Effect of Accelerated Carbonation on the Properties of Cement-solidified Waste Forms. Waste Management, 16, 8, pp. 757–763.

[39] Hills, C.D., Sollars, C.J. and Perry, R. (1993). Ordinary Portland Cement-Based Solidification of Toxic Wastes: The Role of OPC Reviewed. Cement and Concrete Research, 23, pp. 196–212.

[40] Hills, C.D., Sollars, C.J and Perry, R. (1994). Calorimetric and Microstructural Study of Solidified Toxic Wastes: Part 1 -A Classification of OPC/Waste Interference Effects. Waste Management, 14, pp. 589–599.

[41] Antemir, A., Hills, C.D., Carey, P.J., Magnié, M-C.and Polettini, A. (2010). Investigation of 4-year-old stabilised/solidified and accelerated carbonated contaminated soil. Journal of Hazardous Materials, 118, 543–555.

[42] Li, X., Fernández Bertos M., Hills C.D., Simons S.J.R., Carey P.J. (2007). Accelerated Carbonation of Municipal Solid Waste Incineration Fly Ash. Waste Management, 27, pp. 1200–1206

[43] Johnson, D.C., MacLeod, C.L. and Hills, C.D. (2003). Solidification of Stainless Steel Slag by Accelerated Carbonation. Environmental Technology, 24, pp. 671–678.

[44] Ogden M.D., Moon E.M., Wilson A. & Pepper S.E. (2017). Application of chelating weak base resin Dowex M4195 to the recovery of uranium from mixed sulfate/chloride media. Chemical Engineering Journal, 317, 80–89

[45] Gomes, H. I. Jones, A., Rogerson, M, Greenway, G. M., Lisbona, D. F., Burke, I. T. and Mayes, W. M. (2017). Removal and recovery of vanadium from alkaline steel slag leachates with anion exchange resins, Journal of Environmental Management, Volume 187, 1 February 2017, pp. 384-392, ISSN 0301-4797, http://doi.org/10.1016/j.jenvman.2016.10.063.

[46] Quaghebeur, M., Nielsen, P., Horckmans, L. and Mechelen, D.V. (2015). Accelerated carbonation of steel slag compacts: development of high strength construction materials. Frontiers in Energy Research, volume 3, article 52, doi: 10.3389/fenrg.2015.00052.

[47] Rendek, E., Ducom, G., Germain, P. (2006). Carbon dioxide sequestration in municipal solid waste incinerator (MSWI) bottom ash. Journal of Hazardous Materials, 128, 73-79.

[48] Johannesson, B., Utgenannt, P. (2001). Microstructural changes caused by carbonation of cement mortar. Cement and Concrete Research, 31, 925–931.

[49] Padfield, A., Carey, P.J. and Hills, C.D. (2004). Beneficial Re-use of Quarry fines, CO2 Gas and Solid Industrial Waste in the Production of Secondary Aggregates. Engineering Sustainability Journal, 157, pp.149–154.

[50] Bobicki, E.R., Liu, Q., Xu, Z. and Zeng, H. (2012). Carbon capture and storage using alkaline industrial wastes. Prog. Energy Combust. Sci., 38, 302–320.

[51] Power I.M., Harrison, A.L., Dipple, D.M., Wilson, S.A., Kelemen, P.B., Hitch, M., and Southam, G. (2013). Carbon mineralization: from natural analogues to engineered systems, Reviews in Mineralogy and Geochemistry 77 (1): 305-360. https://doi.org/10.2138/rmg.2013.77.9

[52] Teir, S., Eloneva, S., Fogelholm, C.J. and Zevenhoven, R. (2009). Fixation of carbon dioxide by producing hydromagnesite from serpentinite. Appl. Energy, 86, 214–218.

[53] Pan, S.Y., Chung, T.C., Ho, C.C, Hou, C.J., Chen, Y.H. and Chiang, P.C. (2017). CO_2
Mineralization and Utilization using Steel Slag for Establishing a Waste-to-Resource Supply
Chain. Sci Rep. 2017 Dec 8; 7(1):17227. doi: 10.1038/s41598-017-17648-9

[54] Hamilton, J.L., Wilson, S.A., Morgan, B. et al. Turvey, C.C., Paterson, D.J., Jowitt, S.M.,
McCucheon, J. and Southam, G. (2018). Fate of transition metals during passive carbonation
of ultramafic mine tailings via air capture with potential for metal resource recovery.
International Journal of Greenhouse Gas Control, 71. pp. 155-167. https://doi.org/10.1016/j.
ijggc.2018.02.008

[55] Xie, H., Yue, H, Zhu, J, Li, C, Wang, Y, Xie, L and Zhou, X. (2015). Scientific and Engineering
Progress in CO_2 Mineralization Using Industrial Waste and Natural Minerals, Engineering,
1(1): 150–157.

[56] Dri, M., Sanna., A and Maroto-valer, M. (2014). Mineral carbonation from metal wastes: Effect
of solid to liquid ratio on the efficiency and characterization of carbonated products. Applied
Energy, 113, 515–523.

[57] Gerdemann, S.J., O'Connor, W.K., Dahlin, D.C., Penner, L.R. and Rush, H. (2007). Ex Situ
Aqueous Mineral Carbonation. Environ. Sci. Technol., 41, 2587-2593. DOI: 10.102/es0619253

[58] Gunning, P., Hills, C., Antemir, A. and Carey, P. (2011). Novel approaches to the valorisation
of ashes using aggregation by carbonation. In: Proceedings of the 2nd International Slag
Valorisation Symposium, Leuven, Belgium, 2011. pp 18–20.

[59] Carbon8 (2017) Technology- adding carbon dioxide to waste to give it commercial value.
http://c8s.co.uk/technology/.

[60] Morone, M., Costa, G., Polettini, A., Pomi, R. and Baciocchi, R. (2014). Valorization of steel
slag by a combined carbonation and granulation treatment. Minerals Engineering, 59,
82–90. doi: 10.1016/j.mineng.2013.08.009

[61] Salman, M., Cizer, Ö., Pontikes, Y., Santos, R.M., Snellings, R., Vandewalle, L., Blanpain, B.,
Vand Balen, K. (2014). Effect of accelerated carbonation on AOD stainless steel slag for its
valorisation as a CO_2-sequestering construction material. Chemical Engineering Journal, 246,
39–52.

[62] Quaghebeur, M., Nielsen, P., Laenen, B., Nguyen, E. and Van Mechelen, D. (2010). Carbstone:
sustainable valorisation technology for fine grained steel slags and CO_2. Refractories World
Forum 2 (2), 75–79.

[63] Nielsen, P., Baciocchi, R., Costa, G., Quaghebeur, M. and Snellingsa, R. (2017). Carbonate-
bonded construction materials from alkaline residues. RILEM Technical Letters, 2, 53–58.

[64] Evans, K., Nordheim, E. and Tsesmelis, K. (2012). Bauxite residue management. Light Metals.
TMS (The Minerals, Metals and Materials Society, 2012), Edited by: Carlos E. Suarez.

[65] US Dept of Energy, https://www.energy.gov/fe/innovative-concepts-beneficial-reuse-carbon-
dioxide-0; accessed on 28 July, 2018.

[66] Zaelke, D., Young, O. and Andersen, S.O. (2011). Scientific Synthesis of Calera Carbon
Sequestration and Carbonaceous By-Product Applications. Consensus Findings of the
Scientific Synthesis Team January 2011 (http://www.bren.ucsb.edu/news/documents/
Calera_Carbon_Capture.pdf; accessed on 28 July, 2018).

[67] https://hub.globalccsinstitute.com/publications/accelerating-uptake-ccs-industrial-use-
captured-carbondioxide/appendix-f-co2-feedstock

[68] IEA (2009). Cement Technology Roadmap 2009 – Carbon Emissions Reductions up to 2050

[69] http://www.carbicrete.com/; accessed on 05.07.2018

[70] CO_2 Solutions Inc. (2017). CO_2 Solutions Announces Partnership with Carbicrete in the NRG
COSIA Carbon XPrize Competition. Sep 20, 2017 (https://www.prnewswire.com/news-

releases/co2-solutions-announces-partnership-with-carbicrete-in-the-nrg-cosia-carbon-xprize-competition-646022583.html).

[71] Armstrong, S. (2018). These start ups are turning CO2 pollutuon into something useful. http://www.wired.co.uk/article/Xprize-global-warming-climate-change-co2-pollution; 9th April, 2018.

[72] Savage, N. (2017). Carbicrete: Steel in, carbon out. Nature 18 May 2017, vol. 545, S17

[73] https://www.carboncure.com

[74] CSR wire (2018). Solidia Technologies Low-carbon Cement and Concrete Tech Qualifies Under BP's New Advancing Low Carbon Accreditation Programme. CSR News, April 16, 2018. http://www.csrwire.com/press_releases/40934-Solidia-Technologies-Low-carbon-Cement-and-Concrete-Tech-Qualifies-Under-BP-s-New-Advancing-Low-Carbon-Accreditation-Programme

[75] Quartz Media (2017). Concrete Thoughts: The material that built the modern world is also destroying it. Here's a fix (https://qz.com/1320542/regulation-traditionally-lags-behind-innovation-learn-about-the-challenge-in-this-interactive/; accessed on 08 July, 2018).

[76] Power, I.M., Harrison, A.L. and Dipple, G.M. (2016). Accelerating Mineral Carbonation Using Carbonic Anhydrase. Environmental Science and Technology, 50, 5, 2610-2618.

[77] Mirjafari, P., Asghari, K. and Mahinpey, N. (2007). Investigating the application of enzyme carbonic anhydrase for CO_2 sequestration purposes. Industrial Engineering Chemical Research, 46, 3, 921–926.

[78] Lorenzo, F.D., Ruiz-Agudo, C., Ibañez-Velasco, A., Millán R.G., Navarro, J.A.R., Ruiz-Agudo, E. and Rodriguez-Navarro, C (2018). The Carbonation of Wollastonite: A Model Reaction to Test Natural and Biomimetic Catalysts for Enhanced CO_2 Sequestration. Minerals, 8, 5, 209. https://doi.org/10.3390/min8050209, accessed on 01/ 08/2018

[79] Maries. A. and Hills, C.D. (2013). Homogenous catalysis of the accelerated carbonation of Portland cement. 4th International Conference on Accelerated Carbonation for Environmental and Material Engineeirng. 10-12 April, 2013. Leuven, Belgium.

[80] Barbero, R, Carnelli, L, Simon, A., Kao, A, Monforte, A.A., Riccò, M., Bianchi, D. and Belcher, A. Engineered yeast for enhanced CO_2 mineralization. Energy Environ Science, 6, 2, 660–674.

[81] Oliver, T.K., Dlugogorski, B.Z. and Kennedy, E.M. (2014). Biologically enhanced degassing and precipitation of magnesium carbonates derived from bicarbonate solutions. Minerals Engineering, 61, 113-120.

[82] Theulen, J. (2018). Concrete: helping to decarbonise the cement industry. VIEWPOINT. Government Europa Quarterly 26. www.governmenteuropa.eu

[83] Batjes, N.H. (1996). Total Carbon and nitrogen in the soils of the world. European Journal of Soil Science, 47, 151-163.

[84] Goh, K.M. (2011). Greater Mitigation of Climate Change by Organic than Conventional Agriculture: A Review, Biological Agriculture & Horticulture, 27, 2, 205-229, DOI: 10.1080/01448765.2011.9756648

[85] NRC (2015). Climate Intervention: Carbon Dioxide Removal and Reliable Sequestration. National Academies Press, Washington. http://doi.org/10.17226/18805

[86] Antemir, A., Hills, C.D., Carey, P., Magnie M-C (2010). Investigation of 4-year old stabilised/solidifies and accelerated carbonated contaminated soil. Journal of Hazardous Materials 181 (1-3), 543-555.

[87] PASSiFy Project. (2010). Performance Assessment of Solidified/Stabilised Waste-forms, An Examination of the Long-term Stability of Cement-treated Soil and Waste (Final Report), http://www.claire.co.uk/index.php?option=com_resource&controller=article&article=298&category_id=11&Itemid=61

[88] Lake, C.B., Choi, H., Hills, C.D., Gunning, P. and Manaqibwala, I. (2016). Manufactured Aggregate from Cement Kiln Dust, Environmental Geotechnics. http://dx.doi.org/10.1680/jenge.15.00074

[89] Choi. H., Lake, C.B. and Hills, C.D. (2019). Particle size effects on breakage of ACT aggregates under physical and environmental loading. Journal of Hazardous, Toxic and Radioactive Waste (accepted- ref HZENG-724R1).

[90] Hills, C.D., Tripathi, N. and Carey, P.J. (2017). Opportunities for CO_2 mineralisation in Indian waste-based manufactured products. NexGen Technologies for Mining and Fuel Industries. Allied Publishers Pvt Ltd (Pradeep K Singh et al. eds). ISBN 978-93-85926-40-2.

[91] Shipton, B. and Coop, M. R. (2012). On the compression behaviour of reconstituted soils, Soil and Foundation. 52(4),668-681. Available from https://doi.org/10.1016/j.sandf.2012.07.008.

[92] Shahnazari, H. and Rezvani, R. (2013). Effective parameters for the particle breakage of calcareous sands: An experimental study. Engineering Geology, 159, 98-105. Available from http://dx.doi.org/10.1016/j.enggeo.2013.03.005.

[93] Sahu, R.C., Patel, R.K. and Ray, B.C. (2010). Neutralization of red mud using CO_2 sequestration cycle. Journal of Hazardous Materials 79, 28–34.

[94] BRAVO EIPs (2015). A European initiation partnership raw materials commitment (http://bravoiep.eu.bravo-projects/).

[95] GrubeŠa, I.N, BariŠic, I, Fucic, A and Bansode, S.S. (2016). Characteristics and Uses of Steel Slag in Building Construction. Woodhead Publishing Series in Civil and Structural engineering: Number 67. Elsevier ISBN: 978-0-08-100976-5.

[96] Searle, S. and Malins, C. (2013). Availability of cellulosic residues and wastes in the EU, White Paper. The International Council of Clean Transportation (http://www.theicct.org/sites/default/files/publications/icct_eucellulosic-waste-residues_20131022.pdf).

Adrienne Macartney and Pol Knops

11 Mineral carbon sequestration

11.1 Introduction

Mineral carbon sequestration (MCS) is a process where atmospheric CO_2 becomes bound into the chemical structure of minerals and thus is capable of being stably stored over geological time frames. MCS proceeds slowly and naturally on Earth and is an essential component of the global carbon cycle. Significant atmospheric CO_2 removal via silicate weathering partly balances volcanic CO_2 output. Rocks with high olivine content, such as ultra-basic peridotites exhumed from depth, and which are out of chemical equilibrium with the surface environment alter most efficiently. MCS has enormous potential for tackling climate change through geoengineering via acceleration of the mineralisation process. This chapter will introduce the basic chemistry of MCS and the current status of MCS as a geoengineering technique. The commercial viability of reactant products will be explored and an assessment made for the potential role MCS might play in the future of climate change mitigation.

11.2 Basic chemistry

Atmospheric CO_2 dissolves in liquid water to produce a mildly acidified solution through the increase in dissociated H^+ protons (Figure 11.1). This slightly acidic water in contact with silicate rocks causes a release of divalent cations that can combine with aqueous carbonate to precipitate stable carbonate minerals with CO_2 in their structure. The original olivine can also weather into a wide variety of secondary minerals that incorporate CO_2 into their structure.

A carbonate is a salt, possessing the ionic structure CO_3^{2-}, consisting of a strong covalent double bond to one oxygen atom and single bonds to the other two oxygen atoms. Most carbonate minerals consist of the carbonate ion combining with a metal cation, commonly iron (Fe), magnesium (Mg), potassium (K), sodium (Na) and calcium (Ca), but also a wide variety of less common elements. The carbonate molecule can then combine with others to form lattice structures. Sometimes sparsely populated metals can be included within a carbonate lattice sandwiched between non-metal-bearing carbonate lattices, Pb being associated with cerussite, Cu_2 with malachite, Cu_3 with azurite, Zn with smithsonite and so on. In systems where multiple metal elements are available, the metal cations in the

Adrienne Macartney, University of St Andrews, Earth and Environmental Sciences Department, UK
Pol Knops, SCW Systems Diamantweg 36, NL 1812RC Alkmaar

https://doi.org/10.1515/9783110563191-011

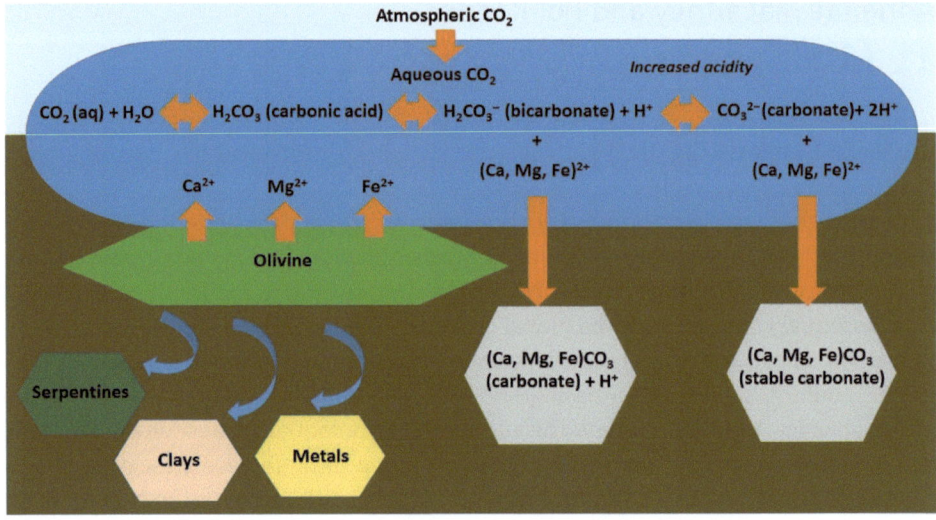

Figure 11.1: Simplified chemical pathways in atmospheric CO_2 weathering the silicate mineral olivine and forming a range of secondary minerals with CO_2 locked into their structure. The mineral pyroxene (not shown) can also weather into stable carbonates, though less efficiently.

carbonate mineral structure can be exchangeable. When CO_2 dissolves in liquid water (eq. 11.1, Figure 11.1), the solution becomes acidified by the increase in dissociated protons (eqs. 11.2, 11.3, Figure 11.1).

$$CO_2(aq) + H_2O \leftrightarrow H_2CO_3(carbonic\ acid) \tag{11.1}$$

$$H_2CO_3(carbonic\ acid) \leftrightarrow HCO_3^-(bicarbonate\ ion) + H^+ \tag{11.2}$$

$$HCO_3^-(bicarbonate\ ion) \leftrightarrow CO_3^{2-}(carbonate\ ion) + H^+ \tag{11.3}$$

When this slightly acidic water comes into contact with silicate rocks, it causes them to release divalent cations such as magnesium, calcium and iron into the solution, which further increase fluid acidity by combining with bicarbonate ions (eq. 11.4, Figure 11.1):

$$(Ca, Mg, Fe)^{2+} + HCO_3^- \rightleftharpoons (Ca, Mg, Fe)CO_3 + H^+ \tag{11.4}$$

Or a neutral combination:

$$(Ca, Mg, Fe)^{2+} + CO_3^{2-} \rightleftharpoons (Ca, Mg, Fe)CO_3 \tag{11.5}$$

This excess of H^+ must be removed from the system if these released cations are to form stable carbonate minerals, and this can occur via Ca-plagioclase dissolution and clay formation:

$$CaAl_2Si_2O_8(\text{Ca-plagioclase}) + 2H^+ + H_2O \rightleftharpoons Ca^{2+} + Al_2Si_2O_5(OH)_4(\text{kaolinite clay})$$

$$(11.6)$$

Incorporation of water into precipitated mineral structures demands volumetric expansion, often >44% [1], which causes cracking [2], and exposure of fresh mineral surfaces for continued reaction [3]:

$$2Mg_2SiO_4(\text{forsterite}) + Mg_2Si_2O_6(\text{Mg}-\text{pyroxene}) + 4H_2O \rightleftharpoons 2Mg_3Si_2O_5(OH)_4(\text{serpentine})$$

$$(11.7)$$

In addition to serpentinisation and hydration of silicates, shown in eqs. 11.6 and 11.7, carbon from CO_2 gets incorporated into the structure of secondary minerals deposited in veins, that is, simple carbonation of olivine (eq. 11.8) and carbonation of a mixture of olivine and pyroxene (eq. 1.9) [2]:

$$Mg_2SiO_4(\text{forsterite}) + 2CO_2 \rightleftharpoons 2MgCO_3(\text{magnesite}) + SiO_2(\text{quartz}) \qquad (11.8)$$

$$Mg_2SiO_4(\text{forsterite}) + CaMgSi_2O_6(\text{CaMg}-\text{pyroxene}) + 2CO_2 + 2H_2O \rightleftharpoons$$
$$Mg_3Si_2O_5(OH)_4(\text{serpentine}) + CaCO_3(\text{calcite}) + MgCO_3(\text{magnesite}) \qquad (11.9)$$

In complex natural systems, the variety available on MCS reactions is nearly endless and the aforementioned equations are merely a few simple examples. The mineral products resulting from abiotic carbonation of basic/mafic rocks will vary depending on the pCO_2 regime of the interacting fluid, with low pCO_2 producing secondary calcites, clays, zeolites [4] and high pCO_2 producing secondary Fe, Mg, and Ca-carbonates; both regimes can produce quartz/chalcedony [5]. When peridotite carbonation proceeds to completion, that is, 100% magnesite and quartz, it is known as a *listvenite* [2]. Neutralisation of water acidity by equation 1.5 commonly proceeds beyond neutral pH on Earth, with hyperalkaline waters (pH>11) reaching the Earth's surface in the form of travertine springs [3, 6–8].

11.3 Natural MCS

In isolation, olivine is the mineral most susceptible to reaction and carbonate formation, and regions where olivine are dominant, such as ophiolites, certainly produce significant atmospheric carbon sequestration. However, some research indicates that carbonation occurs up to 40 times more efficiently in olivine-rich basalt compared with pure olivine [9]. Such basalt regions include large igneous provinces such as the Deccan traps in India, the Siberian flood basalts and the Columbia river basalts in the USA. Ophiolites can also have large regions of reactive basalt [10].

Ophiolites are terrestrial sea floor that has been obducted on to continental crust. Composed dominantly of basalt, gabbro and peridotite, with high-mineral olivine content, ophiolites are strongly out of equilibrium with the surface environment and highly reactive as a consequence; this reactivity is termed *retrograde metamorphism.* Olivine and pyroxenes react with water, forming secondary minerals including carbonate and clays, which tend to form in rock fractures, pore spaces and other lines of weakness in the rocks. Ophiolites, exposed olivine-rich igneous intrusions and global mineral weathering account for ca. 3×10^9 kg draw down of atmospheric carbon, or ca. 1.11×10^{10} kg drawdown of atmospheric CO_2 [11] and are therefore an important part of our planet's carbon budget. The Oman ophiolite alone sequesters ca. 10^7 to ca. 10^8 kg/yr of CO_2 [3]. Flood basalts and ophiolites have the potential to sequester anthropogenic CO_2 on an industrial scale.

The Oman drilling project is an international joint-funded project by the International Continental Scientific Drilling Program. Ophiolites can also have large regions of reactive basalt. Research associated with the project assesses rock reaction dynamics and is useful for both quantifying and understanding natural MCS and assessing geoengineered MCS potential. A workshop was held in 2013 [12] and the first operational drilling commenced in the winter of 2016 [13]. Full research results have not yet been published, yet a host of conference publications relevant to MCS in ophiolite settings are now available [14–20].

11.4 Geoengineered accelerated MCS

Natural MCS is insufficient as a mechanism to offset anthropogenic climate change as it sequesters carbon too slowly (i.e., over geological time scales). Considerable efforts over recent decades to accelerate the reaction process has seen the technology develop from simple laboratory experiments to field site testing. By injecting heated supercritical CO_2 into pre-assessed drill sites, MCS proceeds at a rapid pace. This section briefly examines the larger MCS experiments of CarbFix and Walulla, followed by a review of non-injection MCS utilisation techniques, termed *carbon capture and utilisation* (CCU).

High temperature–pressure reactions proceed faster than low temperature–low pressure reactions (albeit temperature exerts a greater influence over reaction speeds than pressure; pressure dominantly affecting gas solubility and speciation). Using temperature and pressure to increase primary silicate reaction rates can be achieved in industrial/terrestrial MCS scenarios by injecting at depth and using the planet's geothermal gradient. MCS proceeds at a peak rate when the temperature is ca. 185 °C [3]. MCS is exothermic [21] and rocks are naturally heated at depth (to an extent that depends on the local geothermal gradient); as a result, the carbonation reaction can remain at a near optimal temperature subsurface. Furthermore, incorporation of water into the crystal structures of precipitated minerals demands volumetric expansion, often >44% [1], which causes cracking [2] and exposure of fresh mineral surfaces for continued reaction [3].

11.4.1 CarbFix

One of the largest ongoing industrial pilots for geoengineered accelerated MCS is the Hellisheidi Geothermal Power Station CarbFix facility in Southern Iceland. The CarbFix project began in 2007 in the Western Hengill central volcano and fissure swarm, Iceland, injecting geothermal heated water that had been artificially saturated with CO_2 into an olivine tholeiite basaltic terrain at a rate of 0.5 to 0.7 kg/s [22, 23]. The target rock consists of olivine (Fo_{90} to Fo_{80}), plagioclase (An_{90-30}), pyroxene (augite), magnetite, ilmenite, chromium, spinel, and glass [22, 24]. The injected water was mixed with an addition of pure CO_2 equilibrated to 25 bar and another addition of CO_2–H_2S–H_2 equilibrated to 14 bar at a ratio 75:24.2:0.8 [22]. The reason for adding hydrogen sulphide and dihydrogen was to raise the overall acidity of the injected solution, because olivine-rich basaltic rocks release cations at a higher rate if the pH is low. The final equilibrated injection solution has a CO_2 concentration of 0.8 to 0.42 mol/kg and a pH of 3.7–4.0 [22].

For the project, a number of holes (550 m to 800 m in depth) were drilled [25], and the injection site has a geothermal gradient of 80°C/km and an overlying impermeable cap rock of glassy basaltic hyaloclastites [22]. Basalt rocks are rich in divalent cations (Mg^{2+}, Fe^{2+}, Ca^{2+}) and the CarbFix target rock has 6 mol kg^{-1} of cations potentially available for the reaction [22]. Ambient water in the target rocks was ca. 30–60 °C at 300–800 m depth and 180 °C at 1,500 m depth, with pH 8.4 to 9.8, hence relatively alkaline; the injected fluid mix was at ca. 25 °C [22]. Tracer fluids were added into the main injection waters to observe for both leaks in the cap rock and transport fate of the injection fluid. About 1 m^3 water was injected with 50 kg of fluorescent Na dye, and 1 m^3 water was injected with ca. 14 g sulphur hexafluoride SF_6 [23]. The project site has a natural groundwater through flow rate of 25 m/yr [22]. No leakage has yet been discovered.

CarbFix has so far been a ground-breaking success, with 95% of the >170 t of CO_2 injected becoming fully mineralised as calcite in ≤ 2 years [26, 27], demonstrating that industrial-scale MCS can be effective in human, as well as geological, time scales. CarbFix also demonstrates an ideal progression from theoretical calculation and modelling, through to small-scale laboratory experiment into industrial large-scale physical application. Much of the literature around geo-engineered MCS is dominated by the CarbFix project [23, 28–33].

The CarbFix project has now evolved to incorporate the SulFix project. The SulFix project injects hydrogen sulphide. But H2S is instead injected ca. 1350 m [25], where the pressure and temperature is more conducive to sulphide mineral stability and thus long-term storage [34, 35], instead of the CarbFix depth of 550 - 800 m which is aimed at mineral dissolution.

The CarbFix project's thermodynamic database had only 36 mineral reactions described, and the researchers found that many other terrestrial databases lacked the full range of mineral reaction data needed to model the system they were attempting to implement. CarbFix primarily used the EQ3/6 V7.2b database, but found Al-hydroxy complexes particularly lacking; SUPCRT92 was used for computing solubility constants where experimental data were insufficient [36]. One hundred and ten minerals were included in the CarbFix kinetic database; however, they also found the kinetic experimental data lacking and relied on PHREEQC modelling to fill the empirical data gaps. There has since been rapid progress in understanding, testing and developing kinetic databases for minerals associated with geoengineered MCS [37].

Following the CarbFix experiments in Iceland, a great number of laboratory experiments have been conducted, often using simple components such as water, CO_2 gas, basaltic glass or ultramafic rocks, as these react very rapidly. For example, experiments at 40°C for 260 days have produced poorly crystalline carbonates and clays of Ca–Mg–Fe composition, along with Fe-hydroxides and oxy-hydroxides [38]. Other researchers have found that using diluted amounts of CO_2 (18.2 % in a gas stream) reacted with serpentine is more efficient in producing Mg-rich secondary minerals than when 100 % CO_2 is used [39].

11.4.2 Wallula project, Columbia river basalts

In 2009, researchers from the US Department of Energy's Pacific Northwest National Laboratory began the Big Sky Regional Carbon Sequestration Partnership, drilling the first MCS pilot hole into continental flood basalt to a depth of 1253 m [40]. By 2013, 1000 tonnes of supercritical CO_2 were injected into the brecciated Columbia River Basalts at a depth of 828–875 m [41]. The basalt flow tops possessed a permeability of 75 to 150 millidarcies, bounded by low (microdarcy) permeability [40]. Post-experiment site monitoring over numerous years revealed no CO_2 gas in shallow surface soil samples and no site leakage, and subsequent drilled samples showed increased Ca, Mg, Mn and Fe concentrations combined with $^{13}C/^{18}O$ isotopic shifts indicative of subsurface MCS reactions [42].

11.4.3 Flood basalt and ophiolite potential for geoengineered MCS

Continental flood basalts represents less than 10% of continental land mass, but naturally sequesters ca. 33 % of the CO_2 consumed by global silicate weathering [43]. The Siberian flood basalts potentially extend ca. 1.6×10^6 km^2 [44] and the Indian Deccan traps ca. 5×10^5 km^2 [45]; others include the Karoo-Ferrar, Serra Geral-Etendeka, Newark, Antarctica, Ethiopia, Madagascar, as well as smaller potential sites such as the Rajahmundry Traps ca. 15, 000 km^2 [45].

In addition to flood basalts, ophiolites are another geological terrain suitable for geoengineered MSC. Ophiolite is a term used when oceanic crust is obducted on top of continental crust. Ophiolites occur worldwide and retain the same stratified basaltic, gabbroic, peridotite (olivine rich) petrology as the sea floor; albeit deformed, weathered and chemically altered. The Oman drilling project enables research on the viability of using ophiolites as sites of geoengineered MCS [46].

11.5 Green Sand for "passive" industrial MCS

A profoundly different approach compared to in situ MCS is to utilise the mineral olivine (often termed *green sand*) to replace current conventional stone materials in civil or marine applications. Three of the key environments proposed for wide scale use of olivine green sand as a geoengineered technique are as follows:

Re-sanding coastal beaches with olivine and using the surf to mill the olivine to smaller, more reactive sizes that will then chemically alter, transforming into mineral carbonate and thus store CO_2 in stable mineral form over geological time frames on the ocean floor [47–49].

Adding olivine sand to agricultural soils for the dual function of buffering the pH in acidic soils (also called *liming*) and to draw down atmospheric CO_2 via olivine weathering [50]. The addition of olivine sand into agricultural soils causes pH adjustments and changes to the fluid run off chemistry, as well as to the bulk physical composition of soil minerals and clays. This could have important implications for grazing, crop yields and agricultural production. Research suggests adding olivine increases plant growth by up to 15.6%, plant potassium (K) content by up to 16.5%, bioavailability of magnesium (Mg) and nickel (Ni), suppression of Ca and enhanced CO_2 sequestration [51]. Considerable research gaps and opportunities exist in this area.

Marine applications include reacting mineral carbonate, such as limestone, with aqueous carbonic acid produced from adding captured industrial CO_2 to water. Such reactions will form cations and bicarbonate in solution that could then be released into rivers, estuaries and marine environments to further react with natural CO2 to re-form and precipitate as sedimentary marine mineral carbonate [52]. Such proposals have met criticism, however, that they "may merely trade one environmental problem for another" and that underground injection is a safer option [53]. Current dredging technology could place green sand in deeper coastal shelf settings, although further environmental assessments of the positive and negative potential impacts of this currently lack [54].

Other green sand applications include adding green sand to driveways and railway inspection paths [55]. The first green sand projects for roads and railway paths have been implemented in the Netherlands by GreenSand Ltd, supported by Deltares (a Dutch knowledge institute).

CO_2 sequestration efficiency in all these environments is determined by the ambient conditions (moisture, temperature, etc.), residence time and particle size distribution [48, 56]. The ambient reaction conditions in these environments means CO_2 sequestration proceeds at natural rates (rather slowly). However, this slow reaction rate is balanced by the relatively long "residence" times such civil projects are in place for (often >30 years); the weathering and chemical alteration can thus take place over an equivalent time. Quantifying the CO_2 reductions achieved by such projects can be determined by calculating the CO_2 sequestration of the olivine used versus the CO_2 footprint of the conventional material the olivine is replacing. Typically ca. 10 - 30% of the olivine used will react and alter in the lifetime of an average project [48, 57].

11.6 "Active" industrial mineral carbon capture and utilisation

MCS utilisation techniques, sometimes termed *carbon capture and utilisation* (CCU), is the process of taking captured CO_2 and turning it into usable-carbon rich commercial products with negative or neutral carbon footprints. Often this is a circular process as the products created decompose at the end of their life cycle and the carbon can be re-emitted into the atmosphere.

At reaction conditions of ca. 180 °C and 100 Bar using a gravity pressure vessel it was realised that mineral carbonation could occur at a commercial scale in industrial environments [58]. The company Green Minerals and the University of Louvain developed a batch autoclave, whose focus was initially on CO_2 sequestration dynamics, but it was realised the reaction products (carbonate) possess a viable economic value in their own right [59–61]. The focus thus shifted to exploring beneficial uses of carbonate minerals and manufacturing "CO_2 negative materials."

11.6.1 Cement

Using captured CO_2 in the cementation of industrial recycled waste aggregate gained interest around 1995 when carbonation was used to enhance binding quality of waste [62]; this field is being led by companies such as Carbon 8 Aggregates Ltd. (see previous chapter). Captured CO_2 can also be used in the production of new cement, as well as being a waste product of the cement-making process. Manufacturing a metric tonne of average cement produces ca. 0.73–0.99 t of CO_2 as a waste product [63]. However, CO_2 that has been mineralised as carbonate can then be used to produce cement via a number of pathways. An example is magnesium carbonate trihydrate (the mineral nesquehonite), which can be obtained from staged precipitation from desalination brines [64], or alternatively from MgO hydrating into brucite [65]. Nesquehonite, or alternatively dypingite and hydromagnesite, can then be processed into a cement through a variety of pathways such as combining with pulverised fly ash [66]. This cement has a CO_2 content of ca. 30% and are often called *eco bricks* [67]. The formation of nesquehonite from brucite can occur in as little as 30 minutes in a laboratory, using diluted HCl as a catalyst [68]. Companies active in cement CO_2 sequestration include Solidia Technologies Inc., Novacem Ltd., Calera Corporation and CarbonCure Technologies Inc.

Ongoing research, as yet unpublished, on the beneficial use of MCS technology in concrete production is being spearheaded by the CO2MIN project led by HeidelbergCement & Green Minerals', the Institute for Advanced Sustainability Studies and Rheinisch-Westfälische Technische Hochschule at the University of Aachen, and is funded by the German Federal Ministry of Education and Research [69].

11.6.2 Paper

Precipitated calcium carbonate (PCC) is used in the manufacturing of paper. Lime (carbonate) is used as the following:
- a causticizing agent in sulphate (Kraft) plants [70];
- as a reactant to produce calcium bisulphite used to dissolve non-cellulosic parts of wood chips [71];
- is added to chlorine that reacts to produce calcium hypochlorite, which is used to bleach and whiten paper [71];
- can be used for a variety of other roles such as a coagulant in colour removal, a filtration conditioner or as a pH neutraliser [71];
- lime can also be used to recover by-products from process waste such as alcohol, calcium lignosulfonate and yeast [71].

Artificially created carbonate via MCS can replace natural PCC in paper processing and other industries [72]. Initial unpublished results from the University of Darmstadt calculate that such artificial carbonate replacement of PCC could reduce the CO_2 per ton of paper manufactured by ca. 16% and also be cost saving compared with natural PCC.

11.6.3 Polymers

Mineral reaction products of MCS can also be used in industrial commercial polymers. Wollastonite, talc and so on are all MCS reaction products that can be used as functional fillers in industry to manage properties such as shrinkage, strength and thermal conductivity, in the final products [73, 74]. Using MCS reaction products can also reduce process costs and the final product carbon footprint.

Other industrial uses of carbonate include plastics, rubbers, paints and drugs [75].

11.7 Research development

Critical areas of ongoing MCS investigation include CO_2/brine multiphase migration [76] on the pore scale [77, 78], core sample scale [79–81] and large project scale [82–85]. Researchers in this field specifically state that quantitative empirical knowledge of MCS kinetics is not enough, and that more research on mineral surface alteration processes, precipitation controls and reaction pathways are desperately needed [76].

Understanding rate-limiting factors of the MCS process is also important if the technique is to be globally scaled up; the use of acid on serpentine is a good example. Serpentine is an intermediate product of olivine-rich peridotite or dunite undergoing fluid metamorphism, and contains numerous polymorphs. It has been found in laboratory experiments that acid dissolution of serpentine enhances Mg cation release, depending on crystal structure and micro-texture, which can then be incorporated into magnesium carbonate via equation 1.4 [86]. The quantity of Mg released in different serpentinite polymorphs during these experiments were < 5% for Al-bearing polygonal serpentine, ≈ 5% for Al-bearing lizardite 1T, 24–29% for antigorite, ≈ 65% for well-ordered lizardite $2H_1$ ≈ 68% for Al-poor lizardite 1T, ≈ 70% for chrysotile, ≈ 80% for poorly ordered lizardite $2H_1$ and ≈ 85% for nanotubular chrysotile [86].

The use of acid to increase cation production, however, may not be addressing the main rate-limiting factor in MCS. Even with pure olivine experiments, it has been noted that the principal limiting factor in the reaction between olivine and water is usually the precipitation rate of carbonates, not the dissolution rate of olivine and release of cations [87]. Furthermore, olivine dissolution and alteration can be slowed by up to ca. 2 orders of magnitude by the formation of a ca. ≤ 40 nm thick silica passivating layer (PL) on the reaction surface olivine grain [88]. Increasing precipitation rates of carbonate and discovering new techniques to reduce PL layering, such as relying on coastal surf attrition, are all areas where much research potential exists. The truly critical area in MCS research and advancement is in translating academic laboratory and field scale experiments into large-scale industrial pilots and full commercial applications.

11.8 Conclusions

Geoengineered MCS as a negative emissions technology has the potential to significantly reduce atmospheric CO_2; however, considerable investment, research and experimentation is required. There exist important gaps in understanding the chemical and physical processes of natural rock and soil weathering, and even greater lack of understanding about how these natural systems respond to the addition of mineral additives such as olivine. Mineral dissolution kinetics also remain poorly quantified, both in the lab and field, as do plant growth reactions under enhanced olivine conditions. Considerable quantification, costing, modelling and experimentation, mature level pilot projects and technology demonstration of MCS are required before applying the technique on an industrial scale. This research and development time must also be balanced by the need to become CO_2 neutral by 2050.

References

[1] Kelemen P, Matter J, Streit E, Rudge J, Curry W, Blusztajn J. Rates and Mechanisms of Mineral Carbonation in Peridotite: Natural Processes and Recipes for Enhanced, in situ CO_2 Capture and Storage. Annual Review Of Earth And Planetary Sciences 2011, 39,545–576.

[2] Kelemen P. B, Hirth G. Reaction-driven cracking during retrograde metamorphism: Olivine hydration and carbonation. Earth and Planetary Science Letters 2012, 345–348,81–89.

[3] Kelemen P. B, Matter J. In situ carbonation of peridotite for CO_2 storage. Proceedings of the National Academy of Sciences 2008, 105(45),17295–17300.

[4] Neuhoff P. S, Fridriksson T, Arnorsson S, Bird D. K. Porosity evolution and mineral paragenisis of basaltic lavas at Teigarhorn, Eastern Iceland. American Journal of Science 1999, 299, 467–501.

[5] Rogers K. L., Neuhoff P. S., Pedersen A. K., Bird D. K. CO_2 metasomatism in a basalt-hosted petroleum reservoir, Nuussuaq, West Greenland. *Lithos* 2006, 92(1–2), 55–82.

[6] Pentecost A. The quaternary travertine deposits of Europe and Asia Minor. Quaternary Science Reviews 1995, 14(10),1005–1028.

[7] Blank J. G, Green S. J, Blake D *et al*. An alkaline spring system within the Del Puerto Ophiolite (California, USA): A Mars analog site. Planetary and Space Science 2009, 57(5–6), 533–540.

[8] Arcilla C. A, Pascua C. S, Alexander W. R. Hyperalkaline groundwaters and tectonism in the Philippines: significance to natural carbon capture and sequestration. *Energy Procedia* 2011, 4, 5093–5101.

[9] Sissmann O, Brunet F, Martinez I, Guyot F, Verlaguet A, Pinquier Y, Daval D. Enhanced Olivine Carbonation within a Basalt as Compared to Single-Phase Experiments: Reevaluating the Potential of CO_2 Mineral Sequestration. Environmental Science and Technology Letters 2014, 48(10),5512–5519.

[10] McGrail PB, Schaef HT, Ho AM, Chien YJ, Dooley JJ, Davidson CL. Potential for carbon dioxide sequestration in flood basalts. JGR Solid Earth 2006, 111–B12.

[11] IPCC. (2013). *Climate Change 2013: The Physical Science Basis*. Available: http://www.ipcc.ch/report/ar5/wg1/.

[12] Keleman P. Planning the Drilling of the Samail Ophiolite in Oman: Workshop on Scientific Drilling in the Samail Ophiolite. EOS 2013, 94(3), 32.

[13] Website 2: Oman Drilling Project. (Accessed November 11, 2018, at http://www.omandrilling.ac.uk/publications).

[14] de Obeso JC, Keleman PB, Manning CE, Michibayashi K, Harris M. Listvenite formation from peridotite: Insights from Oman Drilling Project hole BT1B and preliminary reaction path model approach. AGU fall meeting 2017, V43G-2941.

[15] Gretchen LFG, Grabowska M, Oyanagi R, Kimura K, Morishita T, Okamoto A, Klein F, Tamura A, Teagle DAH, Takazawa E, Coggon JA, Kelemen PB, Matter JM. Hydrothermal Alteration of the Crust-Mantle Transition and Upper Mantle in the Samail Ophiolite: Insights from Holes CM1A and CM2B of the Oman Drilling Project. AGU fall meeting 2018, V11B-02.

[16] Harris M, Zihlmann B, Mock D, Akitou T, Teagle DAH, Kondo K, Deans JR, Crispini L, Takazawa E, Coggon JA, Keleman PB. Hydrothermal Alteration of the Lower Oceanic Crust: Insight from OmanDP Holes. GT1A and GT2A. AGU fall meeting 2017, V24E-04H.

[17] Kelemen PB, Godard M, Johnson KTM, Okazaki K, Manning CE, Urai JL, Michibayashi K, Harris M, Coggon JA, Teagle DAH, Phase I Science Party TODP. Peridotite carbonation at the leading edge of the mantle wedge: OmDP Site BT1. AGU fall meeting 2017, V24E-06K.

[18] Manning CE, Kelemen PB, Michibayashi K, Harris M, Urai JL, de Obeso JC, Jesus APM, Zeko D. Transformation of Serpentinite to Listvenite as Recorded in the Vein History of Rocks From Oman Drilling Project Hole BT1B. AGU fall meeting 2017, V24E-08M.

[19] Manning CE, Lu S, Kelemen PB, Phase I Science Party TODP. Origin of Serpentinite and Listvenite Near the Basal Thrust of the Samail Ophiolite Recorded in Oman Drilling Project Hole BT1B. AGU fall meeting 2018, V11B-07.

[20] Michibayashi K, Katayama I, Kelemen PB, Okazaki K, Godard M, Takazawa E, Teagle DAH. Quantification of the downhole degree of serpentinization estimated by X-ray CT core imaging (Oman Drilling Project Phase 2, D/V CHIKYU). AGU fall meeting 2018, V12B-01.

[21] Matter JM, Keleman PB. Permanent storage of carbon dioxide in geological reservoirs by mineral carbonation. Nature geoscience 2009, 2, 837–841.

[22] Alfredsson HA, Oelkers EH, Hardarsson BS, Franzson H, Gunnlaugsson E, Gislason SR. The geology and water chemistry of the Hellisheidi, SW-Iceland carbon storage site. International Journal of Greenhouse Gas Control 2013, 12, 399–418.

[23] Matter JM, Broecker WS, Stute M, Gislason SR, Oelkers EH, Stefánsson A, Wolff-Boenisch D, Gunnlaugsson E, Axelsson G, Björnsson G. Permanent Carbon Dioxide Storage into Basalt: The CarbFix Pilot Project, Iceland. *Energy Procedia 2009*, 1, 3641–3646.

[24] Jakobsson SP, Jónsson J, Shido F. Petrology of the Western Reykjanes Peninsula, Iceland. Journal of Petrology 1978, 19, 669–705.

[25] Aradóttir ESP, Gunnarsson I, Sigfússon B, Gunnarsson G, Júliusson BM, Gunnlaugsson E, Sigurdardóttir H, Arnarson MT, Sonnenthal E. Toward Cleaner Geothermal Energy Utilization: Capturing and Sequestering CO_2 and H_2S Emissions from Geothermal Power Plants. Transport in Porous Media 2015, 108(1),61–84.

[26] Sigfusson B, Gislason SR, Matter JM, Stute M, Gunnlaugsson E, Gunnarsson I, Aradottir ES, Sigurdardottir H, Mesfin K, Alfredsson HA, Wolff-Boenisch D, Arnarsson MT, Oelkers EH. (2015). Solving the carbon-dioxide buoyancy challenge: The design and field testing of a dissolved CO_2 injection system. International Journal of Greenhouse Gas Control 2015, 37, 213–219.

[27] Matter JM, Stute M, Snæbjörnsdottir SÓ, Oelkers EH, Gislason SR, Aradottir ES, Sigfusson B, Gunnarsson I, Sigurdardottir H, Gunnlaugsson E, Axelsson G, Alfredsson HA, Wolff-Boenisch D, Mesfin K, Fernandez De La Reguera TD, Hall J, Dideriksen K, Broecker WS. Rapid carbon mineralization for permanent disposal of anthropogenic carbon dioxide emissions. Science 2016, 352(6291),1312–1314.

[28] Alfredsson HA, Wolff-Boenisch D, Stefánsson A. CO_2 sequestration in basaltic rocks in Iceland: Development of a piston-type downhole sampler for CO_2 rich fluids and tracers. Energy Procedia 2011, 4:3510–3517.

[29] Aradóttir ESP, Sigurdardóttir H, Sigfússon B, Gunnlaugsson E. CarbFix: a CCS pilot project imitating and accelerating natural CO_2 sequestration. Greenhouse Gases 2011, 1(2),105–118.

[30] Gislason SR, Wolff-Boenisch D, Stefansson A, Oelkers EH, Gunnlaugsson E, Sigurdardottir H, Sigfusson B, Broecker WS, Matter JM, Stute M, Axelsson G, Fridriksson T. Mineral sequestration of carbon dioxide in basalt: A pre-injection overview of the CarbFix project. International Journal of Greenhouse Gas Control 2010, 4, 537–545.

[31] Gislason SR, Oelkers EH. Carbon Storage in Basalt. Science 2014, 344(6182),373–374.

[32] Johnsson F. Perspectives on CO_2 capture and storage. Greenhouse Gases 2011, 1(2),119–133.

[33] Matter JM, Broecker WS, Gislason SR, Gunnlaugsson E, Oelkers EH, Stute M, Sigurdardóttir H, Stefansson A, Alfreðsson HA, Aradóttir ES., Axelsson G, Sigfússon B, Wolff-Boenisch D. The CarbFix Pilot Project – Storing Carbon Dioxide in Basalt. Energy Procedia 2011, 4:5579–5585.

[34] Juliusson BM, Gunnarsson I, Matthiasdottir KV, Markusson SH, Bjarnason B, Sveinsson OG, Gislason T, Thorsteinsson HH. Tackling the Challenge of H_2S Emissions. Proceedings World Geothermal Congress 2015.

[35] Ragnarsson Á. Geothermal Development in Iceland 2010–2014. Proceedings World Geothermal Congress 2015.

[36] Orkuveita Reykjavíkur.CarbFix Project, 2017. (Accessed November 21, 2018, at https://www.or.is/english/carbfix-project/about-carbfix-0.)

[37] Oelkers EH, Declercq J. CarbFix Report 4 PHREEQC mineral dissolution kinetics database 2014. (Accessed November 21, 2018, at https://www.or.is/sites/or.is/files/kinetic_database.pdf.)

[38] Gysi AP, Stefánsson A. CO_2-water–basalt interaction. Low temperature experiments and implications for CO_2 sequestration into basalts. Geochimica et Cosmochimica Acta 2012, 81, 129–152.

[39] Pasquier LC, Mercier G, Blais JF, Cecchi E, Kentish S. Reaction Mechanism for the Aqueous-Phase Mineral Carbonation of Heat-Activated Serpentine at Low Temperatures and Pressures in Flue Gas Conditions. Environmental Science and Technology Letters 2014, 48(9), 5163–5170.

[40] McCrail BP, Spane A, Sullivan EC, Bacon DH, Hund G. The Wallula basalt sequestration pilot project. Energy Procedia 2009, 4, 5653–5660.

[41] Sullivan EC. The big sky sequestration Wallula basalt pilot: stratigraphy and implications. Geological Society of America Abstracts With Programs 2009, 41(7), 356.

[42] McCrail BP, Spane A, Amonette JE, Thompson CR, Brown CF. Injection and monitoring at the Wallula basalt pilot project. Energy Procedia 2014, 63, 2939–2948.

[43] Dessert C, Dupré B, Gaillardet J, François LM, Allègre CJ. Basalt weathering laws and the impact of basalt weathering on the global carbon cycle. Chemical Geology 2003, 202(3–4), 257–273.

[44] Reichow MK, Saunders AD, White RV, Pringle MS, Al'Mukhamedov AI, Medvedev AI, Kirda NP. $^{40}Ar/^{39}Ar$ Dates from the West Siberian Basin: Siberian Flood Basalt Province Doubled. Science 2002, 296(5574),1846–1849.

[45] Knight KB, Renne PR, Halkett A, White N. (2003). $^{40}Ar/^{39}Ar$ dating of the Rajahmundry Traps, Eastern India and their relationship to the Deccan Traps. Earth and Planetary Science Letters 2003, 208(1–2), 85–99.

[46] Kelemen P, Al Rajhi A, Godard M, Ildefonse B, Köpke J. et al. Scientific drilling and related research in the samail ophiolite, sultanate of Oman. Scientific Drilling 2013, 15, 64–71.

[47] Schuiling RD, de Boer PL. Rolling stones; fast weathering olivine in shallow seas for cost-effective CO_2 capture and mitigation of global warming and ocean acidification. Earth System Dynamics Discussions 2011, 2, 551–568.

[48] Montserrat F, Renforth P, Hartmann J, Leermakers M, Knops PCM,Meysman FJR, Olivine Dissolution in Seawater: Implications for CO_2 Sequestration through Enhanced Weathering in Coastal Environments. Environ. Sci. Technol 2017, 51(7),3960–3972.

[49] Strengers B, Eerens H, Smeets W, van den Born GJ, Ros J. Plan Bureau voor de Leefomgeving (2018) NEGATIEVE EMISSIES Technisch potentieel, realistisch potentieel en kosten voor Nederland. Chapter 10: Silicates

[50] Renforth P, Pogge von Strandmann PAE, Henderson GM. The dissolution of olivine added to soil: implications for enhanced weathering. Applied Geochemistry 2015, 61, 109–118.

[51] Ten Berge H, van der Meer HG, Steenhuizen JW, Goedhart PW, Knops PCM, Verhagen J. Olivine Weathering in Soil, and Its Effects on Growth and Nutrient Uptake in Ryegrass (Lolium perenne L.): A Pot Experiment. PLoS ONE 2012, 7(8), e42098.

[52] Rau GH, Caldeira K. Enhanced carbonate dissolution: a means of sequestering waste CO_2 as ocean bicarbonate. Energy Conversion and Management 1999, 40(17),1803–1813.

[53] Lackner KS. A Guide to CO_2 Sequestration. Science 2003, 300(5626),1677–1678.

[54] Meysman F, Montserrat F. Negative CO_2 emissions via enhanced silicate weathering in coastal environments. Biology Letters 2017, 13(4).

[55] van Helvoort PJ. (2013) "Het Groene Schouwpad" (The Green Inspection Path) https://cdn.mo vares.nl/wp-content/uploads/2013/05/Groene-schouwpad-Movares-eindrapport-2013.pdf

[56] Vink J. 2016, "De klimaatneutrale buitenruimte", Presentation at Dag van de openbare ruimte https://www.vdijk.nl/dagvdor-klimaatneutrale-buitenruimte-met-olivijn-presentatie.pdf

[57] Den Hamer D., Vink J. 2012, Olivijn legt CO_2 vast in de gemeente Rotterdam. Mogelijkheden voor praktijktoepassingen en klimaatdoelstellingen. Deltares report 1206650-000-BGS-0007 Utrecht.

[58] Doucet FJ, South African Centre for Carbon Capture and Storage 2011, "Scoping Study on CO_2 Mineralization Technologies" – Report No CGS- 2011-007.

[59] Santos RM, Verbeeck W, Knops PCM, Rijnsburger KL, Pontikes Y, Van Gerven T. Integrated mineral carbonation reactor technology for sustainable carbon dioxide sequestration: "CO_2 Energy Reactor". Energy Procedia 2013, 37, 5884–5891.

[60] Santos RM, Knops PCM, Rijnsburger KL, Chiang YW. CO_2 Energy Reactor – Integrated Mineral Carbonation: Perspectives on Lab-Scale Investigation and Products Valorization. Front. Energy Res. 2016, 4, 5.

[61] Turri L, Muhr H, Rijnsburger K, Knops P, Lapicque F. CO_2 sequestration by high pressure reaction with olivine in a rocking batch autoclave. Chemical Engineering Science 2017, 171, 27–31.

[62] Lange LC, Hills CD, Poole AB. Preliminary Investigation into the Effects of Carbonation on Cement-Solidified Hazardous Wastes. Environmental Science and Technology 1995, 30(1), 25–30.

[63] Hasanbeigi A, Price L, Lin E. Emerging energy-efficiency and CO_2 emission-reduction technologies for cement and concrete production: A technical review. Renewable and Sustainable Energy Reviews 2012, 16(8),6220–6238.

[64] Galvez-Martos JL, Elhoweris A, Morrison J, Al-horr Y. Conceptual design of a CO_2 capture and utilisation process based on calcium and magnesium rich brines. Journal of CO_2 Utilization 2018, 27, 161–169.

[65] Vandeperre LJ, Al-Tabbaa A. Accelerated carbonation of reactive MgO cements. Advances in Cement Research 2007, 19(2),67–79.

[66] Zhang H, Shen C, Xi P, Chen K, Zhang F. Study on effect of the activated magnesia carbonized building blocks based on the content of fly ash. Construction and Building Materials 2018, 185(10),609–616.

[67] Glasser FP, Jauffret G, Morrison J, Galvez-Martos JL, Patterson N, Imbabi MSE. (2016). Sequestering CO_2 by Mineralization into Useful Nesquehonite-Based Products. Frontiers in Energy Research: part of the Proceedings of the Fifth International Conference on Accelerated

Carbonation for Environmental and Material Engineering (ACEME 2015). (Accessed November 22, 2018, at https://www.frontiersin.org/articles/10.3389/fenrg.2016.00003/full.)

[68] Zhao L, Sang L, Chen J, Ji J, Teng HH. Aqueous Carbonation of Natural Brucite: Relevance to CO_2 Sequestration. Environmental Science and Technology 2010, 44(1),406–411.

[69] HeidelbergCement and RWTH Aachen investigate in binding CO2 in minerals. HeidelbergCement, 2017. (Accessed November 22, 2018, at https://www.heidelbergcement.com/en/pr-29-06-2017.)

[70] Miner R, Upton B. Methods for estimating greenhouse gas emissions from lime kilns at kraft pulp mills. Energy 2002, 27(8),729–738.

[71] Pulp and Paper: National Lime Association, (2018. Accessed November 22, 2018, at https://www.lime.org/lime-basics/uses-of-lime/other-uses-of-lime/pulp-and-paper/.)

[72] Zhao H, Dadap N, Park AHA. 2011. Tailored synthesis of precipitated magnesium carbonates as carbon-neutraler materials during carbon mineral sequestration. 2010 ECI Conference on The 13th International Conference on Fluidization - New Paradigm in Fluidization Engineering. (Accessed November 22, 2018, at http://www.columbia.edu/ca. ap2622/pdf/TAILORED%20SYNTHESIS%20OF.pdf.)

[73] Hadal RS, Dasari A, Rohrmann J, Misra RDK. Effect of wollastonite and talc on the micromechanisms of tensile deformation in polypropylene composites. Materials Science and Engineering: A 2004, 372(1–2), 296–315.

[74] Lee GW, Park M, Kim J, Lee JI, Yoon HG. Enhanced thermal conductivity of polymer compositesed with hybrid filler. Composites Part A: Applied Science and Manufacturing 2006, 37(5),727–734.

[75] Jimoh OA, Otitoju TA, Hussin H, Ariffin KS, Baharun N. Understanding the Precipitated Calcium Carbonate (PCC) Production Mechanism and Its Characteristics in the Liquid-Gas System Using Milk of Lime (MOL) Suspension. South African Journal of Chemistry 2017, 70.

[76] Wang D, Dong B, Breen S, Zhao M, Qiao J, Liu Y, Zhang Y, Song Y. Review: Approaches to research on CO_2/brine two-phase migration in saline aquifers. Hydrogeology Journal 2015, 23, 1–18.

[77] Chen C, Zhang D. Pore-scale simulation of density-driven convection in fractured porous media during geological CO_2 sequestration. Water Resources Research 2010, 46, W11527.

[78] Bandara UC, Tartakovsky AM, Palmer BJ. Pore-scale study of capillary trapping mechanism during CO_2 injection in geological formations. International Journal of Green Gas Control 2011, 5, 1566–1577.

[79] Perrin JC, Krause M, Kuo CW, Miljkovic L, Charoba E, Benson SM. Core-scale experimental study of relative permeability properties of CO_2 and brine in reservoir rocks. Energy Procedia 2009, 1, 3515–3522.

[80] Gunde AC, Bera B, Mitra SK. Investigation of water and CO_2 (carbon dioxide) flooding using micro-CT (microcomputed tomography) images of Berea sandstone core using finite element simulations. Energy 2010, 35, 5209–5216.

[81] Perrin JC, Falta RW, Krevor S, Zuo L, Ellison K, Benson SM. Laboratory experiments on core-scale behaviour of CO_2 exolved from CO_2-saturated brine. Energy Procedia 2011, 4, 3210–3215.

[82] Ghomian Y, Pope GA, Sepehrnoori K. Reservoir simulation of CO_2 sequestration pilot in Frio Brine formation, USA Gulf Coast. Energy 2008, 33, 1055–1067.

[83] Yamamoto H, Zhang K, Karasaki K, Marui A, Uehara H, Nishikawa N. Large-scale numerical simulation of CO_2 geologic storage and its impact on regional groundwater flow: a hypothetical case study at Tokyo Bay, Japan. Energy Procedia 2009, 1, 1871–1878.

[84] Doughty C. Investigation of CO_2 plume behavior for a largescale pilot test of geologic carbon storage in a saline formation. Transport in Porous Media 2010, 82, 49–76.

[85] Chasset C, Jarsjö J, Erlström M, Cvetkovic V, Destouni G. Scenario simulations of CO_2 injection feasibility, plume migration and storage in a saline aquifer, Scania, Sweden. International Journal of Green Gas Control 2011, 5, 1303–1318.

[86] Lacinska AM, Styles MT, Bateman K, Wagner D, Hall MR, Gowing C, Brown PD. Acid-dissolution of antigorite, chrysotile and lizardite for *ex situ* carbon capture and storage by mineralisation. Chemical Geology 2017, 437, 153–169.

[87] Haug TA, Munz IA, Kleiv RA. Importance of dissolution and precipitation kinetics for mineral carbonation. Energy Procedia 2011, 4, 5029–5036.

[88] Béarat H, McKelvy MJ, Chizmeshya AVG, Gormley D, Nunez R, Carpenter RW, Squires K, Wolf GH. Carbon Sequestration via Aqueous Olivine Mineral Carbonation: Role of Passivating Layer Formation. Environmental Science and Technology 2006, 40(15),4802–4808.

[89] Daval D, Sissmann O, Menguy N, Saldi GD, Guyot F, Martinez I, Corvisier J, Garcia B, Machouk I, Knauss JG, Hellmann R. Influence of amorphous silica layer formation on the dissolution rate of olivine at 90 °C and elevated $p\mathrm{CO_2}$. Chemical Geology 2011, 284(1–2), 193–209.

Peter Styring
12 Reactions using impure carbon dioxide

12.1 Introduction

12.1.1 Carbon dioxide sources

If we consider the composition of waste or flue gas, we need to be completely aware of its source. In most cases where CCUS is considered, the flue gas typically comes from a gas or coal-fired power station. The amount of CO_2 present in the waste from gas-generated power is typically 5–10%, while for coal generation it is 10–15% depending on efficiencies. The remainder of the flue gas is primarily nitrogen and water, but there will also be traces of nitrogen oxides (NO_x) and sulphur oxides (SO_x) depending on the purity of the fuel used. There will also be water as the second combustion product. The flue gas can be dehumidified by condensing out the water and NO_x and SO_x removed using acid gas treatment processes. This all costs money and has additional environmental penalties. However, we are still left with a mixed gas stream containing a low concentration of CO_2, which has low reactivity, in nitrogen, which is inert. Increasingly, other sources of CO_2 are being considered, such as iron and steel, ammonia and cement production, as well as fermentation processes where the CO_2 concentration is higher. However, this still leaves us with a problem of how to separate the CO_2 from the N_2. Both are linear, non-polar molecules, but CO_2 does possess a quadrupole moment that means an induced dipole can be created in the proximity of a second species with a strong dipole moment.

In isolation, the CO_2 molecule is linear and symmetrical, as shown in Figure 12.1 (a), and so each internal dipole cancels to give net zero dipole moment for the molecule. However, in the presence of a strongly polar species or an anion (negative charge), the O–C–O bond becomes distorted and no longer linear. In this state, there is a net induced dipole moment on the molecule (b), termed the *electric quadrupole moment*. Once this distortion occurs, the CO_2 molecule gains additional energy and becomes more reactive. This has been proposed as the mechanism by which CO_2 can be separated from N_2 in ionic liquid media [1]. The anion (chloride in Figure 12.2) possesses a negative charge that induces a charge polarisation in the CO_2 molecule. This is stabilised by the partial charge on one of the CO_2 oxygens interacting with the positive charge on the methyl-methylimidazolium cation.

Molecular simulations show that the interaction has a binding energy that can be predicted but importantly shows a deviation of the O–C–O bond angle away from 180 ° [1]. This deviation can also be induced using simple nucleophiles that activate

Peter Styring, UK Centre for Carbon Dioxide Utilisation, Department of Chemical & Biological Engineering, The University of Sheffield, Sheffield, United Kingdom

https://doi.org/10.1515/9783110563191-012

the CO_2 molecule, and so it can be converted to a product. In this simple case, a methyl Grignard reagent can be used, which converts the CO_2 to a carboxylate anion, in this case acetate, which forms the acid on neutralisation with a proton (Figure 12.3).

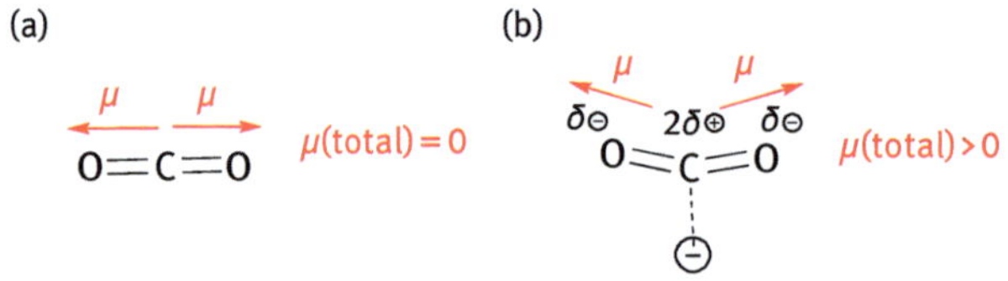

(a)   (b)

Figure 12.1: (a) There is no dipole on a linearly symmetric CO_2 molecule in isolation. (b) The presence of a strong charge induces an electronic quadrupole moment that is non-zero.

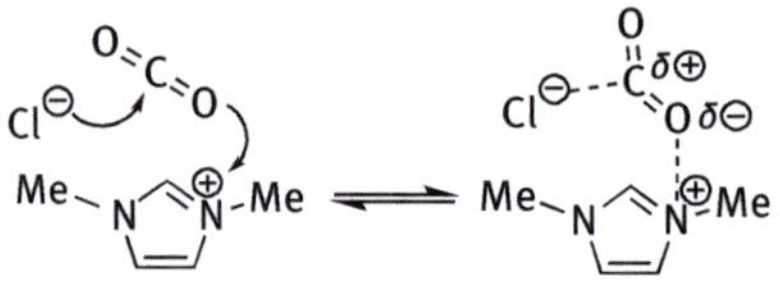

Figure 12.2: Weak chemisorption of CO_2 by the ionic liquid [Mmim][Cl].

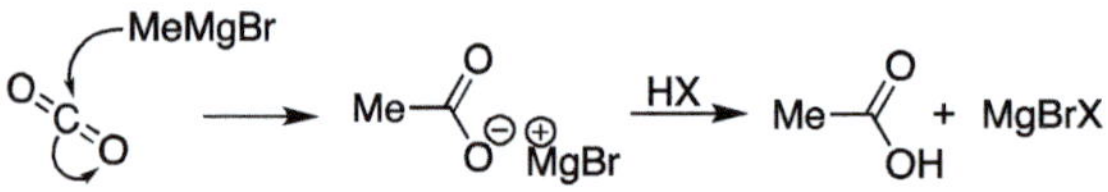

Figure 12.3: Reaction of CO_2 with a nucleophile (methyl) to give a carboxylic acid.

The reason that this unreactive molecule does react is a consequence of the high energy possessed by the Grignard reagent, and in particular the methyl anion acting as a nucleophile. It has been said that CO_2 utilisation reactions should not be considered as a viable technology due to the low reactivity of CO_2 and that efforts should concentrate on the development of capture agents to purify the CO_2 ready for application in carbon capture and storage (CCS) and CO_2-enhanced oil recovery applications. To put this into some kind of perspective, we need to look at the capture process that underpins this concept. Reaction of CO_2 with amines, such as monoethanolamine (MEA), or advanced amines such as piperazines [2] involves the formation of a carbamate intermediate, which then reacts further to produce an ammonium salt (Figure 12.4) in an exothermic process, but one that needs energy input to overcome the activation energy.

The fact that a reaction has taken place means the CO_2 can be separated from the inert nitrogen. Therefore, a product has been formed using the raw flue gas. The problem is that the "product" has no value: it is the pure CO_2 that is the ultimate product. This means that the carbamic acid salt must be thermally decomposed

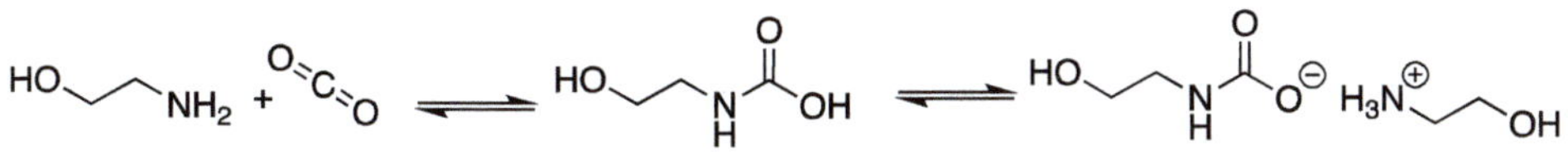

Figure 12.4: Carbon capture process using MEA to chemisorb CO_2.

back to the amine and purified CO_2. There are issues here that must be addressed. The decomposition process also degrades the amine to produce toxic by-products such as nitrosamines and other NO_x species. The MEA needs to be constantly replenished due to this degradation and through evaporative loss of the amine to the atmosphere. Secondly, the reaction gives a product. In this case not a useful one, but it does demonstrate that chemistry can be used to produce a product from the CO_2. As this is the case, why not use the raw flue gas directly to make a product that has value.

If anyone is still under the misconception that CO_2 is inert, they only need to search for "magnesium dry ice reaction" on an internet search engine and look at the images and videos that result. For example, there are some interesting videos on the RSC Advances website (https://bit.ly/2weqFz7). Solid CO_2 reacts with magnesium metal at −78 °C to give the magnesium oxide and carbon according to the equation: $2Mg + CO_2 \rightarrow 2MgO + C$. Of course, this is not an impure gas stream, but it does demonstrate the reactive power of CO_2 under the correct conditions.

12.1.2 The circle of life

Nature offers an insight into CCU processes. Nature does not use concentrated CO_2 as a feedstock. It uses atmospheric concentrations of the gas, let us say 400 ppm. The capture process is a part of the conversion process where the CO_2 is reduced by hydrogen, in the form of protons and not molecular dihydrogen, in the Calvin cycle. The process operates at temperatures below 37 °C and at atmospheric pressure. So, if a biological system can convert CO_2 into products why cannot a chemical system. Of course, it can be achieved using catalysts to promote conversion. However, we do not want to mimic the biological system as it is very inefficient (plants usually are less than 4% efficient using photosynthesis) and incredibly slow. We want to improve on it by reducing conversion times and increasing efficiencies. We will need energy for this: photosynthesis uses direct photochemical conversion. In the chemical domain, we are able to use more diverse sources of sustainable energy, usually from weather-dependent power sources (solar, wind) or geographical sources (hydro, tidal).

Fossil oil was initially atmospheric CO_2 that was used by plants that degraded to give the hydrocarbons that are recovered today. This is a CCU process that occurred on geological timescales. It is the role of the chemist and engineer to achieve similar feats but in minutes and hours. Perhaps the main reason that there is little literature on the direct conversion of impure CO_2 sources is simply that not many people have attempted it, instead considering only the fundamental chemistry of pure reactants.

12.2 Bulk properties

12.2.1 Volumes, masses and moles

The interplay between different technologies is often confused by different, mixed or the misleading use of units. For example, the concentration of CO_2 in air is currently around 410 ppm. If that is volumetric, then 10^6 L of air will contain 410 L of CO_2, that is, 0.041 vol%. However, when we are considering CO_2 chemistry, then the molar concentration is more important. According to NOAA [3], 1 ppm CO_2 in air has a molar concentration of 1 μmol/mole dry air (at 2018 levels that is 410 μmol CO_2/mol air). So, the gravimetric concentration of CO_2 in air is 0.063 wt%. This is worth considering when determining the most efficient CO_2 source to use when using impure gas streams.

12.2.2 Composition of flue gases

The composition of a waste gas is certainly not constant. It varies not only between processes but also between runs of the same process. With a drive to reduce coal combustion in the energy sector, particularly in the United Kingdom, the potential CO_2 concentrations will tend to decrease. For a coal-fired power generator, the concentration of CO_2 in the flue gas is typically 10–14%, while for natural gas combustion this is typically 8% and for natural gas combined cycle only 4%. This makes it more challenging technologically and economically to affect efficient separations, but considerably less challenging than trying to use atmospheric CO_2 at this point in time. If separation is challenging now, it will become increasingly more challenging [4].

If we look at other CO_2 emitters outside the energy sector, the situation is more favourable. Blast furnace gases are in the range 20–30% from pre-combustion to post-combustion. Cement kiln gases are in the range 14–33% and contain fewer other impurities. Of course, other impurities may have a major impact of the subsequent chemistries. In capture processes, using for example amines, impurities can be tolerated as it is a non-catalytic process. However, if catalysts are used in utilisation processes, then impurities such as sulphur and nitrogen oxides need to be removed. In many processes, environmental protection regulations necessitate that these acid gases, as well as particulates and toxic metals, must be removed. The other major impurities are oxygen, as a result of incomplete combustion, and water as the combustion by-product. Water will seriously affect the catalyst and will also have a probably detrimental effect on any reaction. Again, humidity has been suggested as a barrier to CO_2 conversion reactions due to the cost of dehumidification. Here it is worth considering the whole process. In the case of amine capture, the active agent is dissolved in water to form an aqueous solution. Therefore, dehumidification is unnecessary as the water will be a trace when compared to the bulk. However, the resulting purified gas will be highly humid and so will need to be dried before it is further processed. Storage applications require very dry CO_2 to prevent corrosion of pipework when transported under supercritical conditions. For CO_2 utilisation processes directly from flue gases, the dehumidification step is simply shifted to a stage before the reaction (Figure 12.5) and so energy costs are comparable.

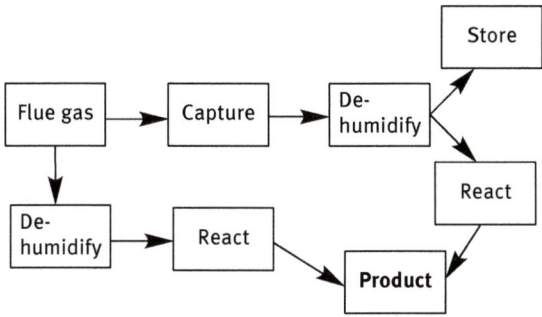

Figure 12.5: Flue gas dehumidification options.

12.3 The challenge

It is not surprising that many research groups choose to use pure CO_2 as the feedstock for reactions. After all, in early stage research, it is about establishing the fundamentals rather than the commercial and industrial practicalities. However, it is suggested that when developing protocols for CO_2 utilisation research a range of gas compositions should be used, much as in the same way that solvents may be varied. Pure gas streams allow the reaction variables to be determined without the added complication of impurities. However, if the reaction is to be transferred to a commercial process, then the obstacles to progress need to be identified. This should be done as early as possible in the reaction development process. So, the challenge will be to develop:
-   reactions that use catalysts composed of Earth crust abundant elements (or no catalysts at all);
-   reactions that use sustainable feedstocks;
-   catalysts that are tolerant to impurities (or which can be easily regenerated in situ) and
-   reactions yielding no by-products if possible.

All of this is a big ask. However, if the capture and purification step can be eliminated, then the cost of product will be reduced significantly as the purification step is often a hot spot on a techno-economic analysis.

Many catalytic reactions are carried out under inert conditions where the reaction vessel is held under an atmosphere of nitrogen or argon to help preserve catalytic activity. By using CO_2/N_2 mixtures, the selectivity should not be adversely affected, but reaction kinetics will be due to dilution. However, this is looking at the reaction again in isolation. The rate of reaction may decrease, but the capture step is eliminated, and so there will be benefits. If the reaction is effective, it will not only produce a valuable product but also concurrently purify the nitrogen, which is actually the process that occurs in carbon capture reactions. So, it is essential to look at the big picture and construct comparative scenarios looking at all the pros and cons.

12.4 Reactions of CO_2 to give value-added products

Xie et al. [5] have eloquently identified the role of CO_2 utilisation in a new economy. They identify many of the problems often overlooked in CCS such as the long-term stability of storage, especially in sites where there is a risk of seismic activity. Of course, costs are identified as a major barrier to CCS deployment while CO_2 utilisation can yield a profit [5]. Much of this is related to the deployment of CO_2 utilisation in the accelerated mineralisation (AM) of pure and waste inorganic feedstocks. Indeed, the roadmap published by the Global CO_2 Initiative includes a report commissioned from the McKinsey Company to look at the mitigation and economic potential of CO_2 utilisation [6]. The report identified three areas that offered the greatest potential. AM through enhanced cementation and the formation of aggregates and construction materials provided the first and third most viable options, while the second was the reduction of CO_2 with hydrogen to give transport fuels. Mineralisation is favourable as it results in permanent storage of CO_2, but in the built environment rather than in the geological arena. Furthermore, mineral carbonation is an exothermic process as carbonate products are below CO_2 in the Gibbs energy diagram (Figure 12.6), and so produce low-grade heat on reaction [7]. Conversely, considerable energy must be supplied to reduce CO_2 to give the fuel product. This is because it is reversing the combustion process that produces energy. The driving force for the production of fuels is our long-term need to maintain a high-energy density fuel transport infrastructure.

The North group have reported extensive studies on the formation of cyclic carbonates from CO_2 and a terminal epoxide. These reactions are catalysed by bimetallic aluminium complexes based on salen-type ligands [8] as shown in Figure 12.7 and work best with a co-catalyst that is often based on a tetraalkylammonium halide salt. Similar results have been reported by us [9], which show that the reaction can occur in the presence of catalyst alone but works best in the co-catalytic system (Figures 12.8 and 12.9). In some cases, a solvent is used in cyclic carbonate formation [9], while in others the epoxide acts in dual roles of solvent and reactant [10].

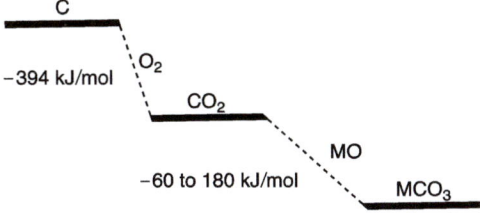

Figure 12.6: Thermodynamics of accelerated mineralisation: enthalpies of formation (not to scale).

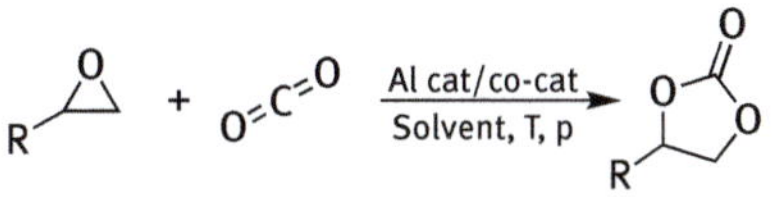

Figure 12.7: General synthesis of cyclic carbonates from CO_2.

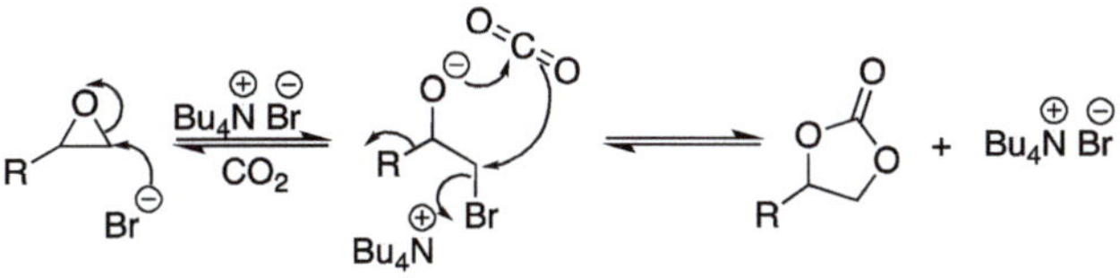

Figure 12.8: Cyclic carbonation formation catalysed by tetrabutylammonium bromide (TBAB).

While most reactions have been carried out using pure CO_2, North has employed two different approaches to testing the aluminium catalysts under more realistic industrial conditions. Reactions have been performed using "simulated glue gas" and real, raw flue gas. It is not possible to buy bottled real flue gas, and so simulated flue gases are used routinely either premixed at 12% CO_2 (BOC-Linde) or by mixing pure gases using mass flow controllers. North has extended this by adding realistic impurities into the simulated flue gas mixture [11]. For example, catalyst deactivation studies were carried out using 19.5% CO_2 in nitrogen, doped with either 1,700 ppm sulphur dioxide (SO_2) or a mixture containing 661 ppm nitrogen monoxide (N_2O) and 36 ppm nitrogen dioxide (NO_2). Reactions were carried out in batch and continuous flow reactors. In the case of the later, a gradual deactivation of catalytic activity was observed. The catalyst could, however, be reactivated irrespective of whether pure CO_2, a CO_2/N_2 mixture or SO_2/NO_x-doped mixtures were used. However, if sulphur trioxide (SO_3), a contaminant often found in flue gas, was used, then irreversible catalyst deactivation was observed.

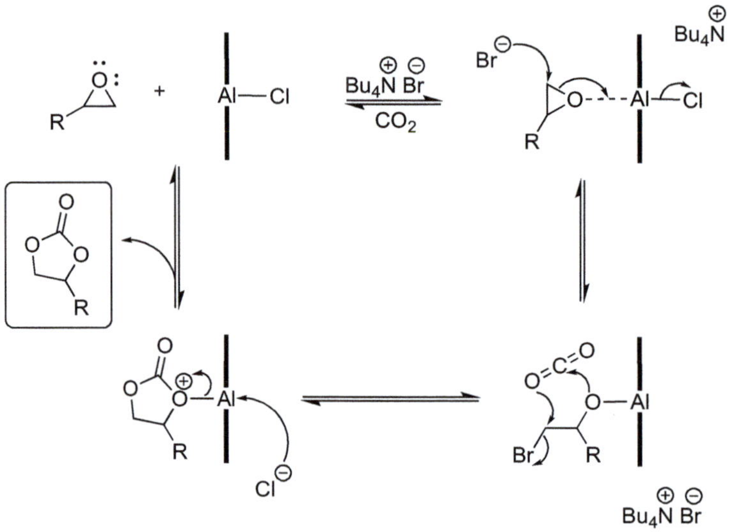

Figure 12.9: Cooperative cyclic carbonation formation catalysed by an aluminium(III)-TBAB co-catalyst system.

North et al. (2011) also reported the cyclic carbonate formation reaction using real flue gas obtained from a coal- and gas-fired power station [12]. A direct sampling of the emissions stack was found to have the following compositions. From gas combustion, the flue gas was captured at 53 °C and had a composition of CO_2 5.1%(v/v), O_2 8.7%(v/v), SO_2 26.1 ppmv, CO 189.3 ppmv and NO_x 33.3 ppmv. Coal combustion flue gas was captured at 52 °C with a composition CO_2 15.1%(v/v), O_2 3.4%(v/v), SO_2 291.4 ppmv, CO 39.9 ppmv and NO_x 443.0 ppmv. Reaction with natural gas flue gas showed conversions and catalytic activity similar to that of the control CO_2/N_2 mixtures with initial deactivation of the catalyst but the ability to regenerate activity. Coal-derived flue gas, however, showed a greater loss in activity, the mechanism for which is unclear. However, ICP measurements showed similar loss of aluminium content in all cases, not just the coal-derived case. It is therefore clear that the catalyst and process can tolerate impurities; however, there is a suggestion that pre-reaction clean-up of coal-derived flue gas is advisable.

The Styring Group have reported a number of stoichiometric processes using organomagnesium complexes (Grignard reagents) to carry out low-temperature carbon–carbon bond-forming reactions with CO_2 [13]. The highly exothermic reaction produces a carboxylate salt, which on acid work-up yields the carboxylic acid. This is a very simple reaction, yet one that is surprisingly effective in the CO_2 utilisation arena. This is a good example of the importance of looking at the whole system rather than the individual process. At face value, a reaction that uses a stoichiometric reagent rather than a catalyst may not seem too efficient. That is compounded by the fact that the metal appears in a waste product, magnesium(II) chloride, which is energy intensive to recycle. However, if the complete supply chain for the process is considered, then magnesium recycling is a not that inefficient process. The manufacture of pure magnesium metal, the precursor of the Grignard reagent, involves the electrolysis of $MgCl_2$ present in sea water. In these cases, the concentration is very low, and so adding excess $MgCl_2$ from the CO_2 utilisation process can actually increase the efficiency of the electrolytic purification process. Alternatively, a thermal process can be used. This typically takes magnesium(II) carbonate that decomposes to MgO and is then further reduced to magnesium metal through reaction with silicon. The magnesium therefore passes around a chemical cycle where it is used and recycled (Figure 12.10). This is a process that has been referred to a metal or chemical looping [7, 14].

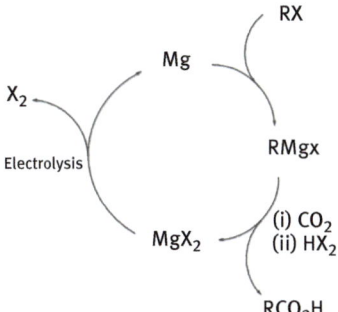

Figure 12.10: Metal looping using magnesium.

The diversity of products that can be obtained using Grignard chemistry makes it particularly valuable as shown in Figure 12.11. Not only can simple Grignard reagents be easily made, with a bit of effort di-Grignard reagents can be formed that offer the opportunity to di-functionalise molecules. Figure 12.11 shows this for the production of terephthalic acid, a monomer in the synthesis of polyethyleneterephthalic acid, extensively used in packaging materials. If this can be made commercially from CO_2, then it represents a route into more sustainable polymers. One of the current drawbacks of this concept is the fact that the aromatic ring is still sourced form petrochemical precursors. If aromatic rings can be produced sustainably and economically from CO_2, then this opens up a whole new domain of potential chemistries. Perhaps an easier immediate target is the di-acid adipic acid that is used in the production of Nylon-6,6 [13].

These materials have been produced by reacting the appropriate Grignard reagent with simulated flue gas with a composition of 12% CO_2 in nitrogen (BOC-Linde), which is dehumidified. However, if impure flue gas is used, then there is a possibility of other impurities resulting in the product stream. SO_x and NO_x will give minor impurities due to their low concentrations, but water will result in debromination of the

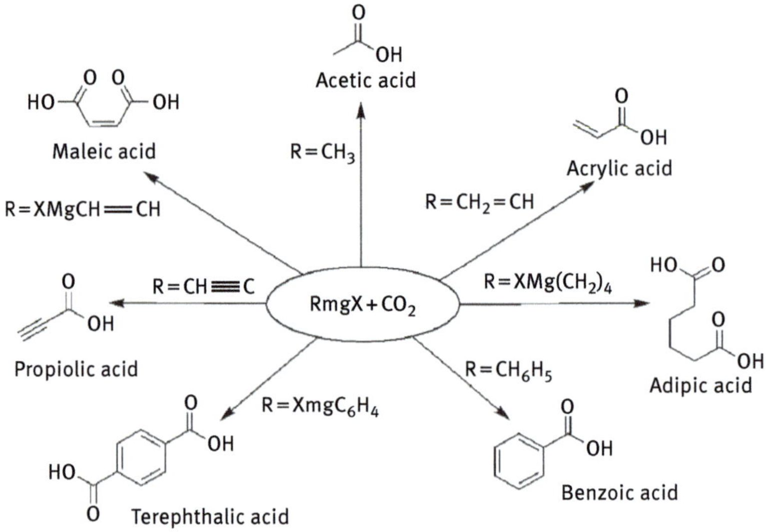

Figure 12.11: Reaction of Grignard reagents with CO_2.

starting material, to give a costly by-product. This is shown in Figure 12.12 for phenyl magnesium bromide (PhMgBr), which produces benzene in a humid atmosphere. Another concern is oxygen that results from incomplete combustion and can occur in the range 3–9% in typical flue gases. The product is phenol (PhOH) that would need to be separated from the reaction mixture or that may react with the benzoic acid produced in the reaction vessel to yield unwanted phenyl benzoate [13].

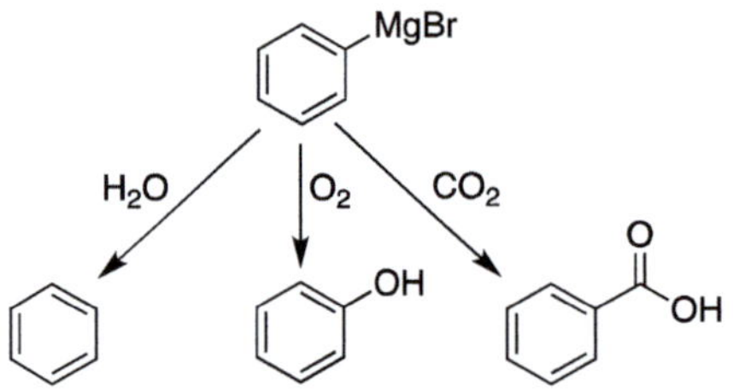

Figure 12.12: Possible competing reactions of a Grignard reagent with humid flue gas.

Studies were carried out where up to 65%(v/v) oxygen was introduced into the simulated flue gas mixture and the coupling reaction carried out. Generally, this is avoided as there is a possibility that explosive alkyl peroxides could be produced in the presence of oxygen in an ethereal solvent. However, careful studies of the reaction of phenyl Grignard with CO_2 in the presence of oxygen yielded some interesting results. It was shown that conversion to the benzoic acid decreased with increasing oxygen concentration; however, there was no evidence of any oxygenated hydrocarbon products [13]. This suggests that the reaction of Grignard reagent with CO_2 is faster than the reaction with oxygen. Furthermore, it suggests that the decrease, which is linear with increasing oxygen concentration (Figure 12.13), is a kinetic effect as a result of the decreasing CO_2 concentration. This leads to the conclusion that oxygen in the flue gas can be tolerated in terms of the chemistry. However, the safe operation of such reactors would need paramount consideration [15].

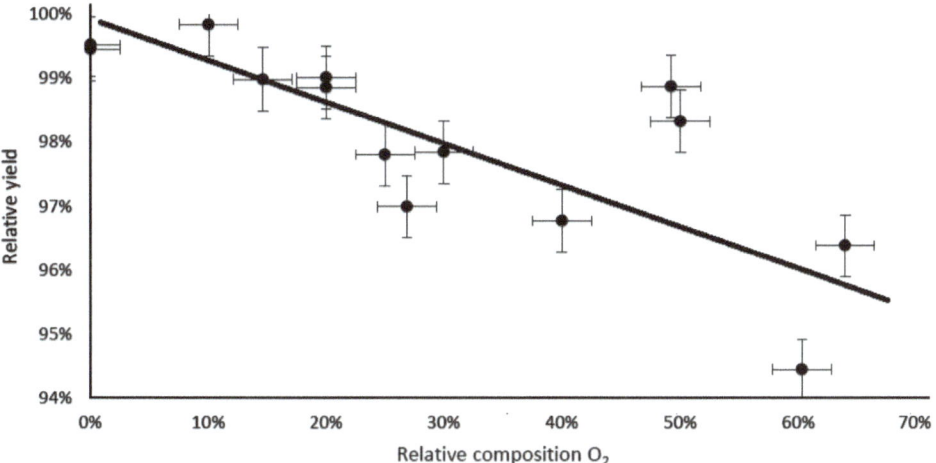

Figure 12.13: Yield of benzoic acid as a function of increasing O_2 concentration in the reaction of phenylmagnesium bromide with CO_2 at 20 °C for 1 h.

Butanol is a direct drop-in replacement fuel for gasoline. They have similar energy densities and octane numbers. Butanol has a low affinity for water and consequently low corrosive properties. This is in direct contrast to ethanol that can cause extensive damage to seals and valves. Consequently, butanol can be used in unmodified gasoline engines as a 100% pure liquid, whereas ethanol is restricted to 15% blends in fossil gasoline. Figure 12.14 shows a reaction sequence for the production of butanol from methanol (CO_2 would be the precursor to this as well) with the Grignard reaction, with flue gas being a key step in the overall transformation process to give a C-2 fragment, acetic acid. This is then esterified and a Claisen condensation performed to give the C-4 product, which is subsequently reduced with hydrogen [14].

If we consider that the first step of the sequence is the reduction of CO_2 to give methanol, then we can see that a carbon cycle is being developed. Looking at this energy profile, it is perhaps surprising to see that each step is exothermic, with the exception of the Claisen condensation that is weakly endothermic. Even the preceding reduction of CO_2 with hydrogen to give methanol is exothermic, with an

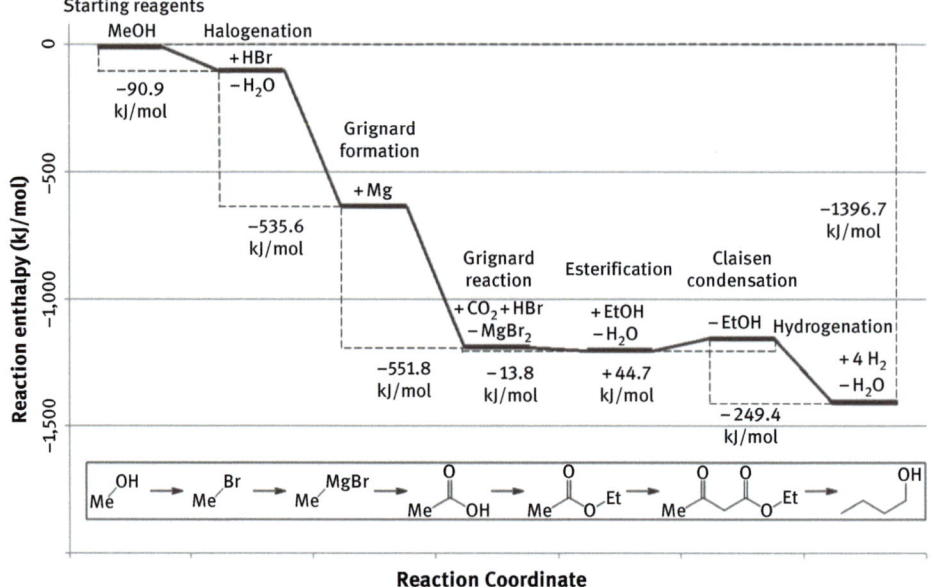

Figure 12.14: Reaction profile for the synthesis of butanol from CO_2-derived methanol.

enthalpy of −49.7 kJ/mol. The energy that drives the overall cyclic process is the electrolytic regeneration of magnesium in the strongly endothermic reaction discussed previously and the renewable hydrogen that is consumed. However, just because the reaction sequence is exothermic overall, it does not mean that the process is free of energy costs as many of the steps have an activation energy that must be overcome.

Figure 12.15 shows the carbon cycle developing. CO_2 is the input and CO_2 is the output on combustion of the fuels product. The key process using impure flue gas is that at the far right of the figure. However, it is not unfeasible that each of the CO_2 utilisation steps could be carried out using flue gas. It is just that people have yet to try and so it is a challenge. The isolated yield of butanol in the laboratory, starting from bromomethane, is 42%, which is good for a multi-step process [14]. However, this is un-optimised and so there is scope for improvement. If the methyl ether is used in place of the ethyl ester described in Figure 12.14, then all carbons in the cycle can be derived from CO_2. These carbons are shown in blue in Figure 12.15.

Tri-reforming of methane using raw flue gas to produce syngas has been extensively studied in China and other regions of the Far East. Each of the three reactions shown in Figure 12.16 yield syngas, each with a different composition. By tuning the different conditions and feedstock ratio, it is therefore possible to refine the ratio selection to give syngas in appropriate ratios to suit the subsequent reaction, such as a Fischer–Tropsch conversion [16].

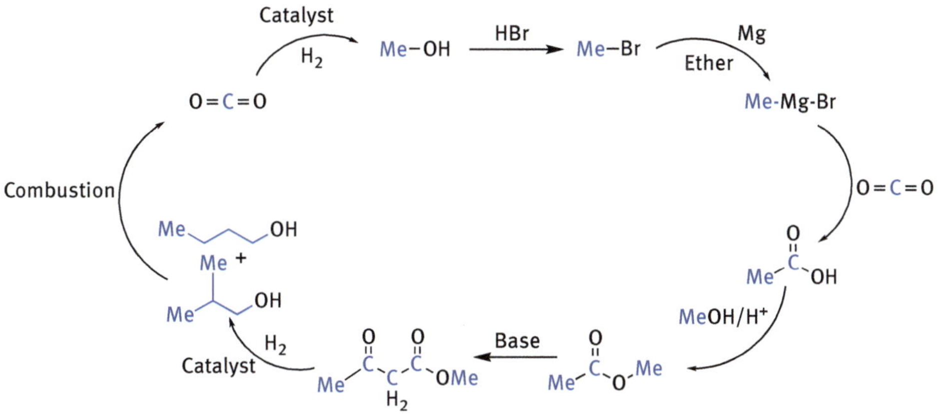

Figure 12.15: Butanol as a combustion fuel in a circular carbon economy. Carbon atoms derived from CO_2 are shown in blue. This is every carbon in the resultant butanol.

$$CO_2 + CH_4 \rightleftharpoons 2CO + 2H_2$$

$$H_2O + CH_4 \rightleftharpoons CO + 3H_2$$

$$0.5O_2 + CH_4 \rightleftharpoons CO + 2H_2$$

Figure 12.16: Tri-reforming of methane to give different syngas compositions.

12.4 Reactions to give inorganic products

Inorganic carbonates have a great benefit when it comes to CO_2 utilisation and mitigation. Carbonates have lower Gibbs energies than CO_2 itself as shown earlier, which means the reaction process is energetically downhill. The case for calcium carbonate formation is shown in Figure 12.17. This means that the reactions are exothermic and so offer the opportunity for heat recovery to feed into other areas of the process. This is a good example of industrial symbiosis in the circular economy. There are many examples where CO_2 is reacted with virgin minerals, such as olivine and serpentine, although these may not be economically and energetically favourable overall due to the process of extracting and grinding the mineral in a suitable form for effective reaction [17, 18].

A novel approach to the reaction of CO_2 with minerals is in the AM of waste mineral residues. The Hills group have reported numerous reactions of CO_2 with waste residues including air pollution control residues and steel slag [19–25]. The advantage of such processes is that the CO_2 is permanently sequestered, at least on geological timescales, in the carbonated product. The reaction is primarily between metal oxides, hydroxides and silicates and CO_2, and mostly those of calcium and magnesium to yield the metal carbonates as shown in Figure 12.18 for the oxides and hydroxides.

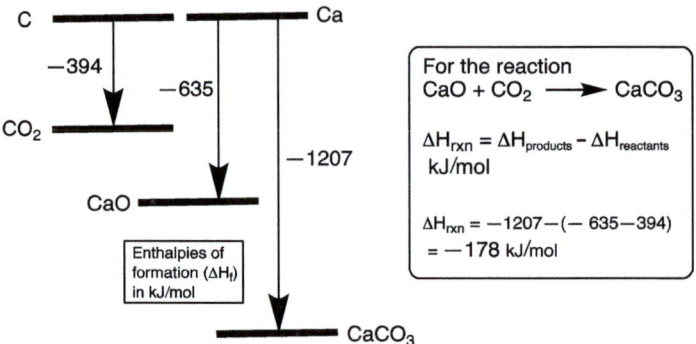

Figure 12.17: Enthalpy of the formation of calcium carbonate showing exothermicity.

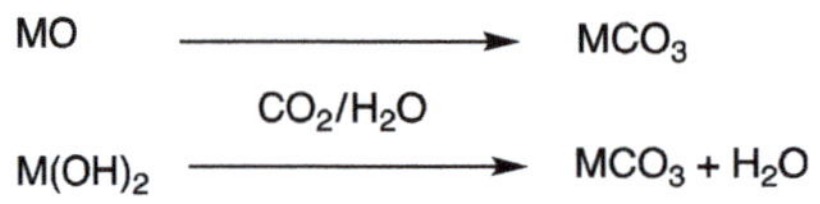

$$MO \xrightarrow{\hspace{2cm}} MCO_3$$

$$M(OH)_2 \xrightarrow{CO_2/H_2O} MCO_3 + H_2O$$

Figure 12.18: Accelerated mineralisation of metal containing mineral waste to carbonates.

Both are essentially atom-efficient reactions in an aqueous medium that help to reduce waste, thereby achieving many of the 12 principles of green chemistry [26]. The environmental credentials of such reactions are enhanced when one considers that this is not only sequestration of CO_2, but also the production of value-added products that help to displace more polluting products from the environment. For example, the work of Hills allows contaminated waste to be converted to building materials and aggregates that have achieved building regulations approval in the UK and can replace existing stocks derived from less sustainable sources. An added benefit is that the materials also have improved mechanical properties over the conventional construction materials. The reported materials use pure CO_2 as the feedstock, but efforts are being made to reduce the CO_2 concentration and so reduce cost, through using more dilute CO_2 sources. The process has now been commercialised on several sites by the spin-out companies Carbon8 Systems and Carbon8 Aggregates.

This approach has been adopted by Xie et al. [27] who have used flue gas CO_2 to treat phosphogypsum waste in a reactive capture process. The cooled flue gas reacts with ammonia to produce ammonium carbonate, as shown schematically in Figure 12.19, which then reacts with phosphogypsum waste in a mineralisation reactor. In addition to CO_2 capture, the process produces two valuable products that are readily separated. These are calcium carbonate and ammonium sulphate, a fertiliser that can simultaneously deliver nitrogen and sulphur nutrients. The process has been economically scaled to produce a demonstrator plant, which has been constructed in Puguang, China.

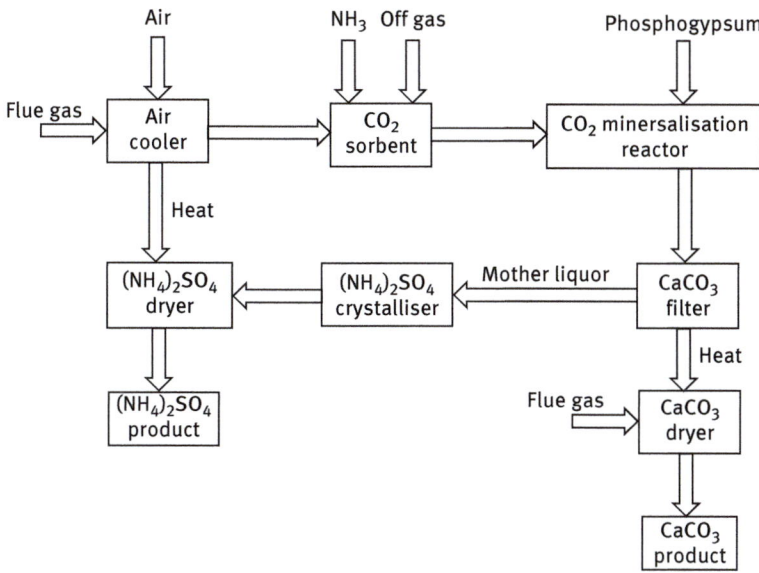

Figure 12.19: Accelerated mineralisation of metal containing mineral waste to carbonates.

The same group have reported a hybrid mineralisation process that can convert low concentrations of CO_2 into valuable carbonates [28]. The first step is the electrolysis of magnesium chloride in aqueous solution using a hydrogen diffusion electrode as the anode and nickel foil as the cathode to give the hydroxide. This is then reacted with CO_2 at different concentrations to give the hydrogen carbonate [$Mg(HCO_3)_2$], which is thermally decomposed to give either magnesium carbonate in the valuable nesquehonite form, or to magnesium metal. This is an interesting process as the latter reaction could be used to regenerate magnesium metal for use in the Grignard reactions discussed earlier in this chapter. The process is seen as being of economic value as one tonne of CO_2 yields 3.16 tonne of nesquehonite, with an energy consumption of 871 kWh. At the time of publication, this equated to a cost of approximately 435.5 RMB (55 €), while the market value of the mined mineral was 3,000 RMB (378 €) for the same quantity. The concentration of CO_2 can be as low as 20% in the simulated flue gas and so has potential in the mitigation of industrial flue gases. This fits in with the widely held belief that while CCS projects are ideally positioned to address the large problem of power sector emissions, CO_2 utilisation projects are ideally suited to address emissions mitigation from the industrial sector. This is in part not only because of the generally higher concentration emissions from industrial processes but also because industrial emitters are often very much remote from possible sequestrations sites. Indeed, this was reinforced by a publication by Pérez-Fortes et al. (2014) that concluded that economically viable CO_2 capture and utilisation processes could be realised by using emissions from the cement and iron and steel industries.

Xie et al. [5] have combined practical experiments on the carbonation of magnesium(II) chloride to give magnesium(II) carbonate with a techno-economic evaluation of CO_2 utilisation with CCS. The report correctly identifies the economic and environmental shortcomings of CCS and proposes that mineral carbonation is a viable route not only to mitigation of CO_2 but also at a profit rather than a loss.

The Styring group have reported an application of AM using impure CO_2 streams to not only cure Portland cement, but to also use toxic mineral waste to produce inert inorganic products [29]. Pure CO_2 and two impure gas supplies were investigated. Although the gases were anhydrous, humidified gas can be used as the AM process is carried out in aqueous slurries. De-humidification is only required at the end of the process to yield the dried product. Initial tests using pure CO_2 were carried out in aqueous slurries of varying composition. Reaction was monitored in terms of the partial pressure of CO_2 consumed using a pressure sensor connected to the reaction flask. The pressure was set at 1 bar of the gas (the process was the same for mixed gas compositions) and the slurry stirred mechanically at 500 rpm. Figure 12.20 shows the reaction of Portland cement with pure CO_2 over a range of slurry compositions. The fastest reactions were observed with 10 g cement mixed with between 20 and 40 mL water. These conditions were good for the formation of powdered calcium carbonate. The reaction proceeded effectively at temperatures ranging from 0 to 50 °C, making it applicable to a wide variety of environments.

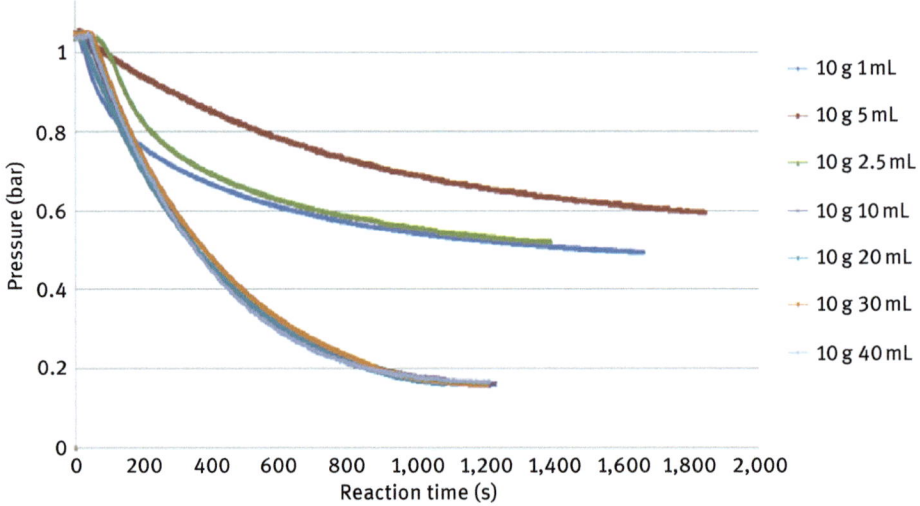

Figure 12.20: Accelerated mineralisation of calcium oxide (Portland cement) in pure CO_2.

The first impure gas used was artificial flue gas (all gases were supplied by BOC/Linde) with a composition of 12% CO_2 in 88% nitrogen. The second gas was perhaps more interesting because of its direct use in localised energy systems: simulated dry biogas was prepared by mixing CO_2 with pure methane in the ratio 0.4 to 0.6 by volume. Reaction with Portland cement is shown in grey in Figure 12.21 over a 19 minute reaction period. The consumption of CO_2 is 24% of the complete gas volume over that period, which means the methane has been concentrated to 80%. If the reaction is continued, consumption plateaus out at 40 minutes to give 98% methane purity. This means that low-grade biogas can be readily upgraded to high-quality biomethane in a simple process that also yields a valuable by-product, calcium carbonate.

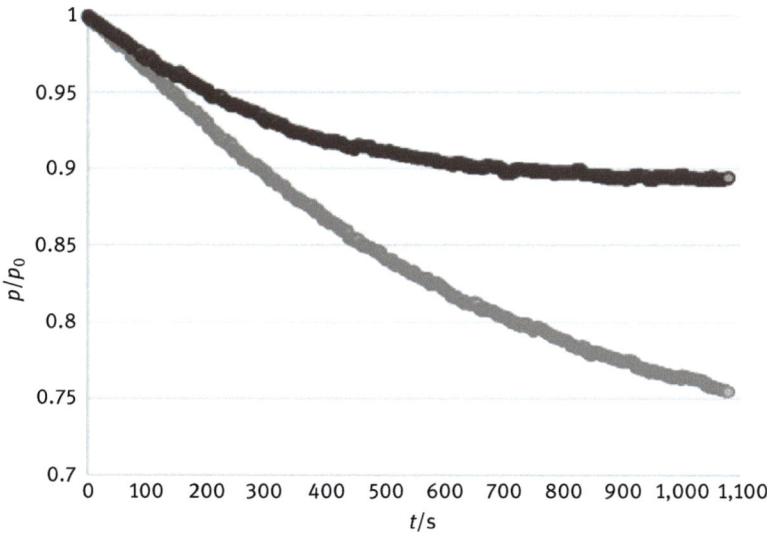

Figure 12.21: Accelerated mineralisation of 10 g Portland cement (grey) and mining tailing waste (black) in water (30 mL) using simulated biogas at 25 °C and 500 rpm.

If mine tailing waste is used instead, then carbonation is also observed but only 12% of the gas volume is consumed after 19 minutes and remains constant until 40 minutes. The limitation is the calcium oxide content of the waste. This produces biogas with 68% purity. However, complete consumption of the CO_2 can be achieved by increasing the mass of waste in the reactor.

12.7 Scope for future studies

The problem is not so much that we cannot use impure gas streams. It is more that we have not really tried. When carrying out the literature review for this chapter, it was genuinely expected that there would be many examples of reactions carried out using raw flue gas streams. Conversely, most reactions are reported using pure CO_2, often food or analytical grade. As with many aspects of the rapidly growing area of CO_2 utilisation or transformations, there is a propagating misconception that impure streams cannot be used. Thus, the myth continues that CO_2 is unreactive and we cannot use impure gas streams. Indeed, the co-reactants also need to be pure. The reader is directed to a video on the Carbon8 Aggregates web page where highly contaminated Air Pollution Control Residues are reacted with an impure CO_2 gas stream at ambient (20 °C) temperature [30]. Hopefully this carbonation reaction will dismiss such myths. A misconception is often regarded as a "rule" when in fact it is only a theory. It is our role as chemists and engineers to test such theories to see if they are valid. In the case of CO_2 reactivity, the theory is clearly invalid. This is based on the evidence of experimental data that shows it clearly is reactive.

Hence, the future perspective should be to go away and test the conditions. Test the catalyst. See if you can use impure and unoptimised conditions. If the reactions do not work well, go back and tune the conditions. A laboratory experiment is often far away from reality. Think of the whole process as it might be deployed on a commercial scale. Challenge convention. Look at micro- and meso-scale reactors. Look at continuous flow conditions rather than batch processing. Go and innovate!

One promising area for the study of impure gas streams is plasmolysis. Moss et al. [31] have shown that plasmolytic reactions of CO_2 can be improved by adding nitrogen: precisely the gas we have been trying to remove in carbon capture processes. Studies show that when a corona discharge reactor is connected to the outlet of a CO_2 refining process, the CO_2 is reduced to the monoxide. The most efficient reaction occurs for a mixture of 80% CO_2 with 20% nitrogen. Although the mechanism has not yet been elucidated, the concurrent oxidation of nitrogen to NO_x species is observed (Figure 12.22). In the figure, negative absorbance represents removal of material while positive absorbance indicated new product formation. This means that the flue gas does need to be concentrated, or refined, but does not need to achieve the grades required for food or pipeline transportation application: hence the cost is significantly lowered.

This was extended to the treatment of simulated biogas (50% CH_4/50% CO_2) using a dielectric barrier discharge plasma reactor [32]. Different packing materials were used, including barium titanate, alumina, magnesium oxide and an unpacked reactor. The rationale was that CO_2 would react in the plasma to give CO with the oxygen radical produced being consumed through reaction with methane to give more CO and also hydrogen as shown by the elementary reactions in Figure 12.23.

The gas mixture was passed through an unpacked plasma reactor and the outlet gas analysed by FTIR spectroscopy. Figure 12.24 shows the resulting spectrum. Again, negative absorbance is removal of starting material; positive absorbance is the product formation.

The removal of methane and CO_2 is clearly observed as is the formation of carbon monoxide. Hydrogen is not detected using the method used. However, the major product is ethane (C_2H_6) and there are other interesting species such as methanol, ethylene (ethene) and acetaldehyde. While the reaction is non-specific, it

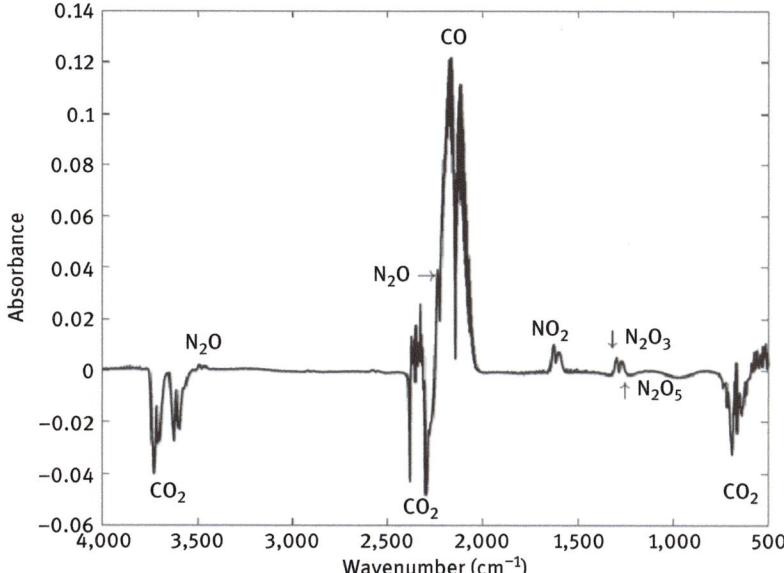

Figure 12.22: FTIR for the products of the reaction of CH_4 and N_2 (80:20) in a pulsed corona plasma reactor.

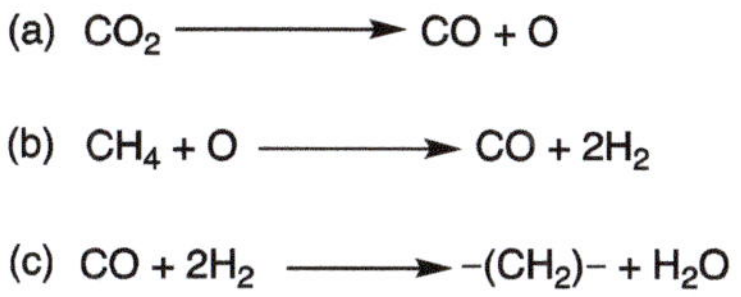

(a) $CO_2 \longrightarrow CO + O$

(b) $CH_4 + O \longrightarrow CO + 2H_2$

(c) $CO + 2H_2 \longrightarrow -(CH_2)- + H_2O$

Figure 12.23: Some of the basic elementary reactions occurring in the co-plasmolysis of CH_4 and CO_2 (50:50) in an unpacked DBD reactor.

does open up the possibility to fine-tune the process to change product distribution, for example, by adding a packed bed containing a specific catalyst. Furthermore, by changing the CO_2 to CH_4 ratio, it is possible to produce different syngas (CO/H_2) ratios in the reactor, which may direct the product formation in different ways. The fact that biogas can be used directly opens up the possibility of using anaerobic digesters to produce highly valued product from waste.

12.8 Conclusions

Many assumptions are challenged, such as CO_2 is inert or unreactive, and that CO_2 must be pure for it to be used in reactions. Amine capture has been used as a method to purify CO_2 in waste gas streams; however, this is a chemical reaction.

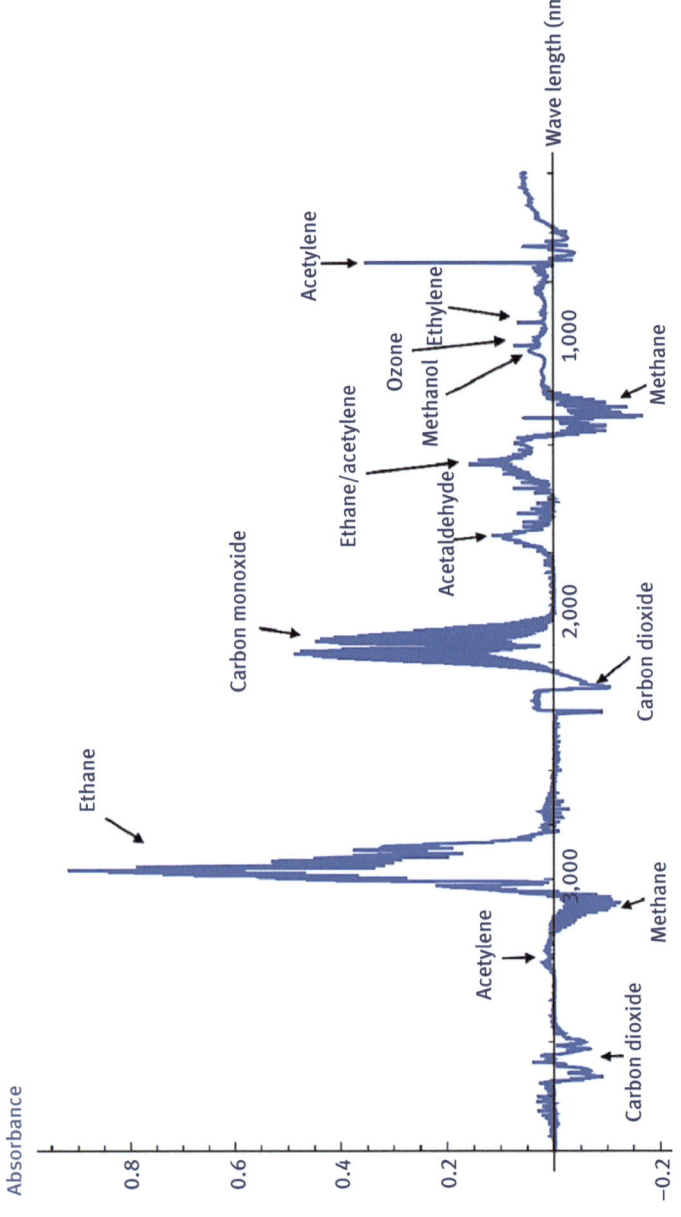

Figure 12.24: FTIR for the products of the reaction of CH_4 and CO_2 (50:50) in an unpacked DBD reactor.

It produces a product that is of no use other than to act as a purification vector. This is a large consumption of energy. Therefore, it seems more efficient to replace this with a reaction that can be of actual value. This chapter demonstrates that the seeds have been sown to develop this technology, whether it is through the production of fuels, chemical feedstocks or inorganic materials such as metal carbonates. It has also been shown that in some cases, such as corona discharge plasmolysis, an impure CO_2 stream can actually be beneficial to the process. The technologies are clearly nascent; however, they do show a potential direction of travel in developing cheap and efficient reaction pathways while also contributing to waste gas purification without the introduction of costly additional process steps. Key to successful development of these concept is the identification of low-cost, sustainable and impurity-tolerant catalyst: or ideally catalyst-free processes. Often, the reason for using pure CO_2 streams is a desire to study the fundamental chemistry of a reaction. However, time is ticking on the path to climate change reduction and so the time has come to take risks. It is time to try using impure waste gas streams to see what is possible. While CO_2 utilisation may not at face value seem large enough to cope with the vast quantities of anthropogenic CO_2 emissions alone, it is a part of the suite of technologies that may contribute in part. When considering this, the reader should consider not only the amount of CO_2 consumed in the reaction but also the amount of indirect CO_2 emissions avoided by not using more fossil oil. Indeed, the Renewable Energy Directive II [33] published by the European Union quite rightly identifies carbon avoided as well as carbon recycled as part of the definition of Renewable Fuels of Non-biological Origin. The scope for reactions using CO_2 is extensive [34], but one must always be wary to ensure that the processes proposed are both environmentally and economically viable. In order to close the carbon cycle, we must eventually consider the case of direct air capture. While direct air capture has been demonstrated in a few cases, it has yet to pass the economic hurdle. This is because it is easier to capture more concentrated CO_2 sources [35, 36]. In a report from the European Commission [35], Pérez-Fortes et al. concluded that emissions from the cement and iron and steel industries showed the greatest promise as CO_2 sources for utilisation reactions.

Acknowledgements: The author is grateful for funding in the field of carbon dioxide utilisation from a number of sources. Engineering and Physical Sciences Research Council (EPSRC) is thanked for funding to the CO2Chem Grand Challenge Network (www.co2chem.com) under three consecutive grants starting in 2010 (EP/H035702/1, EP/K007947/1, EP/P026435/1) and for the 4CU Programme (P/K001329/1). The European Commission are thanked for funding to Smart CO_2 Transformation [SCOT] (FP7, grant agreement No. 319995), CarbonNext (H2020, grant agreement No. 723678).

References

[1]   Dowson GRM, Reed DG, Bellas J-M, Charalambous C and Styring P (2016). Fast and Selective Separation of Carbon Dioxide from Dilute Streams by Pressure Swing Adsorption using Solid Ionic Liquids, *Faraday Discuss*. 192, 511–527.

[2]   Y-J Lin, E Chen and GT Rochelle, Pilot plant test of the advanced flash stripper for CO2 capture, *Faraday Discuss*. **192**, 37–58.

[3]   www.esrl.noaa.gov

[4]   MK Mondal, HK Balsora, P Varshney, Progress and trends in CO_2 capture/separation technologies: A review. *Energy* 46 (2012) 431–441.

[5]   H-P Xie, L-Z Xie, Y-F Wang, J-H Zhu, B Liand and Y Ju, CCU: A More Feasible and Economic Strategy than CCS for Reducing CO_2 Emissions, *Journal of Sichuan University Engineering Science Edition*, 2012, 04, Article 16.

[6] CO_2 Sciences/The Global CO_2 Initiative, Global Roadmap for Implementing CO_2 Utilization (2016).

[7]   Mission Innovation, Accelerating Breakthrough Innovation in Carbon Capture, Utilization, and Storage. https://www.energy.gov/sites/prod/files/2018/05/f51/Accelerating% 20Breakthrough%20Innovation%20in%20Carbon%20Capture%2C%20Utilization%2C% 20and%20Storage%20_0.pdf

[8]   M North and C Young, Bimetallic aluminium(acen) complexes as catalysts for the synthesis of cyclic carbonates from carbon dioxide and epoxides, *Catal. Sci. Technol.*, 2011, 1, 93–99.

[9]   S Supasitmongkol and P Styring, A single centre aluminium(III) catalyst and TBAB as an ionic organo-catalyst for the homogeneous catalytic synthesis of styrene carbonate. *Catal. Sci. Technol.*, 2014, 4, 1622–1630.

[10] M North, Synthesis of Cyclic Carbonates from Epoxides and Carbon Dioxide using bimetallic aluminium(salen) complexes. *ARKIVOC*, 2012, 610–628.

[11] IS Metcalfe, M North and P Villuendas, Influence of reactor design on cyclic carbonate synthesis catalysed by a bimetallic aluminium(salen) complex, *J. CO_2 Utilization*, 2013, 2, 24–28.

[12] M North, B Wang and C Young, Influence of flue gas on the catalytic activity of an immobilised aluminium (salen) complex for cyclic carbonate formation. *Energy Environ. Sci.*, 2011, 4, 4163–4170.

[13] GRM Dowson, I Dimitriou, RE Owen, DG Reed, RWK Allen and P Styring, Kinetic and economic analysis of reactive capture of dilute carbon dioxide with Grignard reagents *Faraday Discuss*. 2015, 183, 47–65.

[14] GRM Dowson and P Styring, Demonstration of CO_2 Conversion to Synthetic Transport Fuel at Flue Gas Concentrations. *Frontiers in Energy Research*, 2017, 5, 26.

[15] P Styring, GRM Dowson and J Cooper, 14th International Conference on Carbon Dioxide Utilization (ICCDU), 2016.

[16] PM Maitlis and A de Klerk, Greener Fischer-Tropsch Processes, Wiley-VCH, Weinheim (2013).

[17] S Madeddu, M Priestnall, E Godoy, RV Kumar, S Raymahasay, M Evans, RF Wang, S Manenye and H Kinoshita, Extraction of $Mg(OH)_2$ from Mg silicate minerals with NaOH assisted with H_2O: implications for CO_2 capture from exhaust flue gas, *Faraday Discuss.*, 2015, 183, 369–387.

[18] H Geerlings and R Zevenhoven, CO_2 Mineralization – Bridge Between Storage and Utilization of CO2. *Annu. Rev. Chem. Miomol. Eng.*, 2013, 4, 103–117.

[19] XM Li, MF Bertos, CD Hills, PJ Carey, S Simon, Accelerated carbonation of municipal solid waste incineration fly ashes. *Waste Management*, 2007, 27, 1200–1206.

[20] G Costa, R Baciocchi, A Polettini, R Pomi, CD Hills, PJ Carey, Current status and perspectives of accelerated carbonation processes on municipal waste combustion residues, *Environmental Monitoring and Assessment*. 2007, 135, 55–75.

[21] PJ Gunning, CD Hills, PJ Carey, Production of lightweight aggregate from industrial waste and carbon dioxide, *Waste Management*, 2009, 29, 10, 2722–2728.

[22] PJ Gunning, CD Hills, PJ Carey, Accelerated carbonation treatment of industrial wastes. *Waste Management*, 2010, 30, 1081–1090.

[23] P Gunning, CD Hills and PJ Carey Accelerated carbonation treatment of industrial wastes, *Waste Management*, 2010, 30, 1081–1090.

[24] K Tota-Haharaj, C Hills and J Monrose, Novel Permeable Pavement Systems Using Carbon-Negative Aggregates, *NexGen Technologies for Mining and Fuel Industries*, 2017, 1, 683–696.

[25] QY Chen, M Tyrer, CD Hills, XM Yang and P Carey, Immobilisation of heavy metal in cement-based solidification/stabilisation: A review, *Waste Management*, 2009, 29,390–403.

[26] PT Anastas and JC Warner, Green Chemistry: Theory and Practice, Oxford University Press: New York, 1998, p.30.

[27] H Xie, H Yue, J Zhu, B Liang, C Li, Y Wang, L Xie and X Zhou, Scientific and Engineering Progress in CO_2 Mineralization Using Industrial Waste and Natural Minerals. *Engineering*, 2015, 1, 150–157.

[28] H Xie, Y Wang, W Chu and Y Ju, Mineralization of flue gas CO_2 with coproduction of valuable magnesium carbonate by means of magnesium chloride. *Chin. Sci. Bull.*, 2014, 59, 2882–2889.

[29] P Styring, DG Reed, L Sicheng and GRM Dowson, Kinetic Analysis of the Accelerated Mineralisation of Lead Mine Tailings, *NexGen Technologies for Mining and Fuel Industries*, 2017, vol. II, 949–956.

[30] Carbon8 Aggregates. https://c8a.co.uk/#foobox-1/1/247982702

[31] M Moss, DG Reed, RWK Allen and P Styring, Integrated CO_2 Capture and Utilisation using Non-Thermal Plasmolysis. *Frontiers in Energy Research*, 2017, 5:20.

[32] O. Huggett-Wilde (Supervisor: P Styring), MEng Thesis, The University of Sheffield, 2016.

[33] https://ec.europa.eu/energy/en/topics/renewable-energy/renewable-energy-directive

[34] P Styring, EA Quadrelli and K Armstrong, Carbon Dioxide Utilisation: Closing the Carbon Cycle, Elsevier, Amsterdam (2015).

[35] M Pérez-Fortes, JA Moya, K Vatopoulos and E Tzimas, CO_2 Capture and Utilization in Cement and Iron and Steel Industries. *Energy Proc.*, 2014, 63, 6534–6543.

[36] P Styring and K Armstrong, CO_2 shortage: why can't we just pull carbon dioxide out of the air?, The Conversation, 2018. https://theconversation.com/co-shortage-why-cant-we-just-pull-carbon-dioxide-out-of-the-air-99255

Shankara Gayathri Radhakrishnan and Emil Roduner

13 Carbon dioxide activation

13.1 Introduction

CO_2 is considered a valuable carbon source in both its free and ligated form. Unfortunately, it is also considered by some to be an unwanted molecule due to its large emission into the atmosphere, leading to climatic changes. With a heat of formation of $\Delta G_f° = -394$ kJ mol^{-1}, it is a thermodynamically stable molecule with a large band gap of ca. 8 eV [1], indicating that a large energy input is required for its chemical transformation. There are exceptions such as reactions that take advantage of the electrophilic character of its carbon atom. Understanding the fundamentals of the catalytic activation of this molecule is the central focus of this chapter, with a particular emphasis on its electrochemical reduction towards valuable products and liquid fuels. This is a central aspect in current concepts of preventing climatic global warming by recycling CO_2 using renewable energies.

13.2 The electronic structure of CO_2

CO_2 is a linear triatomic molecule with two CO bonds of the same length (1.17 Å). It has the highest oxidation state (+4) of carbon, and its molecular orbital (MO) picture reveals that the three atoms of CO_2 provide a total of 16 valence electrons. The ionisation potential amounts to 13.8 eV. The MO energy level of CO_2 can be written as $(K)(K)(K)(3\sigma_g)^2(2\sigma_u)^2(4\sigma_g)^2 (3\sigma_u)^2(1\pi_u)^4(1\pi_g)^4 (2\pi_u)^0$ (Figure 13.1c). The introduction of an electron into the antibonding orbital $2\pi_u$ leads to the formation of the CO_2 radical anion ($CO_2^{•-}$), which is stabilised by the change in structure from its linear to bent form with an electron affinity of -0.67 ± 0.2 eV [2, 3]. Bending of CO_2 from 180 to 135° lowers the energy of the lowest unoccupied molecular orbital (LUMO) by a large amount, especially that of the in-plane $2\pi_u$ orbital (Figure 13.1a), which facilitates the transfer of an electron to the molecule enormously [4, 5]. Furthermore, bending leads to a C=O bond weakening owing to the elongation of up to 0.11 Å (Figure 13.1b) compared to its linear state. Such a bond elongation could lead to the dissociation of CO_2 on the catalyst or electrode surface, thereby forming surface

Shankara Gayathri Radhakrishnan, Chemistry Department, University of Pretoria, Pretoria, Republic of South Africa
Emil Roduner, Institute of Physical Chemistry, University of Stuttgart, Stuttgart, Germany and Chemistry Department, University of Pretoria, Republic of South Africa

https://doi.org/10.1515/9783110563191-013

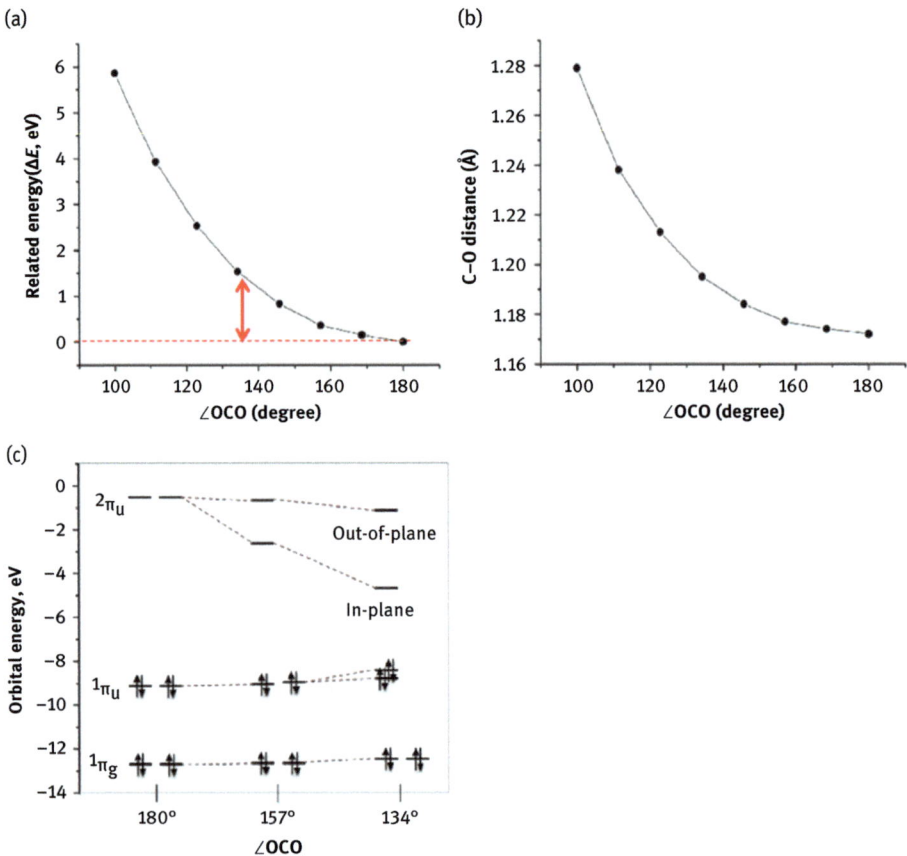

Figure 13.1: The change in the total energy (a) and the C–O bond distances (b) as a function of OCO angles. (c) Walsh diagram for orbital energies upon CO_2 bending [Reprinted with permission from Reference 4].

adsorbed CO and O atoms. All this indicates that the bent state of the CO_2 molecule is the most reactive form and hence electron transfer becomes feasible. $CO_2^{\bullet-}$ is metastable, and in this anionic state, carbon bears a negative charge, making it a nucleophile that leads to an "Umpolung" reactivity [5]. In homogeneous aqueous solution, this bent state is produced electrochemically upon the application of -1.9 V versus NHE [3, 6]. Of course, protonation will considerably stabilise this state. The example of the $CO_2^{\bullet-}$ intermediate illustrates that electrochemical experiments provide electrons and protons via separate and often sequential pathways, while heterogeneous thermal reduction will in general occur by transfer of atomic hydrogen, which eliminates the odd-electron transition states of intermediates.

13.3 Reversible adsorption as a prime requirement for catalysis

13.3.1 Sabatier's principle

Catalysis occurs when the substrate is bound to the catalyst. According to Sabatier's principle, this bonding interaction should be just right, neither too weak nor too strong. If the interaction is too weak, it is insufficient to induce the catalytic step since the concentration of the adsorbed state is low. If it is too strong, the product will remain bound and block or "poison" the catalyst. When the log of the catalytic turnover rate (or assuming Arrhenius behaviour the reaction temperature at a fixed reaction rate) is plotted against the substrate binding energy, this results in a typical volcano-shaped plot where the maximum corresponds to the optimum binding energy.

In a single-step unimolecular reaction, such as the decay of formic acid to CO_2 and H_2, the application of Sabatier's principle is straightforward. In a multi-step reaction, one may identify a rate-determining step with an adsorbed intermediate to which the principle applies, and if there is more than one reactant one may find that the adsorption of one of them dominates the behaviour.

13.3.2 CO_2 adsorption on oxides

CO_2 is often incorrectly regarded as an inert molecule, much like N_2. It reacts readily and reversibly with the non-bonding lone pair of nitrogen in amines, which is the basis for CO_2 capture from the atmosphere of from flue gases. It also binds to the surface of many oxides, with binding energies up to 170 kJ mol^{-1}, as shown in Figure 13.2 [7]. This is understood by the formation of an effective carbonate ion, a well-known and stable species, when CO_2 binds via its electronegative carbon to the oxide dianion (Figure 13.2). This works well for clean oxide surfaces, but one should take into account that under ambient conditions oxide surfaces are often contaminated by adsorbed water molecules or functionalised by surface OH.

When oxides are heated to sufficiently high temperatures, this leads to the loss of O_2, leaving behind two oxygen vacancies at the surface. The electrons of the oxide ions may remain localised in the defect, on reducible neighbouring cations, or become delocalised in the conduction band. It is proposed that such a vacancy is an attractive adsorption site for CO_2 since one of its oxygens can replace the missing oxide ion, and the remaining CO may transfer to the neighbouring metal centre, in particular if it is a transition metal (Figure 13.3 (a) and (b)). We are not aware that CO transfer has been reported; however, a recent density functional theory (DFT) study for CO_2 reduction to methanol on In_2O_3, an extremely successful process [8],

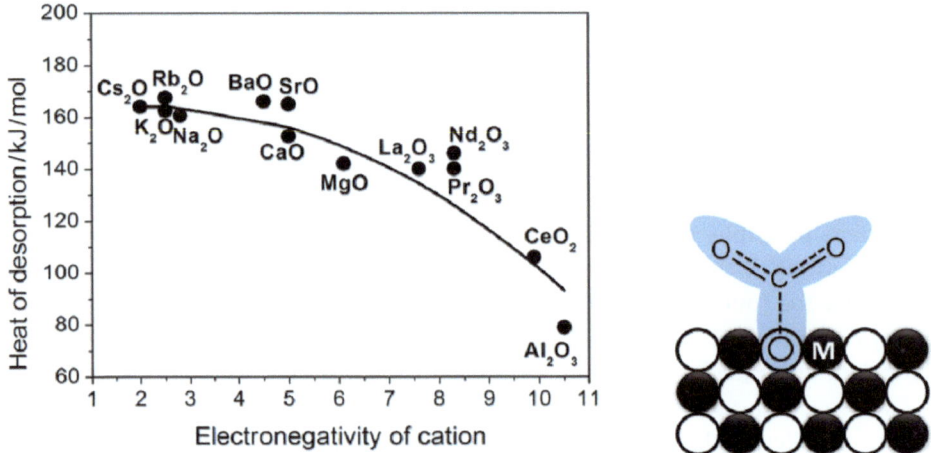

Figure 13.2: Heat of adsorption of CO_2 on oxide surfaces (left) [Redrawn with permission from Reference 7] and suggested mechanism indicating the formation of an effective carbonate ion.

confirmed the importance of oxygen vacancies for the initial adsorption of CO_2 and of subsequent intermediates with evidence of only a transient carbon bond to indium as shown in Figure (13.3c) [9].

Most catalysts used in industrial processes are derived from the metals platinum, palladium, rhodium and ruthenium, which are expensive, toxic and rare. Therefore, more recently, people have started to develop catalysts derived from Earth crust-abundant, less toxic transition metals such as iron, cobalt and nickel, often in the form of organometallic transition metal complexes [10]. Physisorption by van der Waals bonds and covalent chemisorption of CO_2 on single crystal surfaces of various metals has been studied by means of DFT calculations [11]. The results for some important parameters are listed in Table 13.1, and the adsorbed states are depicted schematically in Figure 13.4. There is little variation of the physisorption energy in the range of 21–33 kJ mol^{-1}, and it does not significantly change the structure of the adsorbate. However, the covalent chemisorptive bond is stronger for Ru and Fe, comparable to physisorption for Ir, Rh, Ni and Co, and it is weaker for Pd and Pt where a metastable state is predicted. No energy minimum other than that of the physisorbed state is found for Cu, Ag and Au. Interestingly, chemisorption breaks the symmetry and induces considerable bending of CO_2 that goes along with a transfer of between 0.35 and 1.1 electrons from the metal to the adsorbate. Spontaneous CO_2 bending, covalent binding energy and charge transfer are clearly correlated and largest for a resulting OCO bond angle near 120°, as should be expected based on the electronic structure described in Section 2.

Bending CO_2 to 134° costs about 1.5 eV (144 kJ mol^{-1}), and to 120° costs about 2.9 eV (240 kJ mol^{-1}) as evidenced in Figure 13.1(a) [4]. The chemisorption energies

(a)

(b)

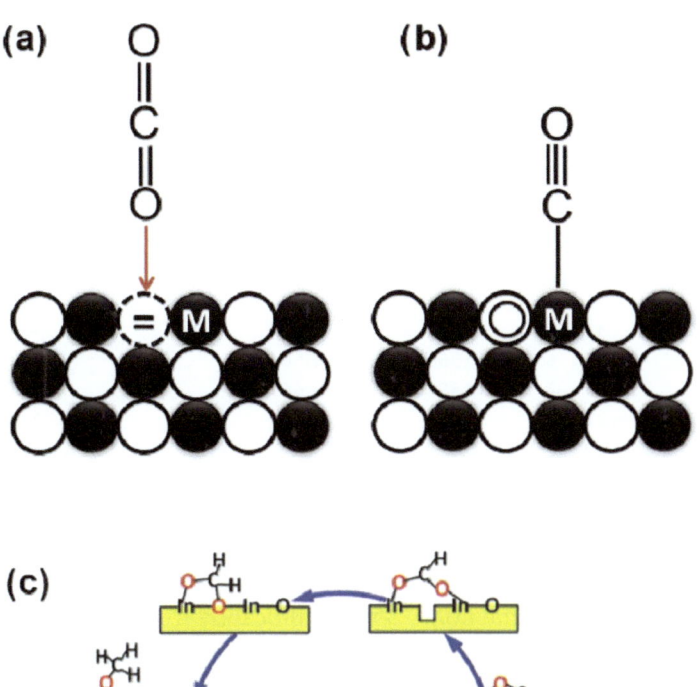

(c)

In$_2$O$_3$

CH$_3$OH

H$_2$O

CO$_2$+3H$_2$

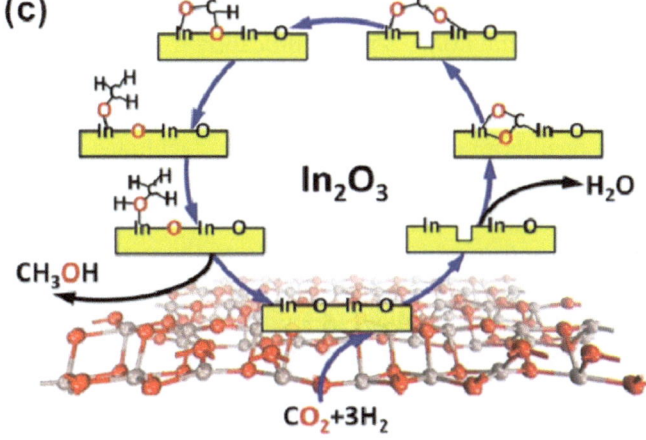

Figure 13.3: Suggested binding of CO$_2$ to a surface oxygen vacancy (a), followed by possible transfer of CO to a neighbouring metal centre M (b), and DFT-based suggestion for thermally catalysed CO$_2$ reduction on In$_2$O$_3$. [(c) Reprinted with permission from Reference 9].

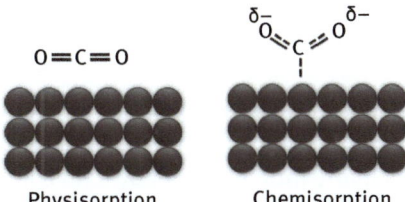

Physisorption
van der Waals bond

Chemisorption
covalent bond

Figure 13.4: Schematic presentation of physisorbed (left) and chemisorbed (right) CO$_2$ on the surface of a metal.

Table 13.1: DFT-calculated van der Waals and covalent adsorption energies, geometries and charge transfer $2\delta^-$ to CO_2 on single crystal surfaces of metals [11].

Metal	Physisorption		Chemisorption		
	Binding energy kJ mol^{-1}	$\angle$ OCO /°	$2\delta^-$ /e	Binding energy kJ mol^{-1}
Fe(110)	−23	121	−1.11	−90
Ir(111)	−33	128	−0.47	−34
Pd(111)	−32	140	−0.35	−17
Ru(0001)	−31	123	−0.83	−61
Rh(111)	−32	134	−0.46	−35
Ni(111)	−26	136	−0.50	−20
Co(0001)	−25	139	−0.64	−30
Pt(111)	−21	131	−0.36	−3
Cu, Ag, Au	−(23−29)	131[*]	−(0.38−0.54)[*]	+(20−40)

* Fixed structural parameters of Pt(111) used for the calculation of the not chemisorbed state.

listed in Table 13.1 are the net energies after bending CO_2 to the indicated angle. Furthermore, bending to 134° lowers the LUMO level significantly from ca. −0.6 eV for the linear molecule down to ca. −2.9 eV [4], which brings it into the range of typical Fermi levels of transition metals and thereby making it accessible for spontaneous electron transfer. The resulting negative charge on CO_2 facilitates protonation at the oxygen as the next step of electroreduction.

The binding energies listed in Table 13.1 [11] are given for the flat geometries of low-index single crystal surfaces, and bonding is directly from the carbon to the metal atom (in most cases on top). The bond strength depends slightly on the type of the facet. Edge, step and corner sites have a lower coordination number and are expected to give rise to stronger bonds, along with a higher degree of bending and electron transfer. These sites are therefore preferred for the reaction, unless the bonding of intermediates or products is sufficiently strong to give rise to poisoning of the active sites.

Alloying is a suitable method to tune the reactivity of the metal to values that can be predicted by interpolating the Fermi levels of the individual components based on their mole fractions [12].

13.3.3 Binding to lone pair moieties

The CO_2 carbon atom is electrophilic and hence reacts with lone pairs of reaction partners. This is well known from the CO_2-capturing mechanism of amines from air, which is subsequently converted directly and efficiently into methanol by a Ru complex catalyst [13]. The simplest reaction is that of ammonia with CO_2 to carbamic acid and further conversion to urea.

$$CO_2 + 2NH_3 \underset{110\,atm}{\overset{160\,°C}{\rightleftharpoons}} H_2NCOONH_4 \overset{160-180\,°C}{\rightleftharpoons} (H_2N)_2C=O + H_2O \qquad (13.1)$$

The pyridinic, pyrrolic and possibly the graphitic nitrogen atoms in N-doped graphene quantum dots have been reported to be active metal-free electrocatalysts in the CO_2 reduction with up to 90% Faradaic efficiency and selectivity for ethylene and ethanol conversion reaching 45% [14]. Coskun et al. reported CO_2 electroreduction catalysed by conductive polydopamine, yielding CO and formate as dominant products [15].

Furthermore, other molecules possess lone pairs of electrons that can react with CO_2. In analogy to ammonia, the water molecule reacts with CO_2 to produce carbonic acid, H_2CO_3, which then reacts further with ammonia to yield ammonium carbamate. Phosphines and sulphur compounds are further candidates. The key criterion is the bond energy, which according to Sabatier's principle should neither be too large nor too small. It is also influenced by the substituents at the atom carrying the lone pair, but more research is required here to identify the most promising compounds for CO_2 conversion to useful products.

13.4 Mechanistic aspects of thermal catalytic versus electrocatalytic activation

As explained in Section 2, CO_2 is a relatively stable molecule and requires large energy input to chemically activate it. Exceptions are the adsorption to oxides (Figure 13.2) and reactions with water to produce carbonic acid and with other molecules containing a lone electron pair (eq. 13.1). Its stability implies that hydrogenation of CO_2 is generally carried out at high temperatures over heterogeneous catalysts. On the other hand, electrochemical methods of CO_2 conversion have various advantages. For example, the conversion of CO_2 to methanol via water electrolysis can be conducted at room temperature. The reaction scheme is as follows:

Anode: $3\,H_2O \rightarrow 6\,H^+ + 6\,e^- + 3/2\,O_2$ $\Delta G° = +711.9\,kJ/mol,\ E° = -1.23\,V$

Cathode: $CO_2 + 6H^+ + 6e^- \rightarrow CH_3OH + H_2O$ $\Delta G° = -15\,kJ/mol,\ E° = +0.026V$

Overall: $2\,H_2O + CO_2 \rightarrow CH_3OH + 3/2\,O_2$ $\Delta G° = +702.5\,kJ/mol,\ E° = -1.21\,V$

$$(13.2)$$

In non-electrochemically catalysed reactions, the activation barriers along the reaction path are overcome by heating to the required temperature. Electrocatalysed reactions have the advantage that they can be conducted at room temperature since the necessary energy is provided by the applied electrical potential [16].

13.4.1 H_2 versus H_2O as hydrogenation agent

Heterogeneously catalysed thermal hydrogenation of CO_2 generally uses H_2, which differs from electrochemical reduction that uses H_2O as a proton source. This is because hydrogenation using H_2O is far too endergonic so that the reaction equilibrium is not on the product side. In electrochemical reactions, the required energy can be supplied electrically, which also makes endergonic reactions accessible.

Hydrogenation of CO_2 via thermal activation primarily produces formic acid, formaldehyde and methanol, while electrochemical reduction follows two-, four-, six- and eight-electron and sometimes even higher-electron reduction pathways to form various hydrocarbons and oxygenated products by tuning the electrochemical potential and the type of electrocatalyst. The major reduction products are carbon monoxide, formic acid, methanol, methane, dimethyl ether, ethylene, as well as other hydrocarbons and fuels [17, 18].

Hydrogenation of CO_2 is an entropically disfavoured process because two different gas phase reactant molecules are consumed. Therefore, CO_2 to formic acid in the gas phase is endergonic ($\Delta G° = +32.8$ kJ mol^{-1}). The formation of methanol is entropically particularly disfavoured, with four gas phase molecules being consumed per reduced CO_2 molecule, as seen in Eqn. 13.3.

$$CO_2(g) + 3H_2(g) \rightleftharpoons CH_3OH(l) + H_2O(l) \tag{13.3}$$

$$\Delta G° = -9.5 \,\text{kJ/mol}; \ \Delta H° = -131 \,\text{kJ/mol}; \ \Delta S° = -409 \,\text{J/mol K}$$

Further disadvantages of thermal hydrogenation, wherein H_2 acts as a hydrogenation agent, include (a) the usage of highly toxic phosphine as a hydrogen source; (b) use of non-aqueous solvents and (c) homogeneous catalysis is difficult to achieve [18].

13.4.2 Gas phase versus liquid phase reaction

Traditionally, the electrochemical reduction of CO_2 is performed in the aqueous phase. The solubility of CO_2 in water is ca. 0.03 M, while it is 10 times larger in acetonitrile. Hence, solubility is a limiting factor in water. Typically, the electrochemical reduction to higher energy products such as methanol, methane and ethanol and higher alcohols is usually inefficient and not selective; the far more common products are CO and formic acid. In this regard, the Latimer–Frost diagram drawn at pH 7 [19] for the multi-electron and multi-proton CO_2 reduction in the aqueous phase can give clues on the preferred reaction intermediates and their stability.

$$CO_2 + e^- \rightleftharpoons CO_2^{•-} \quad E°' = -1.90 \,\text{V} \tag{13.4}$$

$$CO_2 + 2H^+ + 2e^- \rightleftharpoons CO + H_2O \qquad E^{\circ\prime} = -0.53\,V \qquad (13.5)$$

$$CO_2 + 2H^+ + 2e^- \rightleftharpoons HCOOH \qquad E^{\circ\prime} = -0.61\,V \qquad (13.6)$$

$$CO_2 + 4H^+ + 4e^- \rightleftharpoons CH_2O + H_2O \qquad E^{\circ\prime} = -0.48\,V \qquad (13.7)$$

$$CO_2 + 6H^+ + 6e^- \rightleftharpoons CH_3OH + H_2O \quad E^{\circ\prime} = -0.38\,V \qquad (13.8)$$

$$CO_2 + 8H^+ + 8e^- \rightleftharpoons CH_4 + H_2O \qquad E^{\circ\prime} = -0.24\,V \qquad (13.9)$$

$$2H^+ + 2e^- \rightleftharpoons H_2 \qquad E^{\circ\prime} = -0.41\,V \qquad (13.10)$$

A Latimer diagram provides information about the oxidation states of the intermediates, while Frost diagrams connect the relative free energies of these intermediates formed during a redox process. Thus, together, the Latimer–Frost diagram provides insight into the redox stability of the CO_2 electroreduction reaction. In Figure 13.5, the red points are the multi-proton, multi-electron reduction products of CO_2, and the dotted lines correspond to the potentials in eqs. 13.4–13.10. The slope of the line connecting two products/intermediates, that is, the free energy divided by the number of electrons added ($\Delta G/n$) gives the intrinsic thermodynamic activation potential for

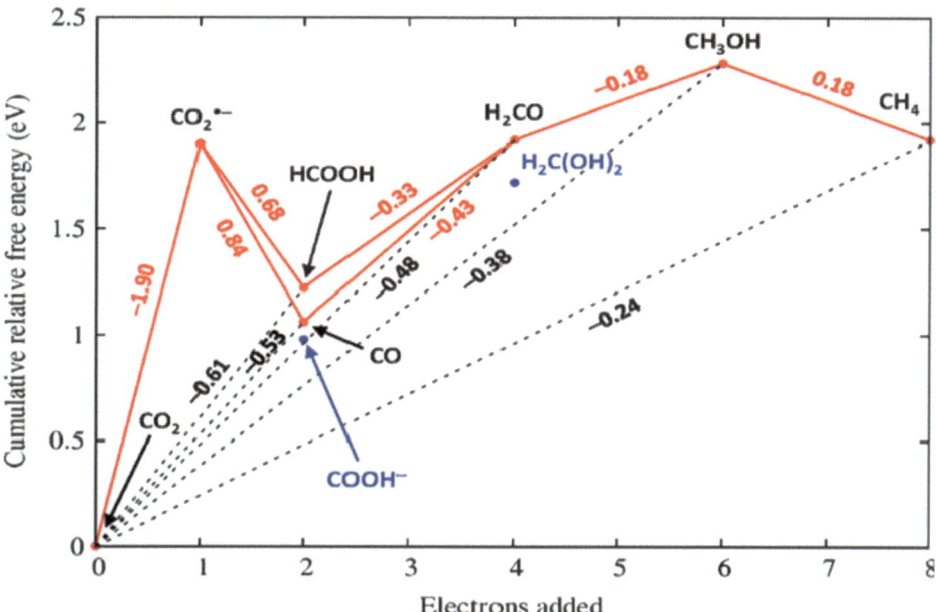

Figure 13.5: Latimer–Frost diagram for the multi-electron, multi-proton reduction of CO_2 in homogeneous aqueous solution at pH 7, based partially on the potentials in eqs. (13.4)–(13.10). Any species lying above the straight line joining two adjacent points (e.g., $CO_2{}^{\bullet-}$ between CO_2 and HCOOH) is thermodynamically unstable with respect to disproportionation. [Reprinted with permission from Reference 19.].

that process. The steeper the slope is, the greater is the energy required for that process. The formation of, for example, $CO_2^{\bullet-}$ from CO_2 is an extremely arduous process. In this diagram, the thermodynamically unstable intermediate that lies above the line connecting two correlated species disproportionates into these two species. For example, formaldehyde, which falls on the line connecting formic acid and methanol, will disproportionate into the latter two as the potential is more positive towards methanol conversion [19].

On the other hand, it is shown that the aqueous phase CO_2 equilibria are less complicated, but the solubility limits the availability of CO_2 at the electrode [20]. Thus, factors that affect the electrochemical reduction of CO_2 in the liquid phase are the solubility and the mass transport effects, which can be overcome by the use of gas phase CO_2 directly as a feedstock.

13.4.3 H transfer versus sequential electron–proton transfer

Thermally activated reactions generally occur by transfer of atomic H. In contrast, water electrolysis produces protons and electrons at the anode at an applied potential of 1.23 V, which then travel through the electrolyte medium and circuit, respectively, to be reduced at the cathode to H_2 or to be attached to CO_2. For example, methanol formation on copper may occur via two reaction pathways (Eqns. 13.11 and 13.12) involving proton and electron transfer, as derived in analogy to Albo et al. [17]:

$$CO_2 + e^- + H^+ \text{ (at cathode surface)} \rightleftharpoons HO^{\bullet}CO \text{ (carboxyl radical)}$$
$$HO^{\bullet}CO + e^- + H^+ \rightleftharpoons H_2O + CO \text{ (carbon monoxide)}$$
$$CO + e^- + H^+ \rightleftharpoons HCO$$
$$HCO + e^- + H^+ \rightleftharpoons CH_2O \text{ (formaldehyde)}$$
$$CH_2O + 2e^- + 2H^+ \rightleftharpoons CH_3OH$$

$$\text{Overall: } CO_2 + 6e^- + 6H^+ \rightleftharpoons CH_3OH + H_2O \quad (13.11)$$

$$CO_2 + e^- + H^+ \rightleftharpoons HCOO^{\bullet} \text{ (formyl radical)}$$
$$HCOO^{\bullet} + e^- + H^+ \rightleftharpoons HCOOH \text{ (formic acid)}$$
$$HCOOH + e^- + H^+ \rightleftharpoons HCO + H_2O$$
$$HCO + e^- + H^+ \rightleftharpoons CH_2O \text{ (formaldehyde)}$$
$$CH_2O + 2e^- + 2H^+ \rightleftharpoons CH_3OH$$

$$\text{Overall: } CO_2 + 6e^- + 6H^+ \rightleftharpoons CH_3OH + H_2O \quad (13.12)$$

Equation 13.11 proceeds via the carboxyl radical (HO$^\bullet$CO) production by the transfer of a proton–electron pair to the oxygen of adsorbed CO_2, which is the rate-determining step [21], while eq 13.12 proceeds via hydrogenation at the carbon under formation of the formyl radical [22].

13.4.4 The effect of an applied bias voltage in electrocatalytic reactions

An electrode is more than just a surface at which electrons are exchanged. Rather, in the same way as in heterogeneous catalysis, details of its structure and electronic properties play a decisive role in how a substrate is bound and what the details of the reaction mechanism are [16].

The application of an electrochemical potential results in electrochemical work done on the system, which helps to overcome the activation barrier of an electron transfer reaction. Basically, this is the analogue of supplying the necessary energy by heating the reaction system under thermal reaction conditions [16]. Increasing the applied potential therefore means that new reaction paths open when their activation energy is reached, which is seen by new products appearing in competition with others. Potential-dependent product distributions have been observed in CO_2 electroreduction experiments on copper electrodes, with hydrogen being observed at low applied potentials, followed by formic acid and CO, and methane and ethene at even higher potentials [23, 24].

Although the Nernst potential for the conversion of CO_2 to its radical anion form is −1.9 V versus NHE, the actual potential required to achieve this, termed the *onset potential*, is higher because of the activation of the electrochemical reaction. Many transition metals have been successfully employed for CO_2 electroreduction, wherein their vacant orbitals and active *d-electrons* facilitate the bonding between the metals and CO_2 [17]. Metallic, non-metallic and molecular catalysts have been the three major categories of catalysts employed for CO_2 electroreduction [25]. Copper and copper-based electrodes have been widely used due to their high probability of obtaining hydrocarbons and alcohols [17, 26]. On the other hand, copper is not selective to one particular product but gives a series of products at low current densities and hence low Faradaic efficiencies [24]. However, it has been reported that the stability of the catalyst is only 4 hours [27, 28]. Until today, CO and HCOOH have been the most common products formed during the CO_2 electroreduction, and there is a large literature present reporting the Faradaic efficiencies of these two products [19, 29]. For example, Au, Ag, Zn, Ni, Pd and Ga are shown to produce CO, while Pd, Hg, In, Sn, Cd, Tl favour HCOOH formation [17].

It is generally known that the bond strength between the metal and CO_2 governs the product selectivity to CO [25]. CO is the major product formed when it binds weakly to the metallic catalyst, so that it is released from the surface before further reduction. It was further suggested that the key to go beyond CO to form other hydrogenated products is that the adsorbed CO must be protonated further to adsorbed CHO. If CHO can be stabilised better than CO, the necessary onset potential is expected to decrease, thereby leading to a more energy-efficient reduction process. Thus, electrocatalysts that bind with CO and CHO with similar tenacity are expected to be better catalysts for CO_2 electroreduction leading to formic acid and higher analogues of carbon [21, 30].

A detailed DFT calculation for CO_2 electroreduction on a copper (211) stepped surface leads to a mechanism that is partly different from those given in equations 13.11 and 13.12 [21]. The applied electrochemical potential is included in the calculation using the computational hydrogen electrode where the electrochemical work $\Delta G°$ of the transfer of an electron across a potential of $E = 1$ V corresponds to 1 eV in each reaction step, or $\Delta G° = -nF \times E = -n \times 96.5$ kJ mol^{-1} (F is the Faraday constant and n the number of transferred electrons), thus effectively lowering the energy of the adsorbed product state of the corresponding step. The calculation finds that the formation of the carboxyl species in the rate-limiting step is the precursor of formic acid. A second pathway that opens simultaneously with HCOOH formation is CO formation, which is in agreement with early experiments [23]. It also goes through carboxyl in the same rate-determining step, but it is followed by cleaving off H_2O to form weakly bound CO, as in eq. 13.11, which if not desorbed evolves further to CH_4 or multiple carbon products (Figure 13.6). Methanol formation was not found from CO_2 electroreduction on copper, neither in the experiment from Hori et al. [23] nor in the

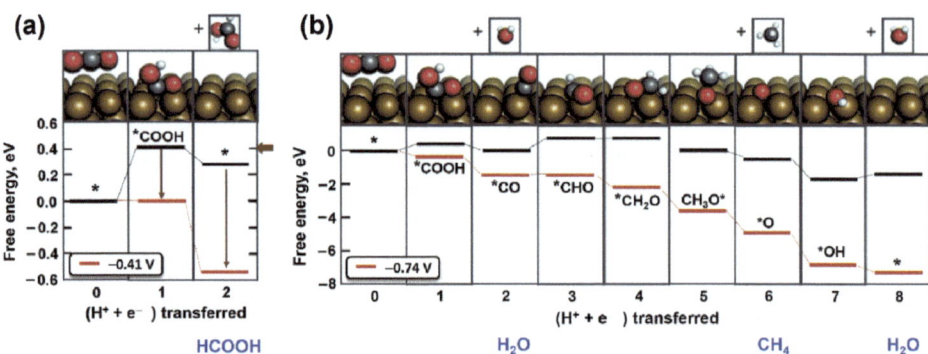

Figure 13.6: DFT calculations of the lowest energy intermediates of CO_2 electroreduction on Cu(211) in the absence of an applied potential (black) and for an applied potential of −0.41 V (red) that drives the reaction towards the HCOOH product (a), and for an applied potential of −0.74 V that drives the reaction to H_2O and CH_4 (b). * represents the energy of the empty adsorption site with the desorbed species, X* that of the adsorbed intermediate. The offset between the black and the red reaction paths represents the accumulated electrochemical work for the transfer of n electrons. Transition states between the intermediates may present additional barriers but were not calculated [adapted from reference 21].

calculations. However, in the non-electrochemical thermal reduction with H_2, methanol is the major product, and the reaction can proceed on Cu(211) via both a formic acid or a CO intermediate [31].

The cumulative electrochemical work conducted by transporting n electrons is shown by the red lines in Figure 13.6. In favourable cases the potential can be chosen such that the profile has a continuously descending slope. The energetics of non-electrochemical steps such as H_2O elimination or dimerisation reactions is not influenced by the applied potential, except that the electrical field polarises the adsorbate and changes the strength of the bond to the catalyst. This may actually prevent desorption and contribute to electrode blocking [16]. Each additional step makes it more tedious to find a single catalyst that can catalyse all steps under the same conditions.

The situation of a continuously descending potential is not normally given; rather, as illustrated in Figure 13.6, it is corrugated [21]. Deep energy minima along a reaction path cause the intermediate to remain bound, blocking the active site and slowing down the process. On the other hand, high-energy intermediates may desorb so that the reduction is terminated before the final product is reached. A typical case is the desorption of CO from metals such as Au, Ag, Zn or Pd to which it is only weakly bound, while on electrodes consisting of Pd, Sn, Hg, Pb, Cd and In the major detected product is formic acid or formate [32]. It is known that CO is formed at low overpotentials, while the reaction proceeds to formic acid at higher overpotentials. This is because in aqueous medium $CO_2^{\bullet-}$ formation is expected to be the rate-determining step, and the metals that can bind stably to $CO_2^{\bullet-}$ produce CO while the metals that weakly bind $CO_2^{\bullet-}$ on the metal surface proceed to formic acid production [33]. Fe, Ni, Co and Pt bind CO strongly and block the catalyst surface, which leads to competing hydrogen evolution reaction (HER). Cu has a special status, with its binding energy to CO strong enough to allow further hydrogenation but not strong enough to cause blocking. It is thus so far the only metal that leads to the formation of hydrocarbons [30].

CO_2 reduction to non-radical products occurs in steps of 2, 4, 6, or 8 electron–proton pairs (hydrogen atoms) to formic acid, CO (after water elimination), formaldehyde, methanol and methane, respectively. Furthermore, besides these C_1 products, various C_2 products like ethane, ethanol, glycolaldehyde, acetaldehyde, acetic acid, ethylene glycol and also C_3 products are observed in minor yields in electrocatalytic reactions [32]. The product selectivity depends on a number of experimental parameters such as the type and morphology of electrocatalyst, its crystal face or edge and corner binding sites, the type of solvent (aqueous, non-aqueous, or none) and the value of the applied electrochemical potential. These parameters determine how strongly a substrate is bound to the catalyst or whether it desorbs, and since the substrate changes during reaction the situation is different for each step. The entire process thus represents a rather complicated system.

In electrocatalytic reactions we should also consider radical intermediates that arise from CO_2 reduction by 1, 3, 5 or 7 electron–proton pairs. As free radicals they would represent a high-energy state because of the unsatisfied valence. However, on the catalyst surface the unpaired electron will actually be used to make the bond to the surface, which stabilises the intermediate and prevents it from desorption. No consistent trend is observed between the intermediates of odd and even electron–proton reduction steps in the energy profile of Figure 13.6.

The detailed mechanism that determines the product selectivity to CO, CH_4 and CH_3OH is still under discussion, as described in a competent recent review by Zhu et al. [32]. In contrast to the standard non-electrochemical high-temperature methanol formation on $Cu/ZnO/Al_2O_3$ system [31], several electrochemical studies found that formic acid (HCOOH) is not further reduced. It was therefore concluded that it cannot be an intermediate of methane and methanol formation on Cu [32], and *CO was proposed to be the key intermediate instead. The following mechanisms were proposed by the groups of Bard [34] and of Koper [35]:

$$^*CO_2 \rightarrow {}^*COOH \rightarrow {}^*CO \rightarrow {}^*CHO \rightarrow {}^*CH_2O \rightarrow {}^*CH_3O \rightarrow CH_3OH \qquad (13.13)$$

$$\rightarrow {}^*COH \rightarrow {}^*C \rightarrow {}^*CH \rightarrow {}^*CH_2 \rightarrow {}^*CH_3 \rightarrow CH_4 \qquad (13.14)$$

$$\rightarrow {}^*CHO \rightarrow {}^*CH_2 \rightarrow CH_3 \rightarrow CH_4 \qquad (13.15)$$

Copper is the most prominent catalyst that leads to the formation of C_2 and higher products. It is clear that C_2 species require the dimerisation of adsorbed intermediates (perhaps *CO, *CHO or *CH_2O); these must therefore possess translational mobility on the catalyst surface.

13.5 CO_2 activation in photosynthesis: The dark reaction in the Calvin cycle

The most extensive CO_2 utilisation occurs in natural photosynthesis. Terrestrial vegetation binds an amount of ca. 120 Gt (gigatons) of carbon annually from the atmosphere, and it emits the same amount when plants degrade. By comparison, burning fossil fuels adds an amount of only 6.4 Gt per year to the atmosphere [36].

Many of the physiological processes are redox reactions, and thus there is a close analogy between photosynthesis and electrochemical CO_2 conversion. Photosynthesis occurs in two separate steps. The first one is the water-splitting reaction, the analogue of the electrochemical anode reaction. It is the energy-intensive step that occurs under absorption of light. The green chlorophyll pigments absorb two photons at an energy near 680 nm for each electron that is liberated in the water oxidation reaction. While the oxygen is dissipated to the atmosphere, the hydrogen is used to reduce nicotinamide adenosine dinucleotide phosphate ($NADP^+$) to NADPH and to synthesise adenosine triphosphate (ATP) from the ADP diphosphate. NADPH and

ATP serve as intermediate energy storage molecules. They are consumed in the second step of photosynthesis, the Calvin cycle, which is light-independent and therefore also called the *dark reaction*. This is the CO_2 fixation and conversion part, the analogue of the electrochemical cathode reaction.

Ribulose 1,5-bisphosphate carboxylase/oxygenase (RuBisCO) is a crucial enzyme in carbon fixation and the most abundant enzyme on earth. [37] The basis for RuBisCO activation is thought to be the carbamylation of the ε-amino group of a lysyl residue on the enzyme's catalytic subunit [38]. This initial step of the Calvin cycle is the CO_2 capture process at lone-pair moieties similar to the description in Section 3.3 (eq. 13.1).

Three five-carbon sugar ribulose bisphosphate (RuBP) units then incorporate one CO_2 each from the carbamate and form a six-carbon molecule. In the next step, the RuBisCO enzyme splits this unit into two identical three-carbon phosphoglycerate units, initiated by the nucleophilic addition of a water molecule and using the energy of six ATP and six NADPH molecules. One of the fragments leaves the cycle and becomes half a glucose sugar molecule; the other five remain in the cycle and regenerate to RuBP by means of three ATP molecules. In summary, one cycle converts three CO_2 molecules to half a glucose molecule using the hydrogens of six NADPH and the energy of nine ATP molecules [39].

13.6 Preventing the formation of H_2 in electrochemical reactions

Hydrogen evolution competes severely during the CO_2 electroreduction reaction and hence must be suppressed to increase selectivity to the desired non-H_2 product. Most of the CO_2 electroreduction reactions are conducted in aqueous medium, and the HER competes with CO_2 reduction because of their comparable standard reduction potentials (Table 13.2).

Table 13.2: Free energies of reaction (ΔG^o) and corresponding standard cell potentials (E^o) for co-electrolysis of water and CO_2 to various products under transfer of n electrons.

Reaction	ΔG^o /kJ mol^{-1}	n	E^o /V
$H_2O_{(l)} \rightarrow H_2 + \frac{1}{2} O_2$	+237	2	−1.229
$CO_2 + H_2O_{(l)} \rightarrow HCOOH_{(l)} + \frac{1}{2} O_2$	+286	2	−1.482
$CO_2 + H_2O_{(l)} \rightarrow CO_{(g)} + \frac{1}{2} O_2 + H_2O_{(l)}$	+257	2	−1.332
$CO_2 + 2H_2O_{(l)} \rightarrow CH_2O_{(g)} + O_2 + H_2O_{(l)}$	+521	4	−1.350
$CO_2 + 3H_2O_{(l)} \rightarrow CH_3OH_{(l)} + 3/2 O_2 + H_2O_{(l)}$	+702	6	−1.213
$CO_2 + 4H_2O_{(l)} \rightarrow CH_{4(g)} + 2O_2 + 2H_2O_{(l)}$	+817	8	−1.058

The thermodynamic standard cell potential for water electrolysis amounts to -1.229 V. However, the reaction is activated, and the onset potential at which current starts to evolve, indicating that electrochemical conversion sets in, is of the order -1.5 V. Technically, the process is carried out under acidic conditions in PEM electrolysers at cell voltages of 1.75–2.20 V and current densities of 0.6–2 A cm^{-2}.

In the co-electrolysis of water with CO_2, the protons and electrons are expected to directly reduce the CO_2 at a suitable cathode. We see from Table 13.2 that $E°$ for the formation of methanol or methane is below that for hydrogen formation. Thus, these reactions are thermodynamically favoured over hydrogen formation. However, the intermediates of these multi-step reactions are of considerably higher energy, and their formation is thermodynamically disfavoured against H_2. There are two corrections to be taken into account to this simple view: (i) The entries for the HCOOH, CO, CH_2O and CH_3OH are for the *free* molecules. Binding to the electrode will lower the energy and thus reduce the activation energies and concomitant potentials. (ii) We have neglected the odd electron intermediates and the possible charged intermediates discussed in Section 4.3. They represent additional complications that may well lead to higher activation energies and more negative onset voltages. The two corrections have opposite sign, and they are of course very significantly electrode dependent.

In this case, we have three options, all of them on the cathode side since the anode reaction remains the same as in standard water electrolysis.

(i) We can try to improve on the cathode catalyst material to reduce the necessary cell voltage. For example, we could try to find a catalyst material that permits the first intermediate to be CO and not HCOOH, implying a different reaction path with non-electrochemical dissociation of CO_2 or electrochemically assisted formation of an unstable transient to HCOOH that loses H_2O. However, CO has to remain bound for further reduction to the desired product. Such catalyst development can be empirical by trial and error, experience and inspiration or it can be realised with the help of quantum chemical calculations.

(ii) We have to find materials in the cathode compartment that inhibit hydrogen evolution by increasing the overpotential to its formation without impeding the desired reaction. It has been proposed that we look for materials in a volcano plot for HER as far as possible to the left or the right from the maximum. Thereby, materials with weak hydrogen binding should be preferred over those with strong binding, as the latter ones would block the surface for the desired catalytic reduction [40]. However, we should keep in mind that HER may occur not only on the surface of the nominal electrocatalysts but prominently also on many carbon support materials and at current collectors or other electrically connected parts in the cathode compartment. Low Faradaic efficiencies for H_2 were generally found for electrodes consisting of Hg, Zn, In, Sn, Pb, Bi and notably also on Cu, Ag and Au. The results also depended on conditions such as pH and crystal facets [various authors in 20].

(iii) We may simply accept the evolved H_2 as a valuable product instead of an unwanted by-product.

In detail, the following further points play a major role in suppression of the HER:

a) It has been found that acid in the reaction media enhances HER, thereby ag-
gressively competing with CO_2 reduction. One method to overcome this has
been proposed by Ogura [40] where the use of halide ions as an electrolyte during CO_2 reduction on copper, via specific adsorption of the halide ions at the
double layer formed at the electrode/electrolyte interface that effectively suppressed the HER (Figure 13.7).

Electrode

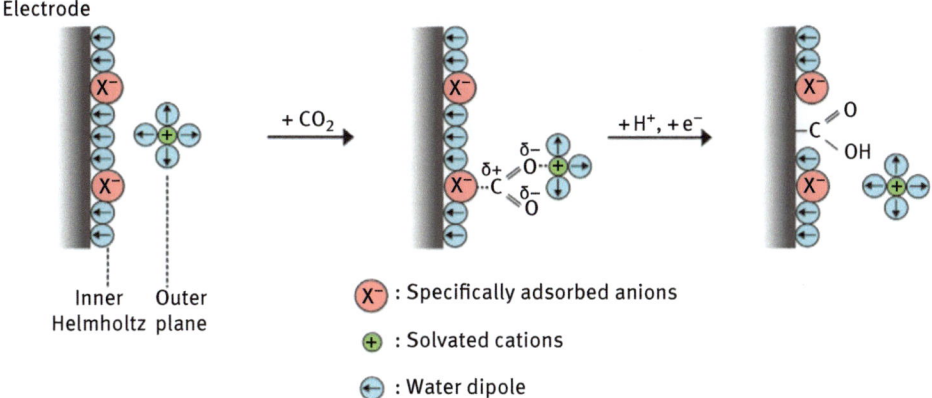

Inner   Outer
Helmholtz plane

X^- : Specifically adsorbed anions

$+$: Solvated cations

$\ominus$: Water dipole

Figure 13.7: CO_2 attraction by electric double layer and formation of the formate radical via an
inner-sphere mechanism [reprinted with permission from Reference 40].

b) On the other hand, basic pH plays an important role as well in the reduction of
CO_2 as it cannot survive in basic media [17], for example, at pH 14 due to the
following processes occurring at the cathode [41]:

$$CO_2(aq) + H_2O + 2e^- \rightarrow HCOO^- + OH^- \qquad E° = -1.02\,V \text{ vs NHE} \qquad (13.16)$$

$$2H_2O + 2e^- \rightarrow 2H_2 + 2OH^- \qquad\qquad E° = -0.83\,V \text{ vs NHE} \qquad (13.17)$$

In basic media, eq. 13.17 is thermodynamically favoured over formate formation
(eq. 13.16), which is only kinetically favoured [20]. Generally, cathode materials of
Hg, In, Pb and Sn favour CO_2 reduction over HER due to their high overpotential
for HER.

c) In general, bimetallic and multi-metallic catalysts are expected to display better
catalytic activity in comparison with their mono-metallic counterparts. For example, Sn supresses HER and hence has been alloyed with Cu [29, 42]. On the

other hand, Pd–Sn was shown to completely suppress CO and H_2 formation, leading to formate being the major CO_2 reduction product [43]. In a different study, ethanol formation was highly enhanced by doping zinc with copper [44]. Thus, a take-home message here is that the $metal_1$–$metal_2$ interaction plays an important role in augmenting a specific catalytic process.

d) Structural and morphological alteration of the electrocatalyst affects competing HER and enhances CO_2ER to form high-energy density products [45]. For example, structurally ordered Cu–Pd alloy favours CH_4 formation where the ordered Cu structure adsorbs CO better to be hydrogenated to CHO. This indicates that Cu may be tuned to create products other than mainly CO and HCOOH, and this with enhanced Faradaic efficiencies (> 80%) [29]. This is also true in the cases where the nanoparticles of molybdenum phosphide supported on In-doped porous carbon gave HCOOH with a Faradaic efficiency reaching 96.5% [46].

e) Consideration must be given to metal-free catalysts owing to their ease of preparation, tailorable structures and high surface area and ease in mixing with other catalysts. For example, N-containing organic catalysts can show preferential adsorption, thereby enhancing the chances for H_2 addition [47, 48].

f) Alternative routes to suppressing the HER could be operating the cell at high CO_2 pressure, temperature and to use a solid polymer electrolyte such as Nafion, which may serve to increase the availability of CO_2 at the electrode.

In a nutshell, the key to reduce hydrogen formation in efficient CO_2 electroreduction is to design a stable electrocatalyst with high overpotential for HER and good affinity for CO_2.

13.7 Concluding remarks

Heterogeneously catalysed CO_2 hydrogenation to methanol using molecular hydrogen at elevated temperature and pressure is a standard technical process. However, CO_2 recycling to liquid fuels or other value products makes no sense when the energy required to make H_2 and to conduct the reaction is obtained by burning fossil fuels. The renewable energies of choice at the required large scale are solar photovoltaic and wind energies, and they are available as electrical energy. Furthermore, strongly endergonic reactions are difficult to carry out thermally because equilibrium disfavours the products, but it is straightforward to supply the energy in the form of electricity, even at room temperature [16]. Therefore, electrochemical CO_2 reduction is a hot and highly competitive subject worldwide.

Currently, most CO_2 electroreduction experiments are conducted in the liquid phase where they are hampered by the CO_2 solubility and the diffusion of reactants and products. The current state of the art for electrochemical hydrocarbon synthesis is still copper, which requires 1 V of overpotential to produce 10 milliamps per cm^2

geometric electrode surface for products further reduced than CO [49]. At this point, the limitation is in most cases the cathode reaction. However, current densities of technical interest are on the order of 1 A cm^{-2} as they are found in PEM fuel cells and PEM electrolysers. To achieve this, the transition will have to be made to cells in which CO_2 is supplied in the gas phase, as it is standard in hydrogen fuel cells. This will eliminate the problems of concentration and diffusion. Furthermore, at high current densities the minimisation of Ohmic resistance in the electrolyte by reducing the anode–cathode distance to 50–100 µm in a membrane setup will be vital.

References

[1]   M. Aresta, Ed., CO_2 Chemistry: Advances in Inorganic Chemistry. Vol. 66, Elsevier Inc., UK, 2014.
[2]   Gutsev GL, Bartlett RJ, Compton RN, Electron affinities of CO_2, OCS and CS_2. J Phys Chem 1998, 108, 6756–6762.
[3]   Koppenol WH, Rush JD, Reduction Potential of the $CO_2/CO_2^{\bullet-}$ Couple. A Comparison with other C_1 Radicals. J Phys Chem 1987, 91, 4430–4431.
[4]   Mondal B, Neese F, Ye S, Bio-inspired insights into CO_2 reduction. Current Opinion Chem Biol 2015, 25, 103–109.
[5]   Álvarez A, Corral Pérez JJ, Olcina JG, Hu L, Cornu D, Huang R, Stoian D, Urakawa A, CO_2 activation over catalytic surfaces. ChemPhysChem, 2017, 18, 3135–3141.
[6]   Ogura K, Ferrell III JR, Cugini AV, Smotkin ES, Salazar-Villalpando MD, CO_2 attraction by specifically adsorbed anions and subsequent accelerated electrochemical reduction. Electrochim Acta 2010, 56 381–386.
[7]   Horiuchi T, Hidaka H, Fukui T et al., Effect of added basic metal oxides on CO_2 adsorption on alumina at elevated temperatures. Applied Catalysis A: General 1998, 167, 195–202.
[8]   Martin O, Martín AJ, Mondelli C, et al. Indium Oxide as a Superior Catalyst for Methanol Synthesis by CO_2 Hydrogenation. Angew Chem Int Ed 2016, 5, 6261–6265.
[9]   Ye J, Liu C, Mei D, Ge Q, Active oxygen vacancy site for methanol synthesis from CO_2 hydrogenation on In_2O_3(110): A DFT study, ACS Catal 3, 2013, 1296–1306.
[10]   Stephan DW, Dogma-breaking catalysis. Nature, 2018, 553, 160–162.
[11]   Ko J, Kim BK, Han JW, Density Functional Theory Study for Catalytic Activation and Dissociation of CO_2 on Bimetallic Alloy Surfaces. J Phys Chem C 2016, 120, 3438–3447.
[12]   Jacobsen CJH, Dahl S, Clausen BS, Bahn S, Logadottir A, Nørskov JK, Catalyst design by interpolation in the periodic table: bimetallic ammonia synthesis catalysts. J Am Chem Soc 2001, 123, 8404–8405.
[13]   Kothandaraman J, Goeppert A, Czaun M, Olah GA, Surya Prakash GK, Conversion of CO_2 from air into methanol using a polyamine and a homogeneous ruthenium catalyst. J Am Chem Soc 2016, 138, 778–781.
[14]   Wu J, Ma S, Sun J, et al. A metal-free electrocatalyst for carbon dioxide reduction to multi-carbon hydrocarbons and oxygenates. Nature Commun 2016, 7, 13869–13875.
[15]   Coskun H, Aljabour A, De Luna P, et al. Biofunctionalized conductive polymers enable efficient CO_2 electroreduction. Sci Adv 2017, 3, *e1700686*.
[16]   Roduner E, Selected fundamentals of catalysis and electrocatalysis in energy conversion reactions - A tutorial. Catalysis Today 2018, 309, 263–268.

[17] Albo J, Alvarez-Guerra M, Castaño P, Irabien A, Towards the electrochemical conversion of carbon dioxide to methanol. Green Chem 2015, 17, 2034.

[18] Wang W-H, Himeda Y, Muckerman JT, Manbeck GF, Fujita E, CO_2 Hydrogenation to Formate and Methanol as an Alternative to Photo- and Electrochemical CO_2 Reduction. Chem Rev 2015, 115, 12936–12973.

[19] Schneider J, Jia H, Muckerman JT, Fujita E, Thermodynamics and kinetics of CO_2, CO, and H^+ binding to the metal centre of CO_2 reduction catalysts. Chem Soc Rev 2012, 41, 2036–2051.

[20] J. Qiao, Y. Liu, J. Zhang, Eds. Electrochemical reduction of carbon dioxide. CRC Press Boca Raton, 2016.

[21] Peterson AA, Abild-Pederson F, Studt F, Rossmeisl J, Norskov JK, How copper catalyzes the electroreduction of carbon dioxide into hydrocarbon fuels. Energy Environ Sci 2010, 3, 1311–1315.

[22] Yang Y, Evans J, Rodriguez JA, White MG, Liu P, Fundamental studies of methanol synthesis from CO_2 hydrogenation on Cu(111), Cu clusters, and Cu/ZnO(0001). Phys Chem Chem Phys 2010, 12, 9909–9917.

[23] Hori Y, Murata A, Takahashi R, Formation of hydrocarbons in the electrochemical reduction of carbon dioxide at a copper electrode in aqueous solution, J Chem Soc. Faraday Trans. 1989, 85, 2309–2326.

[24] Kuhl KP, Cave ER, Abram DN, Jaramillo TF, New insights into the electrochemical reduction of carbon dioxide on metallic copper surfaces. Energy Environ Sci 2012, 5, 7050–7059.

[25] Lu Q, Jiao F, Electrochemical CO_2 reduction: Electrocatalyst, reaction mechanism, and process engineering. Nano Energy 2016, 29, 439.

[26] Le M, Ren M, Zhang Z, Sprunger PT, Kurtz RL, Flake JC, Electrochemical Reduction of CO_2 to CH_3OH at Copper Oxide Surfaces. J Electrochem Soc 2011, 158, E45–E49.

[27] Chang TY, Liang RM, Wu PW, Chen JY, Hsieh YC, Electrochemical reduction of CO_2 by Cu_2O-catalyzed carbon clothes. Mat Lett 2009,63, 1001.

[28] J. Albo, A. Saez, J. Solla-Gullon, V. Montiel, A. Irabien, Production of methanol from CO_2 electroreduction at Cu_2O and Cu_2O/ZnO-based electrodes in aqueous solution. Appl Cat B: Environmental 2015, 176–177, 709–717.

[29] Pander III JE, Ren D, Huang Y, Loo NWX, Hui S, Hong L, Yeo BS, Understanding the Heterogeneous Electrocatalytic Reduction of Carbon Dioxide on Oxide-Derived Catalysts. ChemElectroChem 2018, 5, 219–237.

[30] He J, Johnson NJJ, Huang A, Berlinguette CP, Electrocatalytic alloys for CO_2 reduction, Chem Sus Chem 2018, 11 48–57.

[31] Behrens M, Studt F, Kasatkin I, Synthesis over Cu/ZnO/Al_2O_3 Industrial Catalysts. Science 2012, 336, 893–897.

[32] Zhu DD, Liu JL, Qiao SZ, Recent Advances in Inorganic Heterogeneous Electrocatalysts for Reduction of Carbon Dioxide. Adv Mater 2016, 28, 3423–3452.

[33] Hori Y, Wakebe H, Tsukamoto T, Koga O, Electrocatalytic Process of CO selectivity in Electrochemical Reduction of CO2 at Metal Electrodes in Aqueous Media. Electrochim Acta 1994, 39, 1833–1839.

[34] DeWulf DW, Jin T, Bard AJ, Electrochemical and surface studies of carbon dioxide reaction to methane and ethylene at copper electrodes in aqueous solutions. J Electrochem Soc 1989, 136, 1686–1691.

[35] Schouten KJP, Kwon Y, van der Ham CJM, Qin Z, Koper MTM, A new mechanism for the selectivity to C_1 and C_2 species in the electrochemical reduction of carbon dioxide on copper electrodes. Chem Sci 2011, 2, 1902–1909.

[36] Pachauri RK, Reisinger A, Eds. Intergovernmental Panel on Climate Change, Assessment Report 4 (IPCC AR4), Geneva 2007. (Accessed: 08 August 2018)

[37] Stec B, Structural mechanism of RuBisCo activation by carbamylation of the active site lysine. Proc Natl Acad Sci, 2012, 109, 18785–18790.

[38] Lorimer GH, Miziorko HM, Carbamate formation on the ε-amino group of a lysyl residue as the basis for the activation of Ribulosebisphosphate carboxylase by CO_2 and Mg^{2+}. Biochem 1980, 19, 5321–5328.

[39] https://www.khanacademy.org/science/biology/photosynthesis-in-plants/the-calvin-cycle-reactions/a/calvin-cycle (Accessed 08 August 2018)

[40] Back S, Kim H, Jung Y, Selective heterogeneous CO_2 electroreduction to methanol. ACS Catal 2015, 5, 965–971.

[41] Ogura K, Electrochemical reduction of carbon dioxide to ethylene: Mechanistic approach. J CO_2 Util 2013, 1, 43–49.

[42] Ryu J, Andersen TN, Eyring H, The Electrode Reduction Kinetics of Carbon dioxide in Aqueous Solution. J Phys Chem 1972, 76, 3278–3286.

[43] Sarfraz S, Garcia-Esparza AT, Jedidi A, Cavallo L, Takanabe K, Cu–Sn Bimetallic Catalyst for Selective Aqueous Electroreduction of CO_2 to CO. ACS Catal 2016, 6, 2842–2851.

[44] Bai X, Chen W, Zhao C, et al. Exclusive Formation of Formic Acid from CO_2 Electroreduction by a Tunable Pd-Sn Alloy. Angew Chem Int Ed 2017, 56, 12219–12223.

[45] Ren D, Ang BS-H, Yeo BS, Tuning the Selectivity of Carbon Dioxide Electroreduction toward Ethanol on Oxide-Derived Cu_xZn Catalysts. ACS Catal 2016, 6, 8239–8247.

[46] Gawande MB, Goswami A, Asefa T, et al. Core-shell nanoparticles: synthesis and applications in catalysis and electrocatalysis. Chem Soc Rev 2015, 44, 7540–7590.

[47] Sun X, Lu L, Zhu Q, Wu C, Yang D, Chen C, Han B, MoP Nanoparticles Supported on Indium-Doped Porous Carbon: Outstanding Catalysts for Highly Efficient CO_2 Electroreduction. Angew Chem Int Ed 2018, 57, 2427–2431.

[48] Tornow CE, Thorson MR, Ma S, Gewirth AA, Kenis PJA, Nitrogen-Based Catalysts for the Electrochemical Reduction of CO_2 to CO. J Am Chem Soc 2012, 134, 19520–19523.

[49] X. Duan, J. Xu, Z. Wei, J. Ma, S. Guo, S. Wang, H. Liu, S. Dou, Metal-Free Carbon Materials for CO_2 Electrochemical Reduction. Adv Mater 2017, 29, 1701784.

[50] Montoya JH, Seitz LC, Chakthranont P, Vojvodic A, Jaramillo TF, Nørskov JK, Materials for solar fuels and chemicals. Nature Mat 2017, 16, 70–81.

L. Pastor-Pérez, E. le Saché and T.R. Reina

14 Gas phase reactions for chemical CO_2 upgrading

14.1 CO_2-reforming reactions

Synthetic gas is a building block for the production of chemicals. When fed to further conversion processes, valuable products from diesel and naphthalene to acetic acid and formaldehyde can be obtained. The Fischer–Tropsch synthesis (FTS) in particular allows the production of a variety of hydrocarbon fractions. The FTS is a polymerisation reaction, involving the adsorption of the reactants, the initiation of a hydrocarbon chain and finally the chain growth termination. Therefore depending on the syngas composition, the catalyst used and the reaction conditions, alkanes, alkenes and alcohols of various length are produced [1]. In that sense, greenhouse gases and biomass can be upgraded to gasoline, diesel fuel, jet fuel or waxes through the FTS.

The direct synthesis of methanol is another highly valuable syngas conversion process. Methanol is a versatile chemical that can be easily upgraded to dimethyl ether, acetic acid or formaldehyde. The direct synthesis of methanol from syngas requires a H_2/CO ratio of about 2 [2], while the hydroformylation process needs a H_2/CO ratio of 1. On the other hand, the FTS usually requires a H_2/CO syngas ratio of 2, but lower H_2/CO ratios can improve the selectivity towards long-chain hydrocarbons [3]. Hence depending on the process, the composition of syngas may vary.

The most commonly used technology to produce syngas is the steam reforming of methane (SRM, eq. 14.1) producing a hydrogen-rich syngas with a H_2/CO ratio of about 3. The dry reforming of methane (DRM, eq. 14.2) on the other hand, uses CO_2 as oxidant and leads to syngas with a H_2/CO ratio of maximum 1. DRM has the advantage of utilising two of the most abundant greenhouse gases and hence has been increasingly investigated as a CO_2 recycling strategy [4]. CO_2-rich reforming is then of great interest for flexible syngas production.

Historically, DRM was first described by Fischer and Tropsch in 1928 [5]. They performed the reaction on various base metal catalysts, nickel and cobalt being the most promising ones. Severe deactivation due to carbon deposition was however observed, and researchers have been focused on tackling this problem ever since. On the other hand, SRM was industrially introduced in 1930. The technology went through an industrial break-through in the 1960s when tubular reformers were suc-

L. Pastor-Pérez, E. le Saché, T.R. Reina, Department of Chemical and Process Engineering, University of Surrey, GuildfordUK.

https://doi.org/10.1515/9783110563191-014

cessfully operating at high pressures decreasing significantly the cost and energy consumption of downstream processes such as ammonia synthesis [6]. Figure 14.1 shows a simplified flow sheet of a reforming process. A desulphurisation unit was introduced to protect the reforming catalyst from sulphur poisoning and hence de-activation. The prereformer was introduced in the 1980s and converts, at low temperatures, the higher hydrocarbons before entering the preheater, thus allowing to preheat at high temperatures without risks of pyrolysis [6]. Carbone dioxide can be added as reactant and recycled for steam reforming of CO_2-rich gas. Finally, the SPARG (sulphur passivated reforming) process was introduced in the 1980s to tackle Ni catalyst deactivation. Indeed, Ni-reforming catalysts are prone to deactivation by carbon deposition. In this process, sulphur is injected in the feed stream to poison a portion of the most active sites to prevent carbon formation, while the remaining sites maintain some activity for reforming [7].

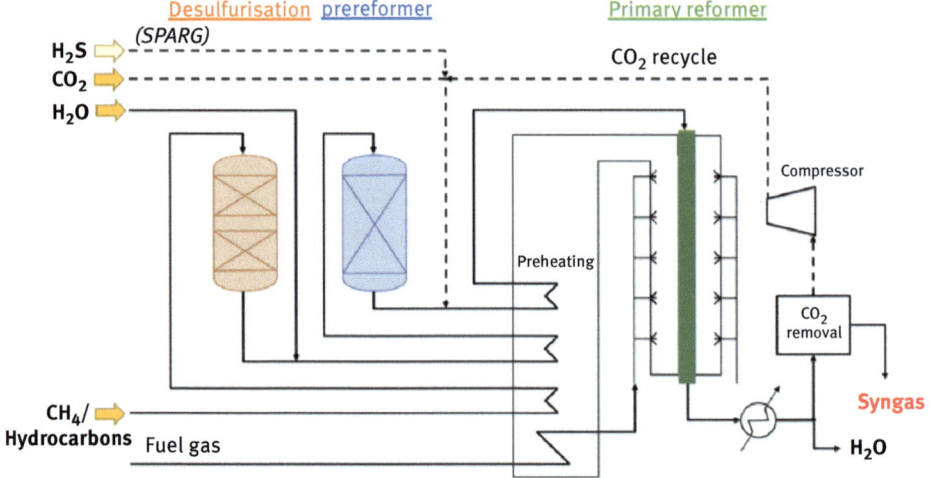

Figure 14.1: Flow sheet of a reforming process adapted from [7] with optional configurations (dotted lines).

14.1.1 Thermodynamic considerations

Thermodynamically, DRM (eq. 14.1) and SRM (eq. 14.2) are highly endothermic reactions. Therefore, elevated temperatures are required to attain high reactant conversions to syngas.

$$CO_2 + CH_4 \leftrightarrow 2CO + 2H_2 \qquad \Delta H°_{298} = 247.3 \text{kJ/mol} \qquad (14.1)$$

$$CH_4 + H_2O \leftrightarrow CO + 3H_2 \qquad \Delta H^\circ_{298} = 206.8 \text{kJ/mol} \qquad (14.2)$$

Alternatively, syngas can be obtained from the partial oxidation of methane (POM, eq. 14.3), also called *oxy reforming*, using O$_2$ as oxidant. POM benefits from fast reaction kinetics, favourable thermodynamics and excellent syngas selectivity at extremely high space velocities. However, catalyst deactivation due to carbon deposition can make the process exothermic and thereby raise safety concerns associated with dangers of explosions [8]. Autothermal reforming (eq. 14.4), the combination of SRM and POM, presents the advantage of not requiring any external source of heat.

$$CH_4 + 1/2\,O_2 \leftrightarrow CO + 2H_2 \qquad \Delta H^\circ_{298} = -35.6 \text{kJ/mol} \qquad (14.3)$$

$$7CH_4 + 3O_2 + H_2O \leftrightarrow 7CO + 15H_2 + CO_2 \qquad \Delta H^\circ_{298} = -6.8 \text{kJ/mol} \qquad (14.4)$$

But with the aim of reducing the impact of greenhouse gases, technologies using CO$_2$ as oxidant are under scrutiny. Since DRM produces a syngas with a H$_2$/CO ratio of maximum 1, the combination of SRM and DRM, known as the *bi-reforming of methane* (BRM, eq. 14.5), is of interest for producing H$_2$-rich syngas. In the same vein, the tri-reforming of methane (TRM, eq. 14.6), is the combination of SRM, DRM and POM and produces a H$_2$-rich syngas and has an improved process efficiency [9].

$$3CH_4 + 2H_2O + CO_2 \leftrightarrow 4CO + 8H_2 \qquad \Delta H^\circ_{298} = 660.9 \text{kJ/mol} \qquad (14.5)$$

$$20CH_4 + H_2O + 3O_2 + CO_2 \leftrightarrow 21CO + 41H_2 \qquad \Delta H^\circ_{298} = 12.9 \text{kJ/mol} \qquad (14.6)$$

Numerous side reactions may occur during reforming processes. In particular, the reverse water gas shift reaction (RWGS, eq. 14.7), consuming part of the H$_2$ produced to reduce CO$_2$ to CO (Figure 14.2). But most importantly, carbon formation reactions are thermally favourable across all the range of temperatures: the Boudouard reaction (eq. 14.8), CO and CO$_2$ reductions (eqs. 14.9 and 14.10) and CH$_4$ decomposition (eq. 14.11). These provoke the coking and deactivation of reforming catalysts.

$$CO_2 + H_2 \leftrightarrow CO + H_2O \qquad \Delta H^\circ_{298} = 41.2 \text{kJ/mol} \qquad (14.7)$$

$$2CO \leftrightarrow C + CO_2 \qquad \Delta H^\circ_{298} = -171 \text{kJ/mol} \qquad (14.8)$$

$$CO + H_2 \leftrightarrow C + H_2O \qquad \Delta H^\circ_{298} = -131 \text{kJ/mol} \qquad (14.9)$$

$$CO_2 + 2H_2 \leftrightarrow C + 2H_2O \qquad \Delta H^\circ_{298} = -90 \text{kJ/mol} \qquad (14.10)$$

$$CH_4 \leftrightarrow C + 2H_2 \qquad \Delta H^\circ_{298} = 75 \text{kJ/mol} \qquad (14.11)$$

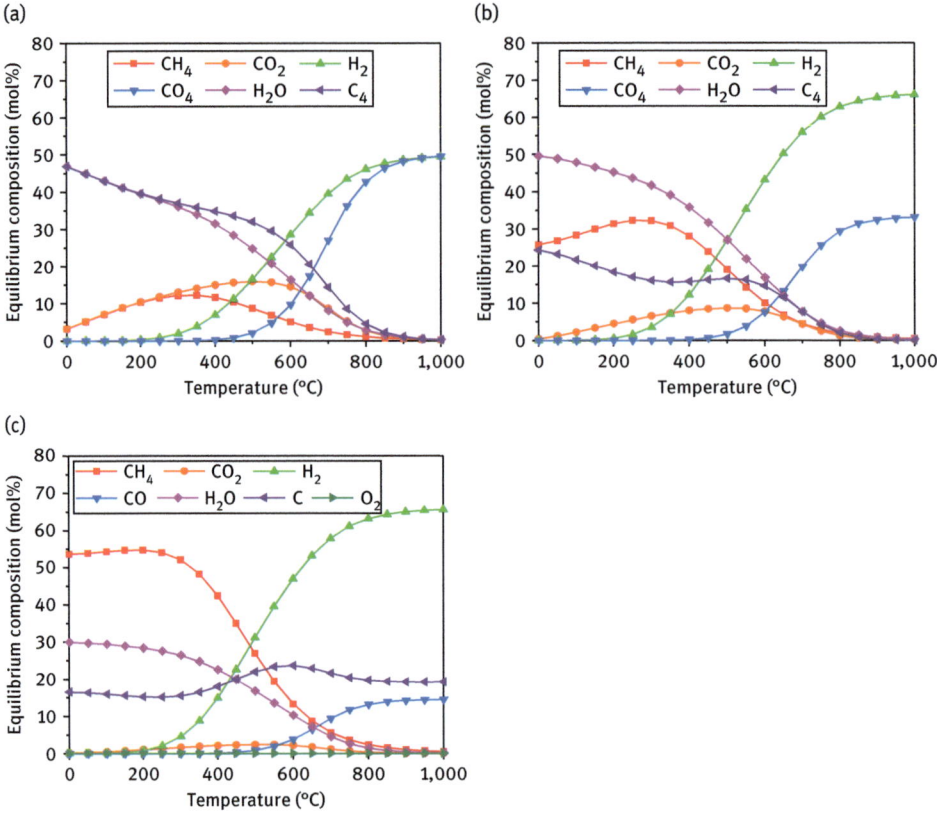

Figure 14.2: Thermodynamic equilibrium plots for (a) DRM, (b) BRM and (c) TRM at 1 bar using stoichiometric feeds (obtained using ChemCad 6.5.5 software).

14.1.2 Reaction mechanism

According to thermodynamics, complete CH_4 and CO_2 conversions are achievable above 800 °C (Figure 14.2). However at such high temperature, methane decomposition is favoured, which results in the formation of carbonaceous species. Hence, a good catalyst needs to kinetically hinder carbon formation. For this reason, the mechanism of the reactions are deeply studied. Although the mechanism may vary with different catalytic systems, it can be summarised in four different steps (Figure 14.3).

First, methane is adsorbed and dissociated from CH_4 to CH_3, CH_3 to CH_2, CH_2 to CH and finally, CH to C and H. Methane activation is believed to be taking place on the metal surface, although Bitter et al. suggested that it occurred at the interfacial sites for Pt/ZrO_2 catalysts [10]. The energy required to dissociate the CH_3–$H_{(g)}$ bond being high (439.3 kJ mol^{-1}), it is agreed that the dissociation of methane on the catalyst is the rate-determining step [11]. CO_2, on the other hand, can adsorb and

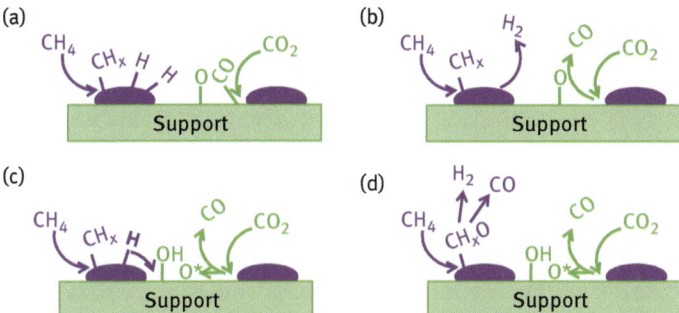

Figure 14.3: Reaction steps for the dry reforming of methane: (a) Dissociative adsorption of CH_4 and CO_2 on the metal and metal–support interface, respectively. (b) Fast desorption of CO and H_2. (c) Formation of surface hydroxyls from hydrogen and oxygen spill over. (d) Surface hydroxyls and oxygen species oxidise CH_x species forming CH_xO and formation of CO and H_2. Adapted from [14].

dissociate on metal surfaces but is activated preferably on the metal–support interface or on the support, forming as a result CO and adsorbed oxygen (Figure 14.3a). The reduction of CO_2 to CO may occur through the formation of carbonate precursors and is facilitated on basic sites [12, 13]. Then, once the reactants are adsorbed, many surface reactions can occur, including side reactions. The desorption of CO and H_2 from the support and the metal respectively are fast reaction steps (Figure 14.3b). Hydrogen spill over from the metal surface to the support is then predicted by most studies, allowing hydrogen to react with adsorbed oxygen species and to form hydroxyl groups (Figure 14.3c). Although oxygen migrates from the support to the metal surface to react with hydrogen-depleted methyl-like CH_x species forming either CH_xO species or direct adsorbed-CO species (Figure 14.3d) [14]. When water is present on the surface, it may migrate to the metal–support interface and help the formation of intermediate CH_xO species [15–17]. Whether the reaction mechanism goes through carbonate or formate intermediates is highly dependent on the catalyst composition; hence, similar to steam reforming, there is no clear agreement about the specific reaction mechanism on the surface of the catalyst.

14.1.3 Metal-based heterogeneous catalysts

As previously stated, to achieve complete conversions, the reaction needs to take place at high temperatures. However under such conditions, metal-supported catalysts are prone to sintering of the metallic phase. Moreover, carbon deposition induced by methane decomposition is also favoured. Therefore, DRM catalysts are not only required to be highly active, but also selective and highly stable. The performance of a catalyst is affected by the nature and particle size of the metal, the structure, texture and nature of the support and by the reaction conditions.

Catalysts composed of noble metals are very active for DRM. They exhibit great carbon resistance compared to transition metals due to lower equilibrium constants for methane decomposition and reduced dissolution of carbon in their lattices [14]. Rostrup-Nielsen and Hansen compared various noble and transition metals supported on $MgAl_2O_4$ spinel and found the following activity order: Ru > Rh, Ni > Ir > Pt > Pd between 500 and 650 °C, while the order for carbon formation was Ni > Pd ≫Ir > Pt > Ru, Rh [18]. The Rh- and Ru-based catalysts only presented negligible amount of carbon formation making them ideal metals for DRM. However, the properties of the support plays a major role in the catalyst performance. Wang and Ruckenstein compared the activity of a rhodium catalyst based on different kinds of oxides supports [19]. The reducible oxides showed very low activity in particular Nb_2O_5 and TiO_2. CeO_2 and ZrO_2 supported catalysts exhibited a very long activation period and lower conversions than irreducible oxide based catalysts like MgO and γ-Al_2O_3 [19]. Although CeO_2 and ZrO_2 do not appear to be suitable supports, Ce and Zr containing binary and ternary supports were found to be more active, due to an improved oxygen mobility and storage capacity as well as an improved dispersion of metal [20].

The low availability and price of noble metals have encouraged researchers to improve the formulation of transition metal catalysts. Nickel has shown good activity for DRM but suffers from severe carbon deposition [21]. As methane adsorption and dissociation on the metal is the rate- limiting step, the focus is on metal dispersion and stability towards sintering. Moreover, carbon deposition occurs preferably on large metal clusters, making Ni dispersion a priority [14]. However, the role of the support alone and its ability to disperse Ni cannot be linearly related to the catalytic performance of a catalyst as various phenomena are induced by metal–support interactions [22]. Ni/Al_2O_3 is the typical state-of-the-art catalyst for DRM. Alumina is an inexpensive material with a high surface area, great thermal stability and the capacity to highly disperse nickel [23]. However at high temperatures, the formation of $NiAl_2O_4$ spinel can affect the catalytic performance [24], plus the acidity of alumina limits its CO_2 adsorption capacity [12]. To facilitate CO_2 adsorption and hence to increase the availability of surface oxygen species or hydroxyl species to prevent the formation of carbonaceous deposits, alumina can be promoted with reducible and basic metal oxides. In particular, nickel on alumina promoted with MgO, La_2O_3, CaO or CeO_2 have been studied by Charisiou et al. and displayed enhanced activity and stability [4, 25, 26]. CeO_2, more specifically, is well known for its high oxygen storage capacity (OSC) [27]. The oxygen vacancies generated by ceria redox mechanisms act as a driving force for CO_2 adsorption and CO formation [28].

Materials originating from well-defined structures like hydrotalcite, fluorite, perovskite or pyrochlore precursors are used in DRM for their ability to stabilise transition metals. Hydrotalcite compounds (HT) have the general formula $[M^{2+}_{1-x}M^{3+}_x$ $(OH)_2]^{x+} (A^{n-}_{x/n})·mH_2O$, where M^{2+} and M^{3+} are metal cations such as Mg^{2+} and Al^{3+}. HT materials have a unique layered structure where the 2+ and 3+ metal cations are randomly distributed. Therefore, Ni ions can be carefully dispersed in the layered structure benefiting from the insulation of Mg and Al ions, minimising nickel aggregation [29, 30]. Perovskites and perovskite-type materials have the general formula ABO_3 and A_2BO_4, respectively, where A is a large cation (rare earth, alkaline earth) and B a smaller cation (transition metal). The partial substitution of A and B ions with other ions is also possible in this type of compounds, thus allowing to adjust their thermal stability and catalytic performance [31]. The most commonly studied perovskite is $LaNiO_3$, partially substituted with various elements (Ce, Ca, Sr, Sm, Nd) in the A site or other metals (Ru, Co) in the B site to ensure small metal crystallites and high oxygen mobility in the metal oxide [31–34]. After an activation pre-treatment, the perovskite are reduced to well-dispersed Ni particles on La_2O_3 [35]. Pyrochlores are materials with the general formula $A_2B_2O_7$, where A is a large trivalent rare earth cation and B a smaller tetravalent transition metal. When Ni is substituted in the B site of the pyrochlore, it becomes active for DRM. Nickel is, in that case, stabilised in an inorganic structure and progressively exsolved during the reaction [36].

Finally, more recently, catalyst design approaches aiming to encapsulate the active metal phase in inorganic cavities show some promising results in preventing sintering of the active sites [37]. Metals or bimetallic nanoparticles encapsulated in oxides with yolk or core shell structures have demonstrated promising results [38]. For instance, Li et al. synthesised a yolk-satellite-shell structured Ni-yolk@Ni@SiO_2 with different shell thicknesses for DRM and obtained high activity and stability, with 11.2 nm shell thickness being the most stable one [37].

In summary, reforming reactions, in general, and DRM, in particular, have regained attention from the catalysis community due their great potential to upgrade CO_2/CH_4 streams (i.e., biogas mixtures). The development of robust catalysts able to overcome sintering and coking is key to facilitate the implementation of this technology in commercial units for CO_2 recycling applications. In response to these needs, new engineered catalysts are under-development using multiple strategies such as multicomponent materials using promoters, stabilisation of Ni on inorganic complex structures or core-shell/yolk-shell configurations.

14.2 Gas phase CO_2 hydrogenations

Although they have been explored for many years for other purposes, nowadays, the reappearance CO_2 hydrogenation reactions (Figure 14.4), a significant representative among chemical CO_2 conversions, is bringing new challenging opportunities for sustainable development in energy and environmental catalysis. Indeed, CO_2 hydrogenation not only reduces the increasing CO_2 build-up but it can also be used to produce added value fuels and chemicals. Nevertheless, the economic perspective of converting CO_2 largely depends on the cost of producing H_2. Hydrogen sources for the

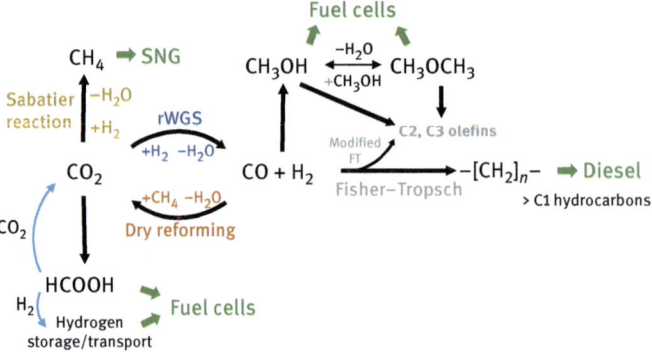

Figure 14.4: Schematic overview of the chemical CO_2 hydrogenation catalytic routes. Adapted from [43].

chemical recycling of CO_2 could be produced either by using still-existing significant sources of fossil fuels (mainly natural gas) or from splitting water (by electrolysis or other cleavage) [39]. Nevertheless, the resources of fossil fuels are diminishing and fuel costs have undergone strong fluctuation in recent years. Therefore, it would be highly desirable to obtain this H_2 from non-fossil fuel sources and processes. The need for this renewable H_2 poses a crucial handicap for using CO_2 as backbone to produce greener fuels and again, the economic perspectives of converting CO_2 will largely depend on producing H_2 in a cost-effective and sustainable manner. Since the sustainable production of H_2 is beyond the scope of this chapter, we will just mention that available (low-cost) sources of renewable H_2 already exist [40, 41]. These technologies validate the viability of CO_2 hydrogenation processes even though it is evident that renewable H_2 synthesis is still a technology under continuous development [42].

Gas phase hydrogenations of CO_2 have been more intensively investigated recently, due to the large demand of added value products like syngas (which is further upgraded to synthetic fuels) or synthetic natural gas (SNG; methane). Syngas along with short-chain olefins are the main raw materials for petrochemistry and thus of the entire value chain of the chemical industry. On the other hand, SNG is a clean energy source with high heating value (HHV = 37.26–28.10 MJ m^{-3}). It not only can be applied into chemical and power industries, but it can also be used as transportation fuel and as a domestic energy resource [44].

Notwithstanding, the implementation of these CO_2 hydrogenation reactions at industrial scale is hindered by the lack of an economically viable active and robust heterogeneous catalysts, able to overcome the CO_2 activation and deliver sufficient amounts of upgraded products at the desired operation conditions. Some advances have been made in the past decade, especially in heterogeneous catalysis; however, effective materials are still under-development. This section provides a summary of the current understanding of reaction mechanisms and catalytic reactivity over a wide variety of catalysts, with an emphasis of practical aspects and two main reactions: the RWGS and CO$_2$ methanation.

14.2.1 CO$_2$ reduction to CO via Reverse Water Gas Shift

The RWGS reaction was first observed by Carl Bosch and Wilhelm Wild in 1914, when they tried (and halfway succeeded) to produce hydrogen from steam and carbon monoxide on an iron oxide catalyst. Presently it is one of the technologies with the highest readiness level for converting CO$_2$ to synthetic fuels or their precursors [45]. Direct CO$_2$ hydrogenation is more thermodynamically favoured than RWGS (due to the thermodynamic constraints related to the RWGS equilibrium reaction), and active catalysts for the RWGS reaction are also active for methanol synthesis or FT reactions [46, 47]. However, the water formed in RWGS decreases the reaction rates of these side processes, thus making convenient to have a process scheme based on two-stage reactors with an intermediate water separation stage (Figure 14.5). Consequently, this step is integrated in the overall process and has been demonstrated on a very promising pilot-scale prototype for industrialised methanol synthesis [48]. CAMERE process (CO$_2$ hydrogenation to form methanol via RWGS reaction) reveals that the two-steps approach (with intermediate water removal) reaches a three times higher productivity than the direct CO$_2$ hydrogenation since the inhibition by water is suppressed and the equilibrium is shifted towards the desired products [49].

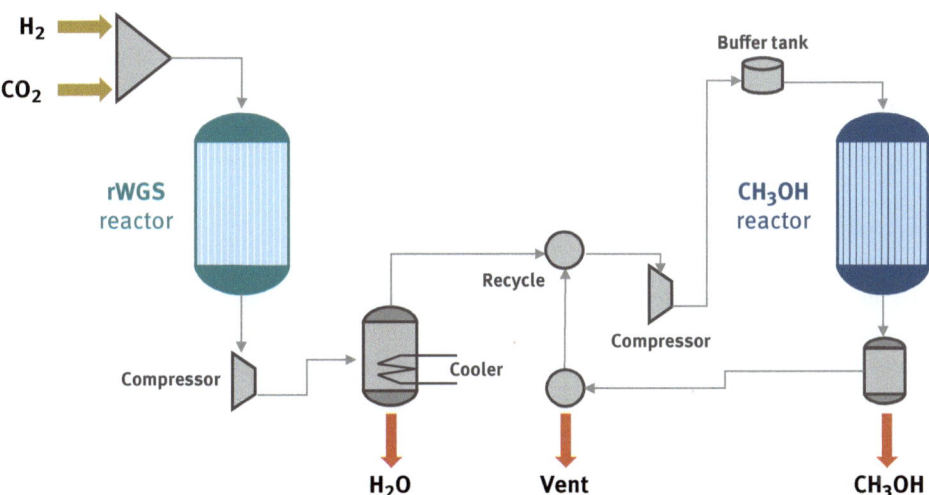

Figure 14.5: Simplified process flow diagram process based on the two-stage reactors (RWGS and methanol synthesis reactor).

14.2.1.1 Thermodynamic considerations

As for thermodynamic aspects, the RWGS reaction (eq. 14.7) is an equilibrium limited process that is thermodynamically favoured at high temperatures due to its endothermic nature.

$$CO_2 + H_2 \leftrightarrow CO + H_2O \qquad \Delta H°_{298} = 41.2\,kJ/mol \qquad (7)$$

Additional side reactions include methanation of CO (eq. 14.12) and the Sabatier reaction (eq. 14.13)

$$CO + 3H_2 \leftrightarrow CH_4 + H_2O \qquad \Delta H°_{298} = -206.5\,kJ/mol \qquad (12)$$

$$CO_2 + 4H_2 \leftrightarrow CH_4 + 2H_2O \qquad \Delta H°_{298} = -165.0\,kJ/mol \qquad (13)$$

The CO_2 conversion is maximised by increasing the H_2/CO_2 ratio and it favours the RWGS reaction [50]. Consequently, when the reaction is carried out at lower temperatures, the equilibrium will increasingly favour the WGS (reverse of eq. 14.7), methanation (eq. 14.12) and Sabatier (eq. 14.13) reactions, as they are exothermic and the most prominent side reactions under these conditions (Figure 14.6). However, the H_2/CO_2 and the temperature must be carefully adjusted to ensure that the conditions applied during the experimental stages are economically favourable for industrial applications. From previous works, it has been demonstrated that there is no significant effect of altering pressure on the reaction activity and position of the equilibrium due to the stoichiometry of the reaction [45]. Moreover, product separation can shift the equilibrium and boost CO_2 conversion [51]. Whitlow and Parrish, from Florida Institute of Technology, built an RWGS demonstration reactor without a catalyst in the system [52]. They incorporated a membrane reactor to separate the products and achieved close to 100% CO_2 conversion (five times the equilibrium conversion at the studied conditions).

14.2.1.2 Reaction mechanism

CO_2 is a thermodynamically stable molecule with a standard formation enthalpy of -393.5 kJ mol^{-1} [12]. The thermodynamic stability of CO_2 provides a major challenge in converting CO_2 to more useful products. It is well established that large-scale production of useful chemicals such as fuels and platform chemicals necessitates continuous operation using heterogeneous catalyst to activate CO_2 over its surface. Thus, first-row transition metals have become a great focus of interest [53–56] since some have been chosen by nature to activate CO_2; for example, the reduction of CO_2 to CO can be mediated by enzymes such as nitrogenase (e.g., MoFeP and FeP) and carbon monoxide dehydrogenase [57].

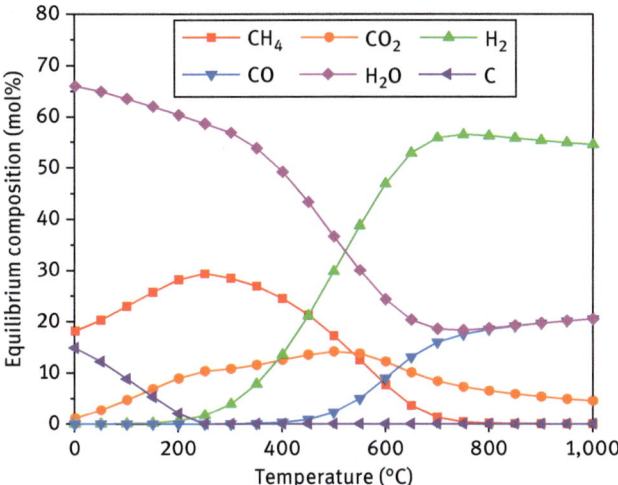

Figure 14.6: Influence of temperature on the thermodynamic equilibrium of the RWGS reaction at 1 bar and H_2/CO_2 molar ratio of 3/1 (obtained using ChemCad 6.5.5 software).

Numerous research has been done employing advanced characterisation tools to unravel the mechanism of RWGS. Cu surfaces [58, 59] and supported Cu/ZnO systems [60] were the most used materials to discuss the mechanism of this reaction, which is, however, still controversial. Currently, there are two major mechanisms proposed: redox mechanism and formate decomposition mechanism.

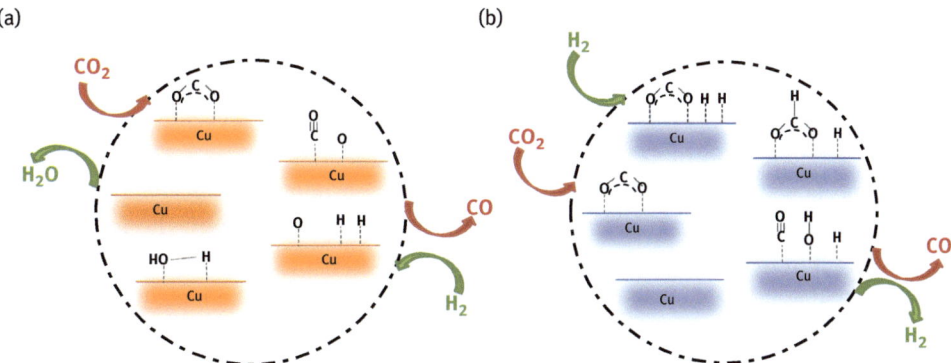

Figure 14.7: RWGS (a) redox and (b) formate proposed reaction mechanism over Cu-based catalysts.

For the redox mechanism (Figure 14.7a) Cu^0 atoms are considered as the active sites to dissociate CO_2, while the reduction of oxidised Cu catalyst has to be faster than the oxidation process to facilitate the reaction [60]. Hydrogen is proposed to be a reducing reagent without direct participation in the formation of intermediates in the RWGS reaction. In the alternative route via formate decomposition (Figure 14.7b), hydrogen atoms are adsorbed on Cu and spill over to the surface associating with CO_2–Cu, leading to the formation of the formate species. The CO is formed from decomposition of the formate intermediate [61, 62]. Even though dissociation of CO_2 on the Cu atoms is considered as the rate-determining step [63], it should be mentioned that the probability for CO_2 dissociation on H_{ads}-rich Cu surfaces is two orders of magnitude larger than on clean Cu surfaces [59]. Moreover, no CO_2 dissociation has been observed in UHV conditions [64].

Ni-based catalysts constitute another commonly studied transition metal formulation. Qin et al. investigated, using density functional theory, the reaction mechanism of the RWGS reaction on an Ni surface and observed that the C–O bond cleavage of CO_2 happened before H_2 dissociation [65]. Hydrogen could promote the charge transfer in the Ni insertion process and ease the dissociation of coordinated CO_2 molecules by reducing the energy barrier. The rate-determining step for the reaction is the migration of hydrogen atom from a Ni centre to an oxygen atom with the formation of water.

Different mechanisms have also been proposed for the reaction over noble metal-based catalysts, specifically over Pd- and Pt-based catalysts [66–68]. Arunajatesan et al. reported that using Pd/Al_2O_3 catalysts and a supercritical mixture of CO_2 and H_2 several surface species including formates, carbonates and CO are unequivocally observed [66]. However, on just the support only surface carbonates are observed. It is likely that the Pd promotes dissociative adsorption of H_2, which spills over onto the support and reacts with the carbonates to form formates and CO. Furthermore, this work suggests that short-residence time continuous reactors are preferred over batch reactors to minimise the effects of possible catalyst deactivation by adsorbed CO. Regarding Pt-based catalyst, Ferri et al. proposed another mechanism [67]. For these catalysts the reaction of CO_2 and hydrogen takes place at the Pt–Al_2O_3 interface, and the formed CO could serve as a probe molecule for the boundary sites. CO_2 adsorbs on oxygen defects of Al_2O_3 thin film to form carbonate-like species, and then reacts with hydrogen to form CO. In another work employing CeO_2 as a support, Goguet et al. proposed that both formates and Pt-bound carbonyls species were observed, but neither of them was the main reaction intermediate, although the formation of CO from formates was likely to occur to a limited extent [68]. They suggest that the RWGS reaction proceeds mainly via surface carbonate intermediates, including reaction between the surface carbonates and oxygen vacancies or the diffusion of the vacancies in the ceria lattice.

14.2.1.3 Metal-based heterogeneous catalysts

Until now, all catalysts tested in the RWGS reaction are based on effective materials for the direct reaction, water gas-shift reaction, adapted to operate with CO_2 and H_2, but not specially developed to work with CO_2. Copper-based catalysts have been the most popularly studied catalytic systems for the WGS reaction in addition to methanol synthesis and, therefore, have also been the most applied to the RWGS reaction.

The RWGS studies of supported metal catalysts consist primarily of Cu and noble metal catalysts (Pt, Pd and Rh) immobilised on a variety supports. Also some promoters like Fe and alkali metals are becoming increasingly used. Studies on these metals and promoters will be highlighted in this section.

The use of Cu for RWGS provides two major advantages: (i) it has been shown to perform RWGS at low temperatures (~165 °C) [63] and (ii) little or no methane is formed as a side product [69]. Hence, the enhancement of Cu activity has been extensively studied using different supports and by incorporation of several promoters into the catalytic system.

Liu et al. have compared a series of bimetallic Cu–Ni/Al_2O_3 catalysts and found that the Cu/Ni ratio has a significant effect on the CO_2 conversion and CO selectivity. As the Cu/Ni increased, the CO selectivity is favoured due to Cu favouring CO formation and, as expected, Ni is more active for CH_4 production [70]. Chen et al. have conducted several studies on the RWGS on Cu nanoparticles supported on different metal oxides [61, 62, 71]. In their first study about the mechanism of CO formation in RWGS reaction they found that the key reaction step is the formation of the formate species from association of H atoms and CO_2. They exposed that CO production can be significantly enhanced by hydrogen over Cu catalyst. TPR profiles showed that Cu_2O is formed during RWGS reaction and/or CO_2 dissociation due to oxygen adatoms reacting with Cu^0 [62]. In their subsequent contribution, they examined the behaviour of Cu nanoparticles supported on SiO_2, concluding also that the RWGS mechanism goes through a formate intermediate and that a high Cu dispersion on SiO_2 enhances CO_2 conversion [61, 71]. This last observation was also found by Gines et al. in a Cu/Zn/Al_2O_3 system [60]. Additionally, Cu/ZnO and Cu–Zn/Al_2O_3 catalysts commonly used for methanol synthesis and WGS reaction have also been tested for the RWGS reaction [72]. A comparison between Cu–ZnO and alumina-supported Cu–ZnO as catalysts for RWGS reaction showed that the presence of Al_2O_3 decreases the crystallite size of the CuO and ZnO particles produced on calcination and at high Cu/Zn ratios increases the dispersion of copper in the final catalyst achieving better catalytic activity. More recently, Cu/ZnO catalysts derived from aurichalcite precursors showed homogeneously distributed CuO particles in ZnO, leading to higher activities and lower apparent activation energies in the RWGS reaction. This ZnO-enriched surface provided an optimised Cu/ZnO interface for the activation and dissociation of CO_2 as the rate-determining step of the RWGS reaction.

Additionally, promoters like alkali metals might alter the catalytic activity of Cu-based materials [73, 74]. K_2O, for example, acts at the interface to Cu by weakening the bonding of surface intermediates, in addition to acting as a promoter for CO_2 adsorption [73].

As was already mentioned, the RWGS is favoured at high temperatures due to its endothermic nature. Nevertheless, copper has poor thermal stability (e.g., sintering of copper) that makes Cu-based catalysts not suitable at this favoured temperature unless these materials are modified by adding a thermal stabiliser. Chem et al. also studied the addition of small amounts of Fe in the Cu/SiO_2 catalyst [75]. Fe prevented Cu NPs sintering by the formation of small particles of Fe around the Cu particles, significantly enhancing the stability and activity of this material. A combination of both strategies, the use of iron acting as a thermal stabiliser and the alkali promotion, was studied by Pastor-Pérez et al. [76]. They found that the addition of Cu to a model Fe/Al_2O_3 catalyst improved the activity and selectivity due to the synergistic Cu–Fe effect and to the enhanced resistance towards active phase sintering. Furthermore, the addition of Cs boosted the CO_2 activation due to its basic character, easing the electron transfer from the catalyst to the reactants and facilitating the CO_2 adsorption on the surface of the catalyst. Additionally, Cs helped to suppress the methanation reaction, which is key to avoid hydrogen consumption in an undesired route.

It is worth mentioning that there are several works based on cerium-based catalysts since it was found that they are also active in both WGS and RWGS reactions. Zhou et al. studied CeCu composite catalysts with different Ce/Cu mole ratios, showing that the synergistic effect of the surface oxygen vacancies of ceria and active Cu^0 species enhanced catalytic activity of the studied CeCu catalysts [77]. The electronic effect between Cu and Ce species boosted the adsorption and activation performances of the reactant CO_2 and H_2 molecules on the CeCu surface. Ni/CeO_2 showed excellent catalytic performance in terms of activity, selectivity and stability for the RWGS reaction. Oxygen vacancies formed in the lattice of ceria and highly dispersed Ni are key active components for the reaction, since bulk Ni favours methane formation [78]. Yang et al. studied the effect of Fe as a promoter in CeO_2–Al_2O_3 supported Ni-based catalysts [79]. In addition to the advantages brought by the presence of CeO_2 in the support, improving Ni reducibility and the overall redox properties due to its excellent oxygen mobility, the iron-promoted system reached an exceptional activity/selectivity balance that was attributed to an electronic enrichment of the Ni surface atoms due to the FeO_x–Ni interaction. However, deactivation of ceria-supported catalysts is a crucial issue that has to be considered. Results obtained with Pt/CeO_2 indicated that carbon deposition took place primarily on the support (ceria) and not on the platinum. This result pointed out that the support was "active" in the reaction and that the deactivation occurred through gradual coverage of the support [80].

Finally, noble metal-based catalysts are also frequently used in the RWGS reaction. These metals (e.g., Pt, Pd and Rh) have high ability towards H_2 dissociation, making them suitable catalysts for this reaction. The most commonly used is Pt, and there are a several recent reports focused on Pt/TiO_2 samples [81–83]. Kim et al. confirmed that the active sites of the Pt/TiO_2 catalyst comprised both the TiO_2 sites and the Pt sites, and the difference in the activity of this catalysts was dependent on the reducibility of the TiO_2 supports. Furthermore, X-ray diffraction analysis showed that the TiO_2 crystallite size was a key factor that dictated the reducibility of the TiO_2 sites [81]. In another contribution, they added that the selectivity of the

Pt/TiO$_2$ catalyst was dependent on the carbonate species formed on the reducible TiO$_2$ sites and on the carbonyl species bonded to the Pt sites [82]. Chen et al. reported that Pt–Ov–Ti^{3+} species formed at the interface between reducible TiO$_2$ support and Pt was identified as the active sites for CO formation, while large Pt particles ease the CO hydrogenation to produce methane at higher temperature [83]. Upon the addition of alkali promoters, potassium has been used in Pt base catalysts. It was found that K enhanced the CO_2 conversion and weakened the strength of CO adsorption on Pt, which hindered the competitive hydrogenation of CO into CH$_4$ through CO bond dissociation [84, 85].

14.2.2 CO_2 methanation: The power-to-gas (P2G) concept

The CO_2 methanation is the hydrogenation of CO_2 to produce methane and water and has been known for over a century as the *Sabatier reaction* [86]. Paul Sabatier revealed how to facilitate the addition of hydrogen to molecules of carbon compounds and he won the Nobel Prize in Chemistry for this contribution in 1912. Nevertheless, this reaction has received renewed interest for its potential applicability for chemical storage of the excess of electrical energy produced by renewable resources recently [87].

Power-to-gas (P2G) processes are considered as an interesting and possible answer to integrate renewable resources, such as solar and wind energy, into the current energy portfolio efficiently. These processes aim at using power to convert H$_2$O into H$_2$ via electrolysis, and storing the obtained fuel to reconvert it later into electricity, when a period of high-power consumption occurs [88]. Thanks to CO_2 methanation, a more practical route that provides energy in a safer and more convenient way can be achieved. Methane is the dominant component of SNG and has several advantages over H$_2$, for example, higher volumetric energy content and safety handling and transportation. In addition, there is no limit for SNG admittance into the gas grid [88]. Moreover, the demand of production of SNG is increasing to overcome the gaps between the supply and demand of natural gas. In this regard, it may be recalled that the common natural gas is a clean energy of HHV (37.26–28.10 MJ m^{-3}) that can be applied in chemical and power industries and also can be used as a transportation fuel [44]. However, it is not a long-term accessible fuel due to the deficient natural gas reservoirs in China [89].

Back to the analysis of SNG, specifically the power-to-methane (P2M) concept is a promising option to absorb and exploit surplus renewable energy in regions where a natural gas infrastructure already exists. A P2M plant basically consists of a H$_2$O electrolyser, a CO_2 separation unit, if CO_2 is not available as pure gas or in a suitable gas mixture, and a methanation module [90, 91]. At periods of power surplus, water splitting is used to produce H$_2$ in the electrolyser. The produced H$_2$ and CO_2 are then converted, in the methanation unit, to a mixture of gases that mainly contains CH_4 and H_2O [92]. This mixture is then treated to produce a methane-rich gas, the so-called synthetic natural gas (SNG). Figure 14.8 illustrates the principle of the P2M concept and its applications.

It is worth mentioning that, although CH_4 produced from CO_2 according to the above routes is considered a way to use off-peak excess renewable energy, this renewable SNG can be also used in the chemical industry to decrease the carbon footprint. However, the direct conversion of CH_4 into chemicals (not via syngas) has considerable difficulty. Hence, the use of CH_4 as raw material for the chemical industry is instead not convenient currently. The scientific community still needs new breakthrough discoveries in the direct methane conversion to chemicals. In addition, an interesting application of Sabatier reaction is also coming to light. CO_2 methanation is conducted for the use of reclaiming oxygen in International Space Station, where the oxygen resource in the exhalant CO_2 from breathing can be transformed into precious water for the astronaut life-support system [93].

14.2.2.1 Thermodynamic considerations

The methanation of CO_2 is a highly exothermic reaction that is expressed as follows (eq. 13):

$$CO_2 + 4H_2 \leftrightarrow CH_4 + 2H_2O \qquad \Delta H°_{298} = -165.0 kJ/mol \qquad (14.13)$$

Although it is an exothermic reaction, it is difficult to achieve because of the high kinetic barriers of the eight-electron reduction process. During the methane synthesis, the carrier of chemical energy is converted, from H_2 that contains low chemical energy density to CH_4 with high chemical energy density. The efficiency of the conversion reaches 83% due to the low heating value at standard conditions, whereby the remaining 17% is released as heat. On the other hand, CO_2 methanation is an exothermic reaction with a negative change of moles; therefore, the synthesis is thermodynamically favoured towards products at low temperature and high pressure as is discussed below.

By-products can be generated during the methanation reaction since there are several competitive reactions for the methanation process, as listed in eqs. 14.7 14.10.

$$CO_2 + H_2 \leftrightarrow CO + H_2O \qquad \Delta H°_{298} = 41.2 kJ/mol \qquad (14.7)$$

$$CO_2 + 2H_2 \leftrightarrow C + 2H_2O \qquad \Delta H°_{298} = -90 kJ/mol \qquad (14.10)$$

The most frequently discussed conditions on the influence of the CO_x conversion, CH_4 selectivity and carbon deposition are the effects caused by the reaction temperature, H_2/CO_x ratio, pressure and the addition of other reactants (H_2O, CH_4, O_2, etc.). Lately some thermodynamic analyses on the methanation process have been attempted. For

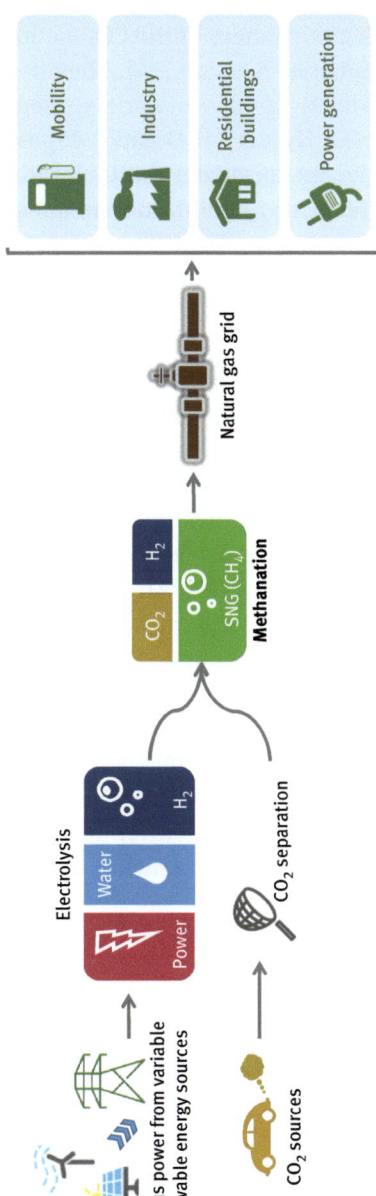

Figure 14.8: Principle of the P2M concept and its applications.

example, Su et al. discussed systematically the thermodynamics of the CO_2 methanation process to gain a better cognition on the Sabatier reaction [94]. They performed the CO_2 methanation starting from CO_2 and H_2 as raw materials, along with six kinds of possible products, namely, CO_2, CO, H_2, CH_4, H_2O and C deposit based on eqs. 14.7 and 14.1. They employed a Gibbs reactor model available from the HSC (enthalpy, entropy and heat capacity) chemistry. Equilibrium composition of different components was obtained in the product under different H_2/CO_2 ratios and temperatures with a constant pressure of 1 bar. Only the major components were taken into consideration and the species presented in a trace amount were neglected. From these studies, it was concluded that, as the reaction temperature increases over 600°C, CO derived from the RWGS reaction is the major product in the CO_2 hydrogenation process, while CH_4 was disfavoured. Meanwhile, the formation of carbon deposits was inhibited to a great extent. The increment of the H_2/CO_2 ratio had a positive effect on the production of CH_4. These trends and calculations are also in agreement with other results reported by Swapnesh et al. [95] and Miguel et al. [96]. Regarding other by-products, Frick et al. reported that carbon and hydrocarbons are also formed but in small amounts [97]. The CO formation as a by-product was suppressed with increasing pressure and decreasing temperature, while that of hydrocarbons (C_nH_{2n} and C_nH_{2n+2} with $2 \leq n \leq 5$) fell with decreasing pressure and temperature.

It can be established that the lower the temperature and higher the pressure, the more favourable the methanation would be from a thermodynamic perspective. Nevertheless, high operating pressure is not economically desirable, and low operating temperature requires a sufficiently active catalyst, which is currently one of the challenges for the commercial implementation of methanation units. A techno-economic compromise must be found.

14.2.2.2 Reaction mechanism

Methanation of CO_2 is a relatively simple reaction; however, its reaction mechanism is still a subject of much debate. There are still open discussions on the type of intermediate compounds involved in the reaction and on the methane formation process. Two main reactions routes have been reported. The first one involves the conversion of CO_2 to CO prior to methanation and the subsequent reaction follows the same mechanism as CO methanation [98, 99]. In contrast, the alternative reaction mechanism proposed (Figure 14.9) involves the formation of formate, carbonate or methanol species (not CO) as the principal intermediate during the reaction [100, 101].

Although there is still no consensus on the kinetics and mechanism, the rate-determining step has been proposed to be either the formation of surface carbon in CO dissociation and its interaction with hydrogen, or the formation of the CH_xO intermediate and its hydrogenation [99, 102].

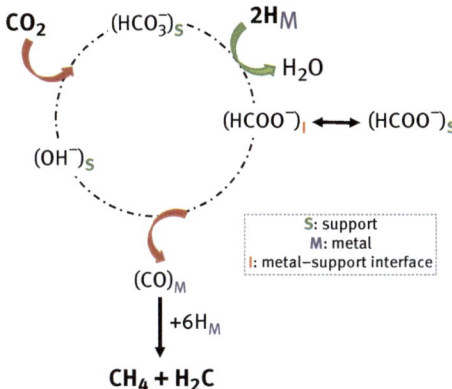

Figure 14.9: Proposed reaction mechanism for CO_2 methanation. Adapted from [98].

On the other hand, two main mechanisms have also been suggested for the transformation of CO_2 to CO_{ads}. The first mechanism includes a RWGS process through a formate intermediate, usually taking place between the support and metal's interface [98, 103–105]. The second mechanism comprises the direct dissociation of CO_2 to CO_{ads} and O_{ads} over a variety of noble metal-based catalysts [106–109].

Karelovic et al. evidenced the direct dissociation of CO_2 [110]. In first place, it was reported that the formation of CO_{ads} species was observed at fairly low temperatures when RWGS does not take place. Secondly, CO was never detected in gas phase even at 200 °C. Moreover, the formation of CO_{ads} was not associated with the formate species until 180 °C was reached. Therefore, it was argued that formates have a minor influence on the reaction path. In contrast, a study of Xu et al. detailed that methane formation was not observed until formate species appeared, proposing that formate may be the critical intermediate [111]. This approach was also suggested by other authors [104, 112, 113].

As mention earlier, the CO_{ads} dissociation is considered as the rate-determining step of the CO_2 methanation, and it has been proposed to proceed by two main pathways. The first one involves direct CO_{ads} dissociation and is suggested to occur over group VIII metal-based catalysts [114–116]; the second one occurs via H-assisted CO_{ads} dissociation [110, 117]. Zhang et al. reported that CO hydrogenation occurs via either a COH or CHO intermediate [118]. That H-assisted CO dissociation is the predominant path in the case of CO hydrogenation over Co and Fe catalysts was also suggested by Ojeda et al. [119]. By conducting a set of spectroscopic studies over supported Rh catalysts, Solymosi et al. found that the dissociation of CO on Rh can be promoted by adsorbed hydrogen [120]. Additionally, some different intermediates were reported by several groups for CO hydrogenation. Specifically, IR experiments were conduct to observe the formation of formyl species over Ru/Al_2O_3 and Rh/TiO_2 catalysts [107, 121]. Mori et al. have captured the existence of H_nCO

intermediates, and observed that the H–CO interaction help to weaken the C–O bond and facilitate its cleavage [122, 123]. It can be concluded that CO dissociation is assisted by H via the formation of Ru carbonyl hydride species. This result is in agreement with the conclusions of Karelovic et al., who proposed the H-assisted CO dissociation process through the formation of Rh carbonyl hydride species [110].

14.2.2.3 Metal-based heterogeneous catalysts

Although catalysts for the Sabatier reaction have been known for a long time, renewed interest in their improvement is promoted by the need to develop catalysts active at lower temperatures and suitable for compact devices (i.e., able to run at high space velocities) to be used in the electrical-to-chemical energy conversion strategy. The need for a fast start and shutdown, the enhanced resistance to deactivation due to these fast cycling modes and the need to be integrated into microreactors are among the additional constraints necessary for the development of compact devices for distributed use.

CO_2 methanation has been investigated using several catalysts based on VIIIB metals, especially Ru, Rh and Pd, supported in different oxides (e.g., SiO_2, TiO_2, Al_2O_3, ZrO_2 and CeO_2). However, the catalytic systems that remain the most used for the methanation reaction are supported Ni-based materials. In these systems the choice of the support is of paramount importance since Ni–support interactions determine the catalytic performance (activity and selectivity) for CO_2 methanation [124].

Ni is a relatively cheap metal and provides high activity and CH_4 selectivity; however, one of the main disadvantages of Ni is its high tendency to get oxidised in oxidising atmospheres – similar to other non-noble metals (i.e., Fe and Co) [92]. In addition, nickel carbonyl can be formed during the reaction, which is very toxic to the human organism [125]. On the other hand, Co-based catalysts show lower activity and Co is more expensive than Ni [126, 127]. Conversely, Fe is very cheap, but it shows low CH_4 selectivity [126, 128]. Indeed, Fe favours the RWGS route as mentioned in the previous section.

Ni/Al_2O_3 catalyst has been the most popular material in CO_2 methanation due to the relatively low price of Al_2O_3 and its ability to finely disperse metal species. Riani et al. compared the catalytic behaviour of Ni nanoparticles with an Ni/Al_2O_3 catalyst with high loading. The best catalytic performance was obtained on the 5:4 (wt_{Ni}/wt_{Al2O3}) Ni/Al_2O_3 catalyst, while bare Ni nanoparticles showed much poorer activity [129]. It has to be recalled that RANEYs nickel, which is a well-known active catalyst for hydrogenation, can also be used in the methanation reaction. Its notable catalytic performance in this reaction is attributed to its unique thermal and structural stability as well as a large BET surface area [130].

A promoting role of the support in this reaction was identified. Amorphous silica was used as a raw material for preparing a series of silica–alumina composites as supports for nickel-based catalysts (Ni/RHA−Al$_2$O$_3$). It was found that hydrogenation activity decreases as the content of alumina increases, indicating that acidic sites are not uniquely responsible for the reaction [131]. Ni/RHA−Al$_2$O$_3$ catalyst prepared by wet-impregnation exhibited favourable catalytic activity owing to the mesoporous structure and high surface area. A strong interaction between metal and oxide along with a high dispersion of NiO and NiAl$_2$O$_4$ nanoparticles was found for this system. In addition, it was reported that the catalytic activity of Ni/RHA−Al$_2$O$_3$ was better than that of Ni/SiO$_2$−Al$_2$O$_3$ due to its better metallic dispersion and higher chemical reaction rate [132]. Zhu et al. studied a series of SiO$_2$, Al$_2$O$_3$ and ZrO$_2$ nanoparticle modified Ni-based bi-modal pore catalysts (20Ni/SiO$_2$−Si, 20Ni/SiO$_2$−Al and 20Ni/SiO$_2$−Zr), The CO$_2$ conversions were in an ascending order of 20Ni/SiO$_2$ (un-modified catalyst) < 20Ni/SiO$_2$−Si < 20Ni/SiO$_2$−Al < 20Ni/SiO$_2$−Zr, corresponding to the increasing order of their surface areas [133].

As a consequence of the significant influence of the support on the dispersion of the active phase, synthesis of highly dispersed metal-supported catalysts has been developed by several research teams. Du et al. studied the dispersion of Ni/MCM-41 catalysts with different amount of Ni [134]. High selectivity (96.0%) and space−time yield (STY, 91.4 g kg^{-1} h^{-1}) were achieved on 3 wt% Ni/MCM-41 at a space velocity of 5760 kg^{-1} h^{-1}, superior to those of Ni/SiO$_2$ catalysts and comparable to Ru/SiO$_2$ catalysts [135, 136]. Reduction at 973 K produced a stable catalyst yielding the best activity and selectivity since most Ni species are reduced to highly dispersed Ni^0 due to the surface anchoring effect.

ZrO$_2$ is another support of interest due to its acidic/basic features and CO$_2$ adsorption abilities on Ni/ZrO$_2$ catalysts [137]. Nevertheless, recently the combination of Ce−Zr oxide has been reported as one of the most promising catalyst supports for CO$_2$ methanation because of its unique redox properties, excellent thermal stability and resistance to sintering. The better performance promoted by the addition of CeO$_2$ is attributed to its high capacity for metal dispersion and to its propensity to create oxygen vacancies that boost oxygen mobility [138−140].

Ocampo et al. found that a 10%Ni/Ce$_{0.72}$Zr$_{0.28}$O$_2$ catalyst for the synthesis of methane exhibited excellent catalytic activity and stability in the reaction during 150 h on stream, achieving 75.9% CO$_2$ conversion and 99.1% CH$_4$ selectivity. The high OSC of the Ce$_{0.72}$Zr$_{0.28}$O$_2$ support and its ability to enhance nickel dispersion (restraining the sintering of the Ni) were the key factors of the high performance obtained [141]. Perkas et al. synthesised Ni/ZrO$_2$ catalysts doped with Ce or Sm cations [139]. They observed a synergistic effect between the surface area and the doping of the rare earth elements. The surface area of the catalyst increased due to the mesoporous structure in the support, which led to the insertion of the Ni particles into the pores. In addition, some recent studies can be found dealing with the effect of Fe and Co addition as promoters on Ni-supported Ce−Zr catalyst. For instance, in a recent work, our team has demonstrated that the use FeO$_x$ and CoO$_x$ boosted the catalytic performance of the reference Ni/CeZr sample [142]. However, the degree of activity promotion was different for each promoter. Fe only improved the activity in the low-temperature range while Co greatly benefited the CO_2 conversion at medium-high temperatures. Overall, the Co–Ni promoted catalysts were the most efficient material within the studied series, showing not only an excellent activity/selectivity balance but also exceptional stability for long-term runs. Furthermore, this catalyst was able to maintain high levels of activity and selectivity at a relatively high space velocity (25,000 mL g^{-1} h^{-1}) and low pressure, which in turn may impact the reactor design facilitating compact reactors configurations [142].

Metal-organic frameworks (MOFs) have emerged recently as a type of catalytic supports due of their advantages in terms of excellent surface areas (typically >1000 m^2 g^{-1}). The so-called MOF-5 (formed from Zn_4O nodes with 1,4-benzodicarboxylic acid struts between the nodes) was used by Zhen et al. as a support for a Ni-based material [143]. A series of Ni@MOF-5 with different Ni loadings for CO_2 methanation have been synthesised by impregnation methods. The 10Ni@MOF-5 showed unexpectedly higher activity for CO_2 methanation than a reference Ni/SiO_2 catalyst at low temperature. It was concluded that the highly uniform dispersion of Ni (dispersion of 41.8%) in the framework of MOF-5 was the main contributor to such a significant activity enhancement. For this catalyst, the CO_2 conversion and CH_4 selectivity were 75 % and over 100 %, respectively, at 320 °C. In addition, this 10Ni@MOF-5 catalyst showed high stability and almost no deactivation in long-term stability tests up to 100 h.

Noble metals, although they are not the most desirable choice for economic reasons, have also been successfully applied in the CO_2 methanation reaction. Ru is believed to be more active and stable in CO_x methanation than Ni. Ni has the problem of deactivation at low temperatures due to the interaction of the metal particles with CO and formation of mobile nickel sub-carbonyls. However, Ru offers positive characteristics such as high activity, CH_4 selectivity (also at low temperatures) and high resistance to oxidising atmospheres [143, 144]. On the contrary, the cost of Ru is its main disadvantage limiting its practical application. Other noble metals like Rh also offer high activity and high selectivity towards CH_4, but its price is also high. Pd has also been tested and the results indicate that it is the least appropriate noble metal for the methanation.

A comparative study was conducted by Garbarino et al. over 20%Ni/Al_2O_3 and 3%Ru/Al_2O_3 commercial catalysts [145]. For the 3% Ru/ Al_2O_3 catalyst (operated at 300 °C, GHSV of 15,000 h^{-1} and in excess H_2) a yield of 96% to methane was achieved without CO co-production. The catalytic activity outperformed the result achieved by a 20% Ni/Al_2O_3 catalyst where a maximum yield of 80% with some CO co-production was observed at 400 °C.

Several works have focused on the comparison of the effect of the support on catalytic properties of Ru nanoparticles in CO_2 hydrogenation. Kowalczyk et al. found that TOFs of Ru-based catalysts were dependent on the Ru dispersion and the type of support [146]. For high metal dispersion, the following order of TOFs ($\times 10^3$ s^{-1}) for the reaction was obtained: Ru/Al$_2$O$_3$ (16.5) > Ru/MgAl$_2$O$_4$ (8.8) > Ru/MgO (7.9) > Ru/C (2.5). It was found that the catalytic activity of Ru nanoparticles was strongly affected by the metal–support interaction. In the case of Ru/C systems, the carbon moiety partially covered the metal surface and reduced the number of active sites. Wang et al. immobilised Ru nanoparticles on the surface of a variety of CeO$_2$ substrates with different morphologies or facets such as nanocubes (NCs), nanorods and nanopolyhedrons with the dominantly exposed {100}, {110} and {111} facets. The highest concentration of oxygen vacancies was reached by Ru/CeO$_2$–NCs catalyst, which resulted in the highest conversion rate of CO$_2$ [147]. A study over a Ru$_{0.05}$Ce$_{0.95}$O$_x$ catalyst reported that this material was sensitive to the surface reduction degree. The CO$_2$ methanation process occurred preferably over the reduced catalyst surface than over an oxidised form [148]. Moreover, Tada et al. investigated the effect caused by the CeO$_2$ content in Ru/CeO$_2$/Al$_2$O$_3$ catalysts on the CO$_2$ methanation catalytic behaviour [149]. The ascending activity order of CO$_2$ methanation at 250 °C follows as Ru/30%-CeO$_2$/Al$_2$O$_3$ > Ru/60%-CeO$_2$/Al$_2$O$_3$ > Ru/CeO$_2$ > Ru/Al$_2$O$_3$. Considering the results of FT-IR experiments, it was perceived that the intermediates of CO$_2$ methanation, including formate and carbonate species, reacted faster with H$_2$ over Ru/CeO$_2$ and Ru/30%-CeO$_2$/Al$_2$O$_3$ catalysts than that on Ru/Al$_2$O$_3$.

From the industrial point of view, the feedstock typically contains traces of sulphur compounds. Therefore, it is very relevant to consider sulphur containing streams to assess the catalytic behaviour of the methanation systems. In this regard, Szailer et al. conducted a careful study on the effect of sulphur on the methanation of CO$_2$. Interestingly, small amounts of H$_2$S (e.g., 22 ppm) promoted the reaction on TiO$_2$- and CeO$_2$-supported metals clusters (e.g., Ru, Rh and Pd), whereas, on other supported catalysts (e.g., ZrO$_2$- and MgO-based) or when the H$_2$S content is high (116 ppm), the reaction rate decreased. The authors explained that when the support is exposed to very low quantities of H$_2$S, the catalyst becomes more active as a result of the formation of new active sites at interfaces between the metal and the support [150]. Perhaps this behaviour resembles the strategy of the well-known SPARG process for reforming reactions.

In summary, a great deal of work has been done on CO$_2$ hydrogenation from both process and catalysis side. Certainly the development of highly active, selective and robust catalysts remains at the edge of research targets for a successful implementation of CO$_2$ hydrogenation units at commercial level. The development of compact units makes necessary high-performing catalysts at very high space velocities, a challenging aspect that opens new avenues for research within the catalysis community. In any case, it is clear that CO$_2$ conversion reactions in the gas phase constitutes a promising route for direct CO$_2$ utilisation. The infrastructure to implement these processes at industrial scale is virtually ready since they are traditional practises in chemical industry and minor adjustments will be required. In this sense, a certain degree of commitment from the industrial chemical sector will be fundamental to validate these approaches beyond academic works.

References

[1]   Jahangiri H, Bennett J, Mahjoubi P, Wilson K, Gu S. A review of advanced catalyst development for Fischer-Tropsch synthesis of hydrocarbons from biomass derived syn-gas. Catal Sci Technol 2014, 4, 2210–29.

[2]   Olah GA, Goeppert A, Czaun M, Prakash GKS. Bi-reforming of methane from any source with steam and carbon dioxide exclusively to metgas (CO-$2H_2$) for methanol and hydrocarbon synthesis. J Am Chem Soc 2013, 135, 648.

[3]   Pakhare D, Spivey J. A review of dry (CO_2) reforming of methane over noble metal catalysts. Chem Soc Rev 2014, 43, 7813–37.

[4]   Charisiou ND, Siakavelas G, Papageridis KN, et al. Syngas production via the biogas dry reforming reaction over nickel supported on modified with CeO_2 and/or La_2O_3 alumina catalysts. J Nat Gas Sci Eng 2016, 31, 164–83.

[5]   Fischer F, Tropsch H. Brennst-Chem 1928, 9,39.

[6]   Rostrup-Nielsen J. Steam reforming of hydrocarbons. A historical perspective. In: Bao X, Xu Y, eds. Stud Surf Sci Catal, Elsevier, 2004, 121–6.

[7]   Mortensen PM, Dybkjær I. Industrial scale experience on steam reforming of CO_2-rich gas. Appl Catal A 2015, 495, 141–51.

[8]   Choudhary Tushar V, Choudhary Vasant R. Energy-Efficient Syngas Production through Catalytic Oxy-Methane Reforming Reactions. Angew Chem Int Ed 2008, 47, 1828–47.

[9]   Arora S, Prasad R. An overview on dry reforming of methane: strategies to reduce carbonaceous deactivation of catalysts. RSC Adv 2016, 6, 108668–88.

[10]   Bitter JH, Seshan K, Lercher JA. On the contribution of X-ray absorption spectroscopy to explore structure and activity relations of Pt/ZrO_2 catalysts for CO_2/CH_4 reforming. Top Catal 2000, 10, 295–305.

[11]   Christian Enger B, Lødeng R, Holmen A. A review of catalytic partial oxidation of methane to synthesis gas with emphasis on reaction mechanisms over transition metal catalysts. Appl Catal A 2008, 346, 1–27.

[12]   Álvarez A, Borges M, Corral-Pérez Juan J, et al. CO_2 Activation over Catalytic Surfaces. ChemPhysChem 2017, 18, 3135–41.

[13]   Bobadilla Luis F, Garcilaso V, Centeno Miguel A, Odriozola José A. Monitoring the Reaction Mechanism in Model Biogas Reforming by In Situ Transient and Steady-State DRIFTS Measurements. ChemSusChem 2016, 10, 1193–201.

[14]   Papadopoulou C, Matralis H, Verykios X. Utilization of Biogas as a Renewable Carbon Source: Dry Reforming of Methane. In: Guczi L, Erdôhelyi A, eds. Catalysis for Alternative Energy Generation. New York, NY, Springer New York, 2012, 57–127.

[15]   Bradford MCJ, Albert Vannice M. The role of metal–support interactions in CO_2 reforming of CH_4. Catal Today 1999, 50, 87–96.

[16]   Osaki T, Mori T. Role of Potassium in Carbon-Free CO_2 Reforming of Methane on K-Promoted Ni/Al_2O_3 Catalysts. J Catal 2001, 204, 89–97.

[17]   Portugal UL, Santos ACSF, Damyanova S, Marques CMP, Bueno JMC. CO_2 reforming of CH_4 over Rh-containing catalysts. J Mol Catal A: Chem 2002, 184, 311–22.

[18]   Rostrupnielsen JR, Hansen JHB. CO_2-Reforming of Methane over Transition Metals. J Catal 1993, 144, 38–49.

[19]   Wang HY, Ruckenstein E. Carbon dioxide reforming of methane to synthesis gas over supported rhodium catalysts: the effect of support. Appl Catal A 2000, 204, 143–52.

[20]   Navarro R, Pawelec B, Alvarez-Galván MC, Guil-Lopez R, Al-Sayari S, Fierro JLG. Renewable Syngas Production via Dry Reforming of Methane. In: Falco MD, Iaquaniello G, Centi G, editors. CO_2: A Valuable Source of Carbon; 2013; London: Springer London, pp. 45–66.

[21] Abdullah B, Abd Ghani NA, Vo D-VN. Recent advances in dry reforming of methane over Ni-based catalysts. J Clean Prod 2017, 162, 170–85.

[22] Bradford MCJ, Vannice MA. Catalytic reforming of methane with carbon dioxide over nickel catalysts I. Catalyst characterization and activity. Appl Catal A 1996, 142, 73–96.

[23] Cheng ZX, Zhao XG, Li JL, Zhu QM. Role of support in CO$_2$ reforming of CH$_4$ over a Ni/γ-Al$_2$O$_3$ catalyst. Appl Catal A 2001, 205, 31–6.

[24] Zhou L, Li L, Wei N, Li J, Basset J-M. Effect of NiAl$_2$O$_4$ Formation on Ni/Al$_2$O$_3$ Stability during Dry Reforming of Methane. ChemCatChem 2015, 7, 2508–16.

[25] Bereketidou OA, Goula MA. Biogas reforming for syngas production over nickel supported on ceria–alumina catalysts. Catal Today 2012, 195, 93–100.

[26] Charisiou ND, Baklavaridis A, Papadakis VG, Goula MA. Synthesis Gas Production via the Biogas Reforming Reaction Over Ni/MgO–Al$_2$O$_3$ and Ni/CaO–Al$_2$O$_3$ Catalysts. Waste Biomass Valorization 2016, 7, 725–36.

[27] le Saché E, Santos JL, Smith TJ, et al. Multicomponent Ni-CeO$_2$ nanocatalysts for syngas production from CO$_2$/CH$_4$ mixtures. J CO2 Util 2018, 25, 68–78.

[28] Leitenburg Cd, Trovarelli A, Kašpar J. A Temperature-Programmed and Transient Kinetic Study of CO$_2$ Activation and Methanation over CeO$_2$ Supported Noble Metals. J Catal 1997, 166, 98–107.

[29] Bhattacharyya A, Chang VW, Schumacher DJ. CO$_2$ reforming of methane to syngas: I: evaluation of hydrotalcite clay-derived catalysts. Appl Clay Sci 1998, 13, 317–28.

[30] Debek R, Motak M, Duraczyska D, et al. Methane dry reforming over hydrotalcite-derived Ni-Mg-Al mixed oxides: the influence of Ni content on catalytic activity, selectivity and stability. Catal Sci Technol 2016, 6, 6705–15.

[31] de Araujo GC, de Lima SM, Assaf JM, Peña MA, Fierro JLG, do Carmo Rangel M. Catalytic evaluation of perovskite-type oxide LaNi$_{1-x}$Ru$_x$O$_3$ in methane dry reforming. Catal Today 2008, 133–135, 129–35.

[32] Dama S, Ghodke SR, Bobade R, Gurav HR, Chilukuri S. Active and durable alkaline earth metal substituted perovskite catalysts for dry reforming of methane. Appl Catal B 2018, 224, 146–58.

[33] de Lima SM, Assaf JM. Synthesis and Characterization of LaNiO$_3$, LaNi$_{(1-x)}$Fe$_x$O$_3$ and LaNi$_{(1-x)}$Co$_x$O$_3$ Perovskite Oxides for Catalysis Application. Mat Res 2002, 5, 329–35.

[34] Evans SE, Staniforth JZ, Darton RJ, Ormerod RM. A nickel doped perovskite catalyst for reforming methane rich biogas with minimal carbon deposition. Green Chem 2014, 16, 4587–94.

[35] Sierra Gallego G, Mondragón F, Tatibouët J-M, Barrault J, Batiot-Dupeyrat C. Carbon dioxide reforming of methane over La$_2$NiO$_4$ as catalyst precursor—Characterization of carbon deposition. Catal Today 2008, 133–135, 200–9.

[36] le Saché E, Pastor-Pérez L, Watson D, Sepúlveda-Escribano A, Reina TR. Ni stabilised on inorganic complex structures: superior catalysts for chemical CO$_2$ recycling via dry reforming of methane. Appl Catal B 2018, 236, 458–65.

[37] Li Z, Mo L, Kathiraser Y, Kawi S. Yolk–Satellite–Shell Structured Ni–Yolk@Ni@SiO$_2$ Nanocomposite: Superb Catalyst toward Methane CO$_2$ Reforming Reaction. ACS Catal 2014, 4, 1526–36.

[38] Price C-AH, Pastor-Perez L, Ramirez Reina T, Liu J. Robust mesoporous bimetallic yolk-shell catalysts for chemical CO$_2$ upgrading via dry reforming of methane. React Chem Eng 2018.

[39] Olah GA, Goeppert A, Prakash GKS. Chemical Recycling of Carbon Dioxide to Methanol and Dimethyl Ether: From Greenhouse Gas to Renewable, Environmentally Carbon Neutral Fuels and Synthetic Hydrocarbons. J Org Chem 2009, 74, 487–98.

[40] Quadrelli EA, Centi G, Duplan JL, Perathoner S. Carbon Dioxide Recycling: Emerging Large-Scale Technologies with Industrial Potential. ChemSusChem 2011, 4, 1194–215.

[41] Barbato L, Iaquaniello G, Mangiapane A. Reuse of CO_2 to Make Methanol Using Renewable Hydrogen. In: Falco MD, Iaquaniello G, Centi G, editors. CO_2: A Valuable Source of Carbon; 2013; London: Springer London, pp. 67–79.

[42] Kusama H, Bando KK, Okabe K, Arakawa H. CO_2 hydrogenation reactivity and structure of Rh/SiO_2 catalysts prepared from acetate, chloride and nitrate precursors. Appl Catal A 2001, 205, 285–94.

[43] Centi G, Perathoner S, Iaquaniello G. Realizing Resource and Energy Efficiency in Chemical Industry by Using CO_2. In: Falco MD, Iaquaniello G, Centi G, editors. CO_2: A Valuable Source of Carbon; 2013; London: Springer London, pp. 27–43.

[44] Koytsoumpa EI, Atsonios K, Panopoulos KD, Karellas S, Kakaras E, Karl J. Modelling and assessment of acid gas removal processes in coal-derived SNG production. Appl Therm Eng 2015, 74, 128–35.

[45] Daza YA, Kuhn JN. CO_2 conversion by reverse water gas shift catalysis: comparison of catalysts, mechanisms and their consequences for CO_2 conversion to liquid fuels. RSC Adv 2016, 6, 49675–91.

[46] Srinivas S, Malik RK, Mahajani SM. Fischer-Tropsch synthesis using bio-syngas and CO_2. Energy Sustain Dev 2007, 11, 66–71.

[47] Saito M, Murata K. Development of high performance Cu/ZnO-based catalysts for methanol synthesis and the water-gas shift reaction. Catal Surv Asia 2004, 8, 285–94.

[48] Skrzypek J, Lachowska M, Serafin D. Methanol synthesis from CO_2 and H_2: dependence of equilibrium conversions and exit equilibrium concentrations of components on the main process variables. Chem Eng Sci 1990, 45, 89–96.

[49] Joo O-S, Jung K-D, Yonsoo J. CAMERE Process for methanol synthesis from CO_2 hydrogenation. In: Park S-E, Chang J-S, Lee K-W, eds. Stud Surf Sci Catal, Elsevier, 2004, 67–72.

[50] Loiland JA, Wulfers MJ, Marinkovic NS, Lobo RF. Fe/γ-Al_2O_3 and Fe-K/γ-Al_2O_3 as reverse water-gas shift catalysts. Catal Sci Technol 2016, 6, 5267–79.

[51] Centi G, Perathoner S. Opportunities and prospects in the chemical recycling of carbon dioxide to fuels. Catal Today 2009, 148, 191–205.

[52] Whitlow JE, Parrish CF. Operation, Modeling and Analysis of the Reverse Water Gas Shift Process. AIP Conf Proc 2003, 654, 1116–23.

[53] Walther D, Ruben M, Rau S. Carbon dioxide and metal centres: from reactions inspired by nature to reactions in compressed carbon dioxide as solvent. Coord Chem Rev 1999, 182, 67–100.

[54] Ohnishi Y-y, Matsunaga T, Nakao Y, Sato H, Sakaki S. Ruthenium(II)-Catalyzed Hydrogenation of Carbon Dioxide to Formic Acid. Theoretical Study of Real Catalyst, Ligand Effects, and Solvation Effects. J Am Chem Soc 2005, 127, 4021–32.

[55] Li J, Jia G, Lin Z. Theoretical Studies on Coupling Reactions of Carbon Dioxide with Alkynes Mediated by Nickel(0) Complexes. Organometallics 2008, 27, 3892–900.

[56] Lu CC, Saouma CT, Day MW, Peters JC. Fe(I)-Mediated Reductive Cleavage and Coupling of CO_2: An $Fe^{II}(\mu$-O,μ-CO)Fe^{II} Core. J Am Chem Soc 2007, 129, 4–5.

[57] Darensbourg DJ. Making Plastics from Carbon Dioxide: Salen Metal Complexes as Catalysts for the Production of Polycarbonates from Epoxides and CO_2. Chem Rev 2007, 107, 2388–410.

[58] Ernst K-H, Campbell CT, Moretti G. Kinetics of the reverse water-gas shift reaction over Cu (110). J Catal 1992, 134, 66–74.

[59] Campbell CT, Ernst K-H. Forward and Reverse Water Gas Shift Reactions on Model Copper Catalysts. Surface Science of Catalysis, American Chemical Society, 1992, 130–42.

[60] Ginés MJL, Marchi AJ, Apesteguía CR. Kinetic study of the reverse water-gas shift reaction over CuO/ZnO/Al$_2$O$_3$ catalysts. Appl Catal A 1997, 154, 155–71.

[61] Chen C-S, Cheng W-H. Study on the Mechanism of CO Formation in Reverse Water Gas Shift Reaction Over Cu/SiO$_2$ Catalyst by Pulse Reaction, TPD and TPR. Catal Lett 2002, 83, 121–6.

[62] Chen C-S, Cheng W-H, Lin S-S. Mechanism of CO formation in reverse water–gas shift reaction over Cu/Al$_2$O$_3$ catalyst. Catal Lett 2000, 68, 45–8.

[63] Fujita S-I, Usui M, Takezawa N. Mechanism of the reverse water gas shift reaction over Cu/ZnO catalyst. J Catal 1992, 134, 220–5.

[64] Nakamura J, Rodriguez JA, Campbell CT. Does CO$_2$ dissociatively adsorb on Cu surfaces? J Phys Condens Matter 1989, 1, SB149.

[65] Qin S, Hu C, Yang H, Su Z. Theoretical Study on the Reaction Mechanism of the Gas-Phase H$_2$/CO$_2$/Ni(3D) System. J Phys Chem A 2005, 109, 6498–502.

[66] Arunajatesan V, Subramaniam B, Hutchenson KW, Herkes FE. In situ FTIR investigations of reverse water gas shift reaction activity at supercritical conditions. Chem Eng Sci 2007, 62, 5062–9.

[67] Ferri D, Burgi T, Baiker A. Probing boundary sites on a Pt/Al$_2$O$_3$ model catalyst by CO$_2$ hydrogenation and in situ ATR-IR spectroscopy of catalytic solid-liquid interfaces. Phys Chem Chem Phys 2002, 4, 2667–72.

[68] Goguet A, Meunier FC, Tibiletti D, Breen JP, Burch R. Spectrokinetic Investigation of Reverse Water-Gas-Shift Reaction Intermediates over a Pt/CeO$_2$ Catalyst. J Phys Chem B 2004, 108, 20240–6.

[69] Rodriguez JA, Evans J, Feria L, et al. CO$_2$ hydrogenation on Au/TiC, Cu/ TiC,and Ni/TiC catalysts: Production of CO, methanol, and methane. J Catal 2013, 307, 162–9.

[70] Liu Y, Liu D. Study of bimetallic Cu–Ni/γ-Al$_2$O$_3$ catalysts for carbon dioxide hydrogenation. Int J Hydrogen Energy 1999, 24, 351–4.

[71] Chen CS, Lin JH, You JH, Chen CR. Properties of Cu(thd)$_2$ as a Precursor to Prepare Cu/SiO$_2$ Catalyst Using the Atomic Layer Epitaxy Technique. J Am Chem Soc 2006, 128, 15950–1.

[72] Stone FS, Waller D. Cu–ZnO and Cu–ZnO/Al$_2$O$_3$ Catalysts for the Reverse Water-Gas Shift Reaction. The Effect of the Cu/Zn Ratio on Precursor Characteristics and on the Activity of the Derived Catalysts. Top Catal 2003, 22, 305–18.

[73] Chen C-S, Cheng W-H, Lin S-S. Study of reverse water gas shift reaction by TPD, TPR and CO$_2$ hydrogenation over potassium-promoted Cu/SiO$_2$ catalyst. Appl Catal A 2003, 238, 55–67.

[74] Figueiredo RT, Santos MS, Andrade HMC, Fierro JLG. Effect of alkali cations on the CuZnOAl$_2$O$_3$ low temperature water gas-shift catalyst. Catal Today 2011, 172, 166–70.

[75] Chen C-S, Cheng W-H, Lin S-S. Study of iron-promoted Cu/SiO$_2$ catalyst on high temperature reverse water gas shift reaction. Appl Catal A 2004, 257, 97–106.

[76] Pastor-Pérez L, Baibars F, le Saché E, Arellano-García H, Gu S, Reina TR. CO$_2$ valorisation via Reverse Water-Gas Shift reaction using advanced Cs doped Fe-Cu/Al$_2$O$_3$ catalysts. J CO2 Util 2017, 21, 423–8.

[77] Zhou G, Dai B, Xie H, Zhang G, Xiong K, Zheng X. CeCu composite catalyst for CO synthesis by reverse water–gas shift reaction: Effect of Ce/Cu mole ratio. J CO2 Util 2017, 21, 292–301.

[78] Wang L, Zhang S, Liu Y. Reverse water gas shift reaction over Co-precipitated Ni-CeO$_2$ catalysts. J Rare Earth 2008, 26, 66–70.

[79] Yang L, Pastor-Pérez L, Gu S, Sepúlveda-Escribano A, Reina TR. Highly efficient Ni/CeO$_2$-Al$_2$O$_3$ catalysts for CO$_2$ upgrading via reverse water-gas shift: Effect of selected transition metal promoters. Appl Catal B 2018, 232, 464–71.

[80] Goguet A, Meunier F, Breen JP, Burch R, Petch MI, Faur Ghenciu A. Study of the origin of the deactivation of a Pt/CeO_2 catalyst during reverse water gas shift (RWGS) reaction. J Catal 2004, 226, 382–92.

[81] Kim SS, Lee HH, Hong SC. The effect of the morphological characteristics of TiO_2 supports on the reverse water–gas shift reaction over Pt/TiO_2 catalysts. Appl Catal B 2012, 119–120, 100–8.

[82] Kim SS, Park KH, Hong SC. A study of the selectivity of the reverse water–gas-shift reaction over Pt/TiO_2 catalysts. Fuel Process Technol 2013, 108, 47–54.

[83] Chen X, Su X, Duan H, Liang B, Huang Y, Zhang T. Catalytic performance of the Pt/TiO_2 catalysts in reverse water gas shift reaction: Controlled product selectivity and a mechanism study. Catal Today 2017, 281, 312–8.

[84] Liang B, Duan H, Su X, et al. Promoting role of potassium in the reverse water gas shift reaction on Pt/mullite catalyst. Catal Today 2017, 281, 319–26.

[85] Yang X, Su X, Chen X, et al. Promotion effects of potassium on the activity and selectivity of Pt/zeolite catalysts for reverse water gas shift reaction. Appl Catal B 2017, 216, 95–105.

[86] Sabatier P, Senderens JB. New synthesis of methane. J Chem Soc 1902, 82, 333.

[87] Centi G, Quadrelli EA, Perathoner S. Catalysis for CO_2 conversion: a key technology for rapid introduction of renewable energy in the value chain of chemical industries. Energy Environ Sci 2013, 6, 1711–31.

[88] De Saint Jean M, Baurens P, Bouallou C, Couturier K. Economic assessment of a power-to-substitute-natural-gas process including high-temperature steam electrolysis. Int J Hydrogen Energy 2015, 40, 6487–500.

[89] Li H, Yang S, Zhang J, Qian Y. Coal-based synthetic natural gas (SNG) for municipal heating in China: analysis of haze pollutants and greenhouse gases (GHGs) emissions. J Clean Prod 2016, 112, 1350–9.

[90] Hoekman SK, Broch A, Robbins C, Purcell R. CO_2 recycling by reaction with renewably-generated hydrogen. Int J Greenh Gas Con 2010, 4, 44–50.

[91] Ghaib K, Ben-Fares F-Z. Power-to-Methane: A state-of-the-art review. Renew Sust Energ Rev 2018, 81, 433–46.

[92] Götz M, Lefebvre J, Mörs F, et al. Renewable Power-to-Gas: A technological and economic review. Renew Energ 2016, 85, 1371–90.

[93] The Sabatier System: Producing Water on the Space Station. NASA, 2011. (Accessed July 17, 2018, at https://www.nasa.gov/mission_pages/station/research/news/sabatier.html.)

[94] Su X, Xu J, Liang B, Duan H, Hou B, Huang Y. Catalytic carbon dioxide hydrogenation to methane: A review of recent studies. J Energy Chem 2016, 25, 553–65.

[95] Swapnesh A, Srivastava Vimal C, Mall Indra D. Comparative Study on Thermodynamic Analysis of CO_2 Utilization Reactions. Chem Eng Technol 2014, 37, 1765–77.

[96] Miguel CV, Soria MA, Mendes A, Madeira LM. Direct CO_2 hydrogenation to methane or methanol from post-combustion exhaust streams – A thermodynamic study. J Nat Gas Sci Eng 2015, 22, 1–8.

[97] Frick V, Brellochs J, Specht M. Application of ternary diagrams in the design of methanation systems. Fuel Process Technol 2014, 118, 156–60.

[98] Marwood M, Doepper R, Renken A. In-situ surface and gas phase analysis for kinetic studies under transient conditions The catalytic hydrogenation of CO_2. Appl Catal A 1997, 151, 223–46.

[99] Lapidus AL, Gaidai NA, Nekrasov NV, Tishkova LA, Agafonov YA, Myshenkova TN. The mechanism of carbon dioxide hydrogenation on copper and nickel catalysts. Pet Chem 2007, 47, 75–82.

[100] Aldana PAU, Ocampo F, Kobl K, et al. Catalytic CO_2 valorization into CH_4 on Ni-based ceria-zirconia. Reaction mechanism by operando IR spectroscopy. Catal Today 2013, 215, 201–7.

[101] Beaumont SK, Alayoglu S, Specht C, Kruse N, Somorjai GA. A Nanoscale Demonstration of Hydrogen Atom Spillover and Surface Diffusion Across Silica Using the Kinetics of CO_2 Methanation Catalyzed on Spatially Separate Pt and Co Nanoparticles. Nano Lett 2014, 14, 4792–6.

[102] Sehested J, Dahl S, Jacobsen J, Rostrup-Nielsen JR. Methanation of CO over Nickel: Mechanism and Kinetics at High H_2/CO Ratios. J Phys Chem B 2005, 109, 2432–8.

[103] Panagiotopoulou P, Kondarides DI, Verykios XE. Mechanistic aspects of the selective methanation of CO over Ru/TiO_2 catalyst. Catal Today 2012, 181, 138–47.

[104] Schild C, Wokaun A, Koeppel RA, Baiker A. Carbon dioxide hydrogenation over nickel/zirconia catalysts from amorphous precursors: on the mechanism of methane formation. J Phys Chem 1991, 95, 6341–6.

[105] Panagiotopoulou P, Kondarides DI, Verykios XE. Mechanistic Study of the Selective Methanation of CO over Ru/TiO_2 Catalyst: Identification of Active Surface Species and Reaction Pathways. J Phys Chem C 2011, 115, 1220–30.

[106] Fisher IA, Bell AT. A Comparative Study of CO and CO_2 Hydrogenation over Rh/SiO_2. J Catal 1996, 162, 54–65.

[107] Eckle S, Anfang H-G, Behm RJ. Reaction Intermediates and Side Products in the Methanation of CO and CO_2 over Supported Ru Catalysts in H_2-Rich Reformate Gases. J Phys Chem C 2011, 115, 1361–7.

[108] Beuls A, Swalus C, Jacquemin M, Heyen G, Karelovic A, Ruiz P. Methanation of CO_2: Further insight into the mechanism over $Rh/γ-Al_2O_3$ catalyst. Appl Catal B 2012, 113–114, 2–10.

[109] Jacquemin M, Beuls A, Ruiz P. Catalytic production of methane from CO_2 and H_2 at low temperature: Insight on the reaction mechanism. Catal Today 2010, 157, 462–6.

[110] Karelovic A, Ruiz P. Mechanistic study of low temperature CO_2 methanation over Rh/TiO_2 catalysts. J Catal 2013, 301, 141–53.

[111] Xu J, Su X, Duan H, et al. Influence of pretreatment temperature on catalytic performance of rutile TiO_2-supported ruthenium catalyst in CO_2 methanation. J Catal 2016, 333, 227–37.

[112] Kustov LM, Tarasov AL. Hydrogenation of carbon dioxide: a comparison of different types of active catalysts. Mendeleev Commun 2014, 24, 349–50.

[113] Wang X, Shi H, Kwak JH, Szanyi J. Mechanism of CO_2 Hydrogenation on Pd/Al_2O_3 Catalysts: Kinetics and Transient DRIFTS-MS Studies. ACS Catal 2015, 5, 6337–49.

[114] Zhen W, Li B, Lu G, Ma J. Enhancing catalytic activity and stability for CO_2 methanation on Ni-Ru/--Al_2O_3 via modulating impregnation sequence and controlling surface active species. RSC Adv 2014, 4, 16472–9.

[115] Solymosi F, Erdöhelyi A, Bánsági T. Methanation of CO_2 on supported rhodium catalyst. J Catal 1981, 68, 371–82.

[116] Weatherbee GD, Bartholomew CH. Hydrogenation of CO_2 on group VIII metals: II. Kinetics and mechanism of CO_2 hydrogenation on nickel. J Catal 1982, 77, 460–72.

[117] Andersson MP, Abild-Pedersen F, Remediakis IN, et al. Structure sensitivity of the methanation reaction: H_2-induced CO dissociation on nickel surfaces. J Catal 2008, 255, 6–19.

[118] Zhang S-T, Yan H, Wei M, Evans DG, Duan X. Hydrogenation mechanism of carbon dioxide and carbon monoxide on Ru(0001) surface: a density functional theory study. RSC Adv 2014, 4, 30241–9.

[119] Ojeda M, Nabar R, Nilekar AU, Ishikawa A, Mavrikakis M, Iglesia E. CO activation pathways and the mechanism of Fischer–Tropsch synthesis. J Catal 2010, 272, 287–97.

[120] Solymosi F, Pásztor M. Analysis of the IR-spectral behavior of adsorbed CO formed in $H_2 + CO_2$ surface interaction over supported rhodium. J Catal 1987, 104, 312–22.

[121] Novák É, Fodor K, Szailer T, Oszkó A, Erdöhelyi A. CO_2 Hydrogenation on Rh/TiO_2 Previously Reduced at Different Temperatures. Top Catal 2002, 20, 107–17.

[122] Mori Y, Mori T, Hattori T, Murakami Y. Electronic effect on surface reaction rates in the carbon monoxide hydrogenation over rhodium catalysts. J Phys Chem 1990, 94, 4575–9.

[123] Mori Y, Mori T, Miyamoto A, Takahashi N, Hattori T, Murakami Y. Support effect on surface reaction rates in carbon monoxide hydrogenation over supported rhodium catalysts. J Phys Chem 1989, 93, 2039–43.

[124] Wang W, Wang S, Ma X, Gong J. Recent advances in catalytic hydrogenation of carbon dioxide. Chem Soc Rev 2011, 40, 3703–27.

[125] Kao Y-L, Lee P-H, Tseng Y-T, Chien IL, Ward JD. Design, control and comparison of fixed-bed methanation reactor systems for the production of substitute natural gas. J Taiwan Inst Chem Eng 2014, 45, 2346–57.

[126] Rönsch S, Schneider J, Matthischke S, et al. Review on methanation – From fundamentals to current projects. Fuel 2016, 166, 276–96.

[127] Habazaki H, Yamasaki M, Zhang B-P, et al. Co-methanation of carbon monoxide and carbon dioxide on supported nickel and cobalt catalysts prepared from amorphous alloys. Appl Catal A 1998, 172, 131–40.

[128] Gao J, Liu Q, Gu F, Liu B, Zhong Z, Su F. Recent advances in methanation catalysts for the production of synthetic natural gas. RSC Adv 2015, 5, 22759–76.

[129] Riani P, Garbarino G, Lucchini MA, Canepa F, Busca G. Unsupported versus alumina-supported Ni nanoparticles as catalysts for steam/ethanol conversion and CO_2 methanation. J Mol Catal A: Chem 2014, 383–384, 10–6.

[130] Sane S, Bonnier JM, Damon JP, Masson J. Raney metal catalysts: I. comparative properties of raney nickel proceeding from Ni-Ai intermetallic phases. Appl Catal 1984, 9, 69–83.

[131] Chang F-W, Tsay M-T, Kuo M-S. Effect of thermal treatments on catalyst reducibility and activity in nickel supported on RHA–Al_2O_3 systems. Thermochim Acta 2002, 386, 161–72.

[132] Chang F-W, Kuo M-S, Tsay M-T, Hsieh M-C. Hydrogenation of CO_2 over nickel catalysts on rice husk ash-alumina prepared by incipient wetness impregnation. Appl Catal A 2003, 247, 309–20.

[133] Zhu P, Chen Q, Yoneyama Y, Tsubaki N. Nanoparticle modified Ni-based bimodal pore catalysts for enhanced CO_2 methanation. RSC Adv 2014, 4, 64617–24.

[134] Du G, Lim S, Yang Y, Wang C, Pfefferle L, Haller GL. Methanation of carbon dioxide on Ni-incorporated MCM-41 catalysts: The influence of catalyst pretreatment and study of steady-state reaction. J Catal 2007, 249, 370–9.

[135] Weatherbee GD, Bartholomew CH. Hydrogenation of CO_2 on group VIII metals: I. Specific activity of $NiSiO_2$. J Catal 1981, 68, 67–76.

[136] Vance CK, Bartholomew CH. Hydrogenation of carbon dioxide on group viii metals: III, Effects of support on activity/selectivity and adsorption properties of nickel. Appl Catal 1983, 7, 169–77.

[137] Yamasaki M, Habazaki H, Asami K, Izumiya K, Hashimoto K. Effect of tetragonal ZrO_2 on the catalytic activity of Ni/ZrO_2 catalyst prepared from amorphous Ni–Zr alloys. Catal Commun 2006, 7, 24–8.

[138] Liu H, Zou X, Wang X, Lu X, Ding W. Effect of CeO_2 addition on Ni/Al_2O_3 catalysts for methanation of carbon dioxide with hydrogen. J Nat Gas Chem 2012, 21, 703–7.

[139] Perkas N, Amirian G, Zhong Z, Teo J, Gofer Y, Gedanken A. Methanation of Carbon Dioxide on Ni Catalysts on Mesoporous ZrO_2 Doped with Rare Earth Oxides. Catal Lett 2009, 130, 455–62.

[140] Zhou L, Wang Q, Ma L, Chen J, Ma J, Zi Z. CeO_2 Promoted Mesoporous Ni/γ-Al_2O_3 Catalyst and its Reaction Conditions For CO_2 Methanation. Catal Lett 2015, 145, 612–9.

[141] Ocampo F, Louis B, Roger A-C. Methanation of carbon dioxide over nickel-based $Ce_{0.72}Zr_{0.28}O_2$ mixed oxide catalysts prepared by sol–gel method. Appl Catal A 2009, 369, 90–6.

[142] Pastor-Pérez L, le Saché E, Jones C, Gu S, Arellano-Garcia H, Reina TR. Synthetic natural gas production from CO_2 over Ni-x/CeO_2·ZrO_2 (x = Fe, Co) catalysts: Influence of promoters and space velocity. Catal Today 2017.

[143] Mills GA, Steffgen FW. Catalytic Methanation. Catal Rev 1974, 8, 159–210.

[144] Brooks KP, Hu J, Zhu H, Kee RJ. Methanation of carbon dioxide by hydrogen reduction using the Sabatier process in microchannel reactors. Chem Eng Sci 2007, 62, 1161–70.

[145] Garbarino G, Bellotti D, Riani P, Magistri L, Busca G. Methanation of carbon dioxide on Ru/Al_2O_3 and Ni/Al_2O_3 catalysts at atmospheric pressure: Catalysts activation, behaviour and stability. Int J Hydrogen Energy 2015, 40, 9171–82.

[146] Kowalczyk Z, Stołecki K, Raróg-Pilecka W, Miśkiewicz E, Wilczkowska E, Karpiński Z. Supported ruthenium catalysts for selective methanation of carbon oxides at very low COx/H_2 ratios. Appl Catal A 2008, 342, 35–9.

[147] Wang F, Li C, Zhang X, Wei M, Evans DG, Duan X. Catalytic behavior of supported Ru nanoparticles on the {100}, {110}, and {111} facet of CeO_2. J Catal 2015, 329, 177–86.

[148] Upham DC, Derk AR, Sharma S, Metiu H, McFarland EW. CO_2 methanation by Ru-doped ceria: the role of the oxidation state of the surface. Catal Sci Technol 2015, 5, 1783–91.

[149] Tada S, Ochieng OJ, Kikuchi R, Haneda T, Kameyama H. Promotion of CO_2 methanation activity and CH_4 selectivity at low temperatures over Ru/CeO_2/Al_2O_3 catalysts. Int J Hydrogen Energy 2014, 39, 10090–100.

[150] Szailer T, Novák É, Oszkó A, Erdőhelyi A. Effect of H_2S on the hydrogenation of carbon dioxide over supported Rh catalysts. Top Catal 2007, 46, 79–86.

Index

https://doi.org/10.1515/9783110563191-015